A PROJECT SQUID WORKSHOP

Combustion Measurements
modern techniques and instrumentation

Edited by

R. GOULARD

School of Engineering and Applied Sciences
The George Washington University
Washington, D.C.

ACADEMIC PRESS

New York   San Francisco   London

A SUBSIDIARY OF HARCOURT BRACE JOVANOVICH, PUBLISHERS

1976

HEMISPHERE PUBLISHING CORPORATION

Washington   London

Library of Congress Cataloging in Publication Data

Project Squid Workshop on Combustion Measurements in Jet
    Propulsion Systems, Purdue University, 1975.
    Combustion measurements.

    At head of title: A Project Squid workshop.
    Workshop held May 22–23, 1975, and sponsored by the
Office of Naval Research (Project Squid), and the Air
Force Office of Scientific Research.
    Bibliography: p.
    Includes index.
    1. Combustion—Congresses. 2. Combustion engineering
—Congresses. I. Goulard, Robert J., 1926–
II. Project Squid. III. United States. Air Force.
Office of Scientific Research. IV. Title.
QD516.P73   1975   621.4'023'028   76-25999
ISBN 0-12-294150-0 (Academic Press)
ISBN 0-89116-037-X (Hemisphere)

Proceedings of the Project SQUID WORKSHOP ON COMBUSTION MEASUREMENTS IN JET
PROPULSION SYSTEMS, held at Purdue University, West Lafayette, Indiana, May 22–23, 1975,
and sponsored by the Office of Naval Research (Project SQUID), and the Air Force Office of
Scientific Research.

1 2 3 4 5 6 7 8 9 0   D O D O   7 8 3 2 1 0 9 8 7 6

This work relates to the Department of the Navy Contract No. N00014-67-A-0226-0005 issued by the
Office of Naval Research under contract authority No. NR 098-038. However, the content does not necessarily reflect the position or policy of the Department of the Navy or the Government, and no official
endorsement should be inferred. The United States Government has a royalty-free, nonexclusive and
irrevocable license throughout the world for Government purposes to publish, translate, reproduce, deliver,
perform, dispose of, and to authorize others to do so, all or any portion of this work.

CONTENTS

iii

PART II
Optical Measurement Techniques

SESSION 7
Interferometry
Chairman: M. M. El Wakil

PART IV
Forum
Chairman: S. S. Penner

PART V
Review and Suggested Experiments
R. Goulard, A. M. Mellor, and R. W. Bilger

PART VI
Conclusions and Recommendations

PREFACE

Rapid advances in the knowledge of combustion processes are the necessary condition to a satisfactory solution of our energy and environment problems. As better measurement techniques become available, a new range of combustion phenomena can be studied with the accuracy necessary for meaningful interpretation. Research progress depends on a better understanding of what are the most promising combinations of new research interests and new measurement techniques. Specifically, the potential of optical methods as compared to that of classical probes should be evaluated in the context of the phenomenology needed for the design of current and future combustors.

This workshop was conceived primarily as an encounter between combustor technologists and instrumentation specialists. Within each group a wide range of interests was in evidence. Measurement needs and opportunities are quite different in small laboratory flames or in large combustors buried in test engines, in experimental work or in computer modeling, in systems analysis or in kinetics rates determination. Similarly, proven inexpensive techniques were discussed as well as sophisticated new ones, a few of which appear to be mere promising concepts at this time.

It seems in this regard that the workshop has been successful in identifying—in broad terms at least—the different sorts of instrumentation likely to be useful at the various levels of combustor design, from basic phenomenology to real engines. At each level, this matching of needs and available techniques was conducted as quantitatively as possible (see in particular the forum, pp. 389–416). However, a healthy self-consciousness kept the constraining environment of prototype combustor testing very much at the center of most discussions. The more sophisticated laboratory techniques were evaluated for their "hardening" potential, as well as for the cost and lead times involved in their possible test cell applications.

Ninety-eight participants met for two days. The first morning was devoted mostly to a presentation of what combustion specialists felt they needed in terms of measurements, from a general, modeling, and prototype view. The next two half-days were made up of short presentations on the various optical and immersed probe techniques, followed by discussions. On the last afternoon, a forum tried to establish the consensus of—and the differences among—the participants' views.

A review paper was assembled after the workshop (pp. 417-470) with a view to summarizing and ordering its salient conclusions, and to expanding on the specific measurements that can be considered usefully at this time. Finally, at the sponsors' request, a brief statement of conclusions and recommendations (pp. 471-474) was formulated.

* * *

The workshop was held under the sponsorship of the Squid Project (Office of Naval Research) at Purdue University, May 22-23, 1975. The entire workshop was recorded, later edited, and now appears in the form of these proceedings.

Special mention should be made of the continuous support and encouragement given by Mr. J. Patton, Power Branch, ONR, whose interest in the Squid Workshop program has been unflagging from the start. Dr. R. Roberts, former Power Branch manager, ONR, and Dr. S. N. B. Murthy, Squid Project director, contributed their ideas and support in the early formulation of the workshop objectives.

Two more agencies gave support and guidance: the Air Force Office of Scientific Research (Dr. T. Wolfson) and the National Science Foundation (Dr. J. Belding, now at ERDA, and Dr. D. Senich). Another strong source of support was Dr. W. M. Roquemore, Aero Propulsion Lab, AFSC, Wright Patterson AFB, who provided a vital input, especially in regard to the participation of the engine manufacturers community. He also encouraged several participants in another Air Force program at Purdue to become significant contributors to the workshop program.

The support of the Purdue University staff was at its usual level of excellence. Mr. Carl Jenks and the Stewart Center attended with quiet efficiency to the needs of our 98 visitors. Our gratitude goes to the staff of the Thermal Sciences and Propulsion Center who helped diligently with the arduous task of transcribing the tapes. Dr. F. P. Chen played a very helpful role in editing the discussions and in integrating the artwork into the manuscript.

R. Goulard
Editor and Workshop Chairman

THIRTY YEARS OF RESEARCH

The primary objective of the workshop on Measurement Techniques in Combustors was to evaluate the application of new optical, unobtrusive measurement techniques to practical combustor design and development. A secondary goal was to bring instrumentation specialists together with engineers and scientists versed with fundamental combustion problems, as well as with designers actively involved with conceptual and operational problems of engines.

There has been a resurgence of interest in recent years in basic problems relating to air-breathing engines due to increasing demands for higher performance, for smaller and lighter-weight power plants, and for operation over wider ranges of operating conditions. This interest is exemplified in the current need to develop deeper fundamental understanding of the physical phenomena involved in all aspects of engine design and development.

To satisfy this need, the Power Program of the Office of Naval Research, through Project SQUID, initiated a series of workshops to consider selected topics from the standpoint of (1) a critical evaluation of current efforts, (2) determining the extent of agreement in explaining various phenomena associated with the subject, and (3) discussion of possible new approaches to solution of problem areas. These workshops have been held:

- Research in Gas Dynamics of Jet Engines, ONR/Chicago, December 4–5, 1969, Project SQUID Report
- Fluid Dynamics of Unsteady 3-D Separated Flows, Georgia Tech., June 10–11, 1971, AD736248
- Laser Doppler Velocimetry Flow Measurements, Purdue University, March 9–10, 1972, AD753243
- Aeroelasticity in Turbomachines, Detroit Diesel Allison, June 1–2, 1972, AD749680
- Laser Raman Diagnostics, G.E. Research & Development Center, May 10–11, 1973, GE-2-PU, Project SQUID Report
- Laser Doppler Velocimetry Flow Measurements II, Purdue University, March 1974, PU-R1-75, Project SQUID Report

- Turbulent Mixing: Non-Reactive and Reactive Flows, Purdue University, May 20–21, 1974, Plenum Press, New York
- Unsteady Flow in Jet Engines, United Aircraft Research Laboratory, (UARL) now (UTRC), July 11–12, 1974, UARL-3-PU, Project SQUID Report
- Measurement Techniques in Combustors, Purdue University, May 22–23, 1975, ADA020386, Project SQUID Report PU-R1-76 (*this volume*)
- Transonic Flows in Turbomachines, Naval Postgraduate School, February 11–13, 1976
- Turbulence in Internal Flows, Airlie House, June 1976

At the time of this workshop, the Office of Naval Research was actively planning its thirtieth anniversary for 1976. The year 1976 also marks the thirtieth anniversary of Project SQUID. This volume is part of a set of scientific publications released by ONR in recognition of, and on the occasion of, its thirtieth anniversary.

James R. Patton, Jr.
Power Program
Office of Naval Research
U.S. Department of the Navy

Anniversary Theme:

Exploring new horizons to protect our heritage

PART I
Needed Measurements in Combustor Technology

NEW DIRECTIONS IN COMBUSTION RESEARCH
as related to jet propulsion systems

IRWIN GLASSMAN
Princeton University

I. INTRODUCTION

Recent workshops have directed their attention to various aspects of problems that could be placed in that broad discipline that we generally call combustion. This report naturally draws most heavily on the NSF Workshop on Energy-Related Combustion Research (Glassman and Sirignano, 1974) held at Princeton University. But the theme of this SQUID Workshop has a sharper focus - jet propulsion systems. Thus rather than review the results of the Princeton/NSF Workshop, I have chosen to discuss only those elements which I thought most relevant to jet propulsion devices and include some subjects not discussed at Princeton. Considering the limitations on the length of this report it is most apparent that this effort cannot be a comprehensive essay. Indeed the references have been chosen so that they are mostly review works and more recent publications not included in these reviews; thus the reference list is not extensive. The reader is encouraged to explore the references in the review articles for a far greater listing.

II. FACTORS GOVERNING NEW DIRECTIONS IN COMBUSTION RESEARCH

Considering the possible enormous breadth of the subject, I have felt it necessary to choose what some would consider an unorthodox approach.

3

I have made no effort to blanket the field. Yet my experiences have taught me that prevailing concerns stimulate thought and generally form the background for generating new research. Thus I have concluded the following underlying concerns will exert the dominant forces which determine the major new propulsion combustion efforts in the next decade:

     1.   the non-availability of standard fuels
     2.   emissions from aircraft
     3.   aircraft fire safety
     4.   limited wars and their strategic implications as exemplified by the recent events in Southeast Asia and the Middle East.

Thus I believe that the major combustion challenges facing the jet propulsion field are:

(a) The understanding associated with the burning of liquid fuels, particularly "new" liquid fuels, which could contain mixtures of hydrocarbons of higher and more varied molecular weight, higher percentages of aromatics, concentrations of bound nitrogen and sulfur, water in emulsified form, antimisting agents, hydrogenated components, etc.

(b) The design of new low emission gas turbine combustion systems, particularly lean pre-mixed and catalytic systems, and the convenient determination of the temperature, velocity and specie concentration leaving the combustor can.

(c) Elements related to fire safety - the effectiveness and mechanisms associated with extinguishing and inerting agents, particularly the gaseous halons and dry powders, the characteristics of flame spread across liquid and solids, simple inexpensive ignition systems for low volability (Jet A type) fuels, associated misting and ignition problems, simple test methods for evaluating which materials are fire safe, etc.

(d) Problems associated with small ramjets. The Middle East conflagrations have shown the utility of longer range missile devices and that a sophisticated enemy could successfully counter subsonic devices. These factors support the need for further development of supersonic ramjet technology.

All the above areas are important and must be investigated. Too often a field concentrates on one or two popular elements and forgets the need elsewhere. Frequently I get the uneasy feeling that from the point of view of jet devices we are giving too much emphasis improving our predictive capability for turbulent reacting flows although this subject is one of the major recommendations of the Princeton/NSF Workshop. In the next paragraphs some of the important fundamental research and instrumentation required in each of the areas listed above will be discussed. Indeed many fundamental problems arise in a few of the areas. An effort will be made to discuss some of the more important ones, particularly those which lead to specific instrumentation problems.

III. "NEW" LIQUID FUELS AND ASSOCIATE PROBLEMS

Irrespective of whether one is dealing with pure, conventional multi-component or "new" fuels, it must be recognized that the ability to measure the droplet size distribution in sprays is still one of the most challenging instrumentation problems in the field (Harrje and Reardon, 1972). Whether one uses impinging jets, swirl or air assisted nozzles - whatever - he is faced with the problem of measuring a droplet range from sub-micron to sizes of the order of one hundred microns not only during the atomization process, but droplets under the very adverse conditions of actual combustion. Laser holography is making great inroads in this area, but digital scanning equipment necessary to reduce the photographic data is unique and expensive. A good, simple inexpensive instrument (or technique) to evaluate the quality of a fuel injector is still very necessary.

Similarly there is always the desire to be able to determine both liquid and vapor concentrations in a spray. It would appear that these measurements would be accomplished only by immersion instruments. To my knowledge a suitable isokinetic probe to perform this task has never evolved.

It is obvious that initial studies of multicomponent and emulsified fuel combustion will focus around single droplets rather than sprays. However, it is already apparent that such particles suspended on fibers are affected by the presence of the fibers themselves. The fibers affect circulation within the droplet and cause agglomeration of the micro-dispersed phase in emulsified fuels. Instrumentation which will permit the study of single droplets is most needed. Two areas that look particularly attractive are the "gating" of a stream of charged particles generated by various means (Williams, 1973) and the use of a modified Millikan cell together with laser optics to hold the charged particle specimen precisely within the laser beam (Davis and Chorojean, 1974). Problems of charging non-polar fuels particles exist in both schemes and the Millikan device has not been adapted for (large) particles other than aerosols (sub-micron).

Recent results obtained in England and at Princeton (Dryer, 1975) give indication that "good" emulsified fuels can lead to droplet explosions, decreased heterogeneity, and reduced emissions, particularly of NO_x and smoke. Instruments are required to help define "good" emulsions; that is, to evaluate the size (order of microns) of the dispersed phase particles (water) within the major droplet (fuel) whose size is the order of tens of microns.

Fuels that will provide our needs decades from now will undoubtedly come from secondary and tertiary resource forms or be derived synthetically from coal. Although the constitution of such fuels cannot be predicted explicitly, it is safe to speculate that they will contain the higher order aliphatic hydrocarbons and aromatics. Particularly from the viewpoint of understanding over-all heat release rates in ramjets, emissions and possible radiation problems in all propulsion devices, it becomes necessary to understand the homogenous gas phase oxidation mechanisms and kinetic rates of higher order paraffins, olefins and benzene. Some instrumentation problems arise in this area. They are particularly concerned with the ability to analyze for specific products. The ability to measure concentrations of formaldehyde in the 100 ppm and less range is particularly difficult and crucial (Colket, et al., 1974). Formaldehyde is the major aldehyde in most emissions and the most easily to be photolyzed by sunlight.

There is still difficulty in estimating hydrocarbon emissions, even by flame ionization detectors (FID). There are problems relating to the FID response being more biased towards paraffins than to partially oxygenated compounds and unsaturates. FID can record in the parts per billion range. Means of identifying and classifying odorants will become more urgently needed.

I would be remiss in discussing emissions if I were to omit comment on NO_x formation. Indeed environmental arguments with respect to NO_x have been the principal tool of the opponents of the SST. But in the context of "new directions" it is not necessary to reflect on NO_x formation from atmospheric nitrogen. Undoubtedly our newer fuels, such as those derived from shale and coal, will contain sufficient quantities of fuel bound nitrogen to be of concern. The detailed study of the mechanisms of NO_x formation from fuel bound nitrogen appears to be academic. If the fuel bound nitrogen is there, it will form NO_x and there does not appear anything one could do about it. Aerodynamic quenching of its formation would also quench the energy release reaction. It would appear much more profitable to concentrate on the unknown reduction steps for NO_x (Bowman, 1975). The fact that greater than equilibrium concentrations of NO_x form from fuel-bound nitrogen emphasizes the need to know more about its reduction steps and any means for accomplishing its reduction. I am not sure what new instrumentation techniques such studies would impose, but this problem will be one of primary concern in the future.

Our present means of burning aromatic fuels has always led to carbon formulation problems. Condensed phase particles form in all jet propulsion devices (smoke in turbojets and ramjets, metal oxides in rockets and ramjets), yet there is still no clear understanding of the nucleation-condensation phenomenon. My intuition, from work with lasers on aerosols

I have seen, leads me to speculate that laser scattering techniques may be the key to our understanding of this problem. It is difficult to be specific about what has to be measured, but let me report that carbon found in flames can be given an approximate empirical formula of C_8H. When examined under an electron microscope, the deposited carbon appears to consist of a number of roughly spherical particles, strung together like pearls on a necklace. The diameters of these particles vary from 100 to 2,000 Å and most commonly lie between 100 to 500 Å. Each particle is made up of a large number (10^4) of what are called crystallites. Each crystallite consists of 5 to 10 sheets of carbon atoms, each containing 100 carbon atoms. Thus each spherical particle contains about 10^5 to 10^6 carbon atoms.

The ability to study complex kinetic mechanisms such as those which exist in hydrocarbon oxidation may be aided by some new developing laser techniques. These techniques employ tunable dye lasers to photo-inhibit a particular elementary reaction step in order that the rate of another elementary step by measured (Lotokhov, 1973). The requirements on such lasers are more severe than those which are used to photo-induce chemical reactions. Response requirements can be thought to be related with the characteristic times of the two processes. The time associated to photo-inhibiting a reaction should be related to the characteristic frequency of the bond to be formed and thus is in the order of 10^{-13} sec. The time associated with photo-inducing is related to the lifetime of the initially excited species and is many orders of magnitude longer than the photo-inhibiting time. The ability to control one or more key elementary reaction steps in a complex reaction mechanism is an exciting prospect.

IV.   TURBOJET COMBUSTOR DESIGN AND ASSOCIATED PROBLEMS

Two of the most promising techniques for reducing emissions from turbojet combustors would appear to be by the use of pre-mixed lean fuel-air mixtures and catalytic combustors (National Academy of Science, 1975). Much work remains to be done in both areas, but they have shown substantial reduction in emission indices in laboratory tests.

The research required to help perfect pre-mixed combustions for aircraft gas turbines include such areas of lean flammability limits, lean propagation velocities, burning of lean, wet mixtures, and quenching (in order to prevent flashback). These areas have been discussed in the report on the Princeton/NSF Workshop. Improved instrumentation to facilitate measurement of lean (wet and completely homogenous) flame speeds and flammability limits is needed. It is unfortunate that there is so little effort in this field.

There appears to be no doubt that the catalytic combustion field will develop as one of the most significant new areas not only for jet propulsion systems, but for all systems that have the facility to control emissions by lean operation. In certain furnace type applications, decreased stack losses are also possible with catlytic schemes.

Catalytic reactions have been studied for a very long time and the instrumentation techniques used are well documented. It would appear that here too, however, lasers would play an important role in the heterogenous kinetics of catalytic surfaces. My intuition is that a tunable dye laser exciting only specific species absorbed on a catalytic surface would be a valuable tool in such studies.

Without a doubt one of the greatest contributions to turbojet design would be the development of non-immersion instrumentation to measure the temperature, velocity and composition profiles of a turbojet combustor exhaust. Accurate (few percent) knowledge of temperature and velocity profiles could lead to combustor modification and also much easier turbine design. Laser Raman gas thermometry, laser doppler velocimetry and the various laser methods for detection of pollutants need the active development attention they are currently getting (APS, 1974). There are still numerous problems in the practical utilization of these techniques - particularly noise, overall signal to noise levels for low concentrations, laser beam quality and reliability, spatial resolution, signal processing, etc. These problems and the whole area, rate as the number one next (or current) generation challenge in the combustion field.

V.   AIRCRAFT FIRE SAFETY

No one cocerned with jet propulsion systems and combustion can ignore the fire hazard introduced by the presence of the engine's fuel or the overall fire vulnerability of the aircraft of which the jet engines are a part. (AGARD, 1971, 1975).

One of the most fundamental problems in this area is related to flame spreading across liquid fuel spills and solid combustible surfaces. The greatest controversy is concerned with the induced air flows as the flame spreads. Which of the analytical models proposed to describe the flame spreading is correct, can only be resolved by velocity measurements close to the moving flame front. In particular one needs a spatial resolution of the order of a millimeter and the ability to measure 1 to 50 cm/sec flows in a two dimensional eddy of the order of one centimeter. It would appear that this difficult task can be accomplished only by laser-doppler velocimetry.

Optical instrumentation is necessary to measure the ignition characteristics of misting fuels (and fuels containing anti-misting agents) when
a projectile enters a fuel tank.

Halon and dry powders are being proposed more and more as inerting and
extinguishing agents. Surprisingly it is still not clear whether these
agents are effective due to their altering the kinetics of the combustion
reaction or simply through their thermal capacity.

Reliable and meaningful test methods for evaluating the flammability
characteristics of the light weight polymeric material currently in use in
today's commercial jet aircraft are urgently needed. Flammability characteristics in this context refers to ignition, flame spreading, smoke generation, flash-over, smoldering, toxic gas generation, etc.

It is unfortunate that the basic combustion fraternity gives so little
of its attention to the aircraft fire safety problem.

VI. COMBUSTION PROBLEMS IN RAMJETS

In some ways one could say that it is political considerations
which will finally bring ramjets to prominence in the propulsion scene. The
need for certain tactical systems make both liquid and solid fueled ramjets
the logical devices to examine. Certain recent military operations and tests
have sharpened the focus on ramjet systems. Ramjets that will be considered
will pose some interesting and challenging combustion problems. The necessity in tactical systems to stabilize over a wide range of air-fuel ratios
reawakens us to the fact that not much is really known about bluff-body stabilization in liquid fuel systems. For that matter, aerodynamic stabilization similar to that in present turbojet combustor cans, or that accomplished
by abrupt changes (steps) in the flow channel, should not be ignored in ramjet considerations. It is my feeling that the recirculating zone that ac-
complishes the stabilization in all these systems will become the subject
of intensive study. Fuel, air and heat addition to the recirculating zone
are all supplemental tools to extend the blow-out range of any stabilizing
device (Fetting et al. 1959). These three dimensional zones pose a real
challenge to laser techniques, since it will be necessary to measure temperatures, species concentrations and velocities. The problem of aerodynamic stabilization in ramjets is no different than that of the primary
zone of a turbojet combustor can. The difference here is that the stabilization problem will be approached without concern about pollutant emissions.

Ramjet inlets are never tested in full scale until they are flown,
thus though it is possible to measure the mean air velocity in test there
appears to be no really suitable instrumentation to measure velocity (particularly profiles) in flight. Most tactical ramjet missiles will use high

viscosity fuels. Unfortunately, standard turbine meters are very sensitive to viscosity and therefore are unsatisfactory. Thus while we concern
ourselves so extensively with advanced laser and other optical instrumentation, we have overlooked the important need for high response flow meter
for viscous fuels. Indeed developmental ramjets suffer from low frequency
(feed system 150 cps) and high frequency (combustion process or acoustic
1500 cps) instabilities. Thus the need for high response instrumentation
both for the viscous fuel and large volume compressible air systems in this
device should be most apparent.

VII. TURBULENT REACTING FLOWS

These final remarks, I know, will be most controversial Many times
in this paper I have pointed out the need to make measurements of temperature, concentration and velocity. It was taken for granted that the experimental situations discussed were mostly turbulent, reacting systems.
However, most of the measurements referred to where essentially mean measurements. I must express concern that great, expensive instrumentation
efforts will be diverted to "turbulence" measurements in reacting flows.
The tremendous efforts in aerodynamic turbulence, though providing much
fundamental information, have never contributed extensively to the solution of most practical aerodynamical problems. Turbulence is just too complex. Turbulence with chemical reaction is even more complex. I seriously doubt whether extensive efforts in this area will have any greater success for combustion than non-reactive turbulence studies gave for practical aerodynamic studies.

While one must never lose sight of the long range importance of the
more academic virtues of establishing actual predictive capability for
turbulent reacting flows, it is my opinion that complexity of most practical combustion systems, particularly those with recirculation, will prevent realization of actual improvements in practical systems from such research in the time frame necessary for impacting the short range problems
discussed here.

REFERENCES

1.   AGARD (1971). "Aircraft Fuels, Lubricants and Fire Safety," AGARD
     Conf. Proc. No. 84.

2.   AGARD (1975). "Aircraft Fire Safety," AGARD Conf. Pre-print No. 166.

3. American Physical Society (1974). "APS Study on the Technical Aspects of Efficient Energy Utilization - The Role of Physics in Combustion," Princeton University School of Engineering, August, 1974.

4. Blazowski, W. S., and Walsh, D. E. (1975). "Catalytic Combustion: An Important Consideration for Future Applications," to appear in Comb. Sci. and Tech.

5. Bowman, C. T. (1975). "Kinetics of Pollutant Formation and Destruction in Combustion," to appear in Progress in Energy and Combustion Science, N. Chigier, Ed., Pergammon, London.

6. Colket, M. B., Naegeli, D. W., Dryer, F. L., Glassman, I. (1974). "Flame Ionization Detection of Carbon Oxides and Hydrocarbon Oxygenates," Env. Sci. and Tech. $\underline{8}$, 43.

7. Davis, E. J. and Chorbajean, E. (1974). " The Measurement of Evaporation Rates of Submersion Aerosol Droplets," Ind. Eng. Chem. Fundam. $\underline{13}$, 272.

8. Dryer, F. L.,(1975). "Fundamental Concepts on the Use of Emulsions on Fuels," Princeton University Aerospace and Mechanical Sciences Report No. 1224.

9. Fetting, F., Choudhury, A. P. R., and Wilhelm, R. H. (1959). "Turbulent Flame Blow-off Stability Effect of Auxiliary Gas Addition into Separation Zone," Swinth (Int'l) Combustion Symposium, Butterworths, London.

10. Glassman, I., and Sirignano, W. A. (1974). "Summary Report of (NSF) Workshop on Energy-Related Basic Combustion Research," Princeton Universoty Aerospace and Mechanical Sciences Report No. 1177.

11. Harrje, D., and Reardon, F. H., Eds. (1972). "Liquid Propellant Rocket Combustion Instability," NASA SP-194.

12. Lotokhov, V. S. (1973). "The Use of Lasers to Control Selective Chemical Reactions," Science $\underline{180}$, 451.

13. National Academy of Sciences (1975). "Environmental Impact of Stratospheric Flight," Washington, D. C.

14. Williams, A. (1973). "Combustion of Droplets of Liquid Fuels: A Review," Comb. and Flame $\underline{21}$, 1.

DISCUSSION

CHIGIER, Univ. of Sheffield - I would like to take up the challenge which you gave about the importance of turbulent studies in combustion research. I believe that the key to many of the problems in combustion research lies in the aerodynamics and in the understanding of turbulence. A few very important developments have taken place in turbulence which will have a tremendous impact on combustion. We no longer consider turbulent flow to be a random isotropic phenomenon, but we believe on the basis of evidence that is coming forward, that many turbulent flows are made up of what we now refer to as coherent structures, i.e. large eddies which move throughout the combustion system. Many of the problems which have been raised, such as the problems of unmixedness, or the high rates of formation of nitrogen oxide in a system which is basically lean, etc., are all basically tied up with the knowledge of the structure of turbulent flow. And I would suggest that we have to place a great deal of emphasis on learning to understand this problem.

For the understanding of these coherent structures, combustion offers an ideal flow visualization technique because we find in turbulent diffusion flames, that combustion only takes place at the interface of these eddies, where burning takes place within the limits of flamability. We see these large eddies moving through the system and, instead of taking indiscriminate time-averaging which many people have done in the past, we now use conditional sampling, and by this means, we can determine what is happening locally with these large eddies. We are finding that there is no burning taking place inside the large eddies which are mainly made up of fuel, also there is no burning taking place in the surrounding air - combustion takes place at the interface. Therefore I would like to disagree very strongly with your final conclusion that we should not emphasize fluid dynamics.

GLASSMAN - I still would guess that you won't find anything very different in flame burners. The height of a turbulent flame for instance, is still proportional, in any practical condition that I will warrant, to the jet diameter. I could study the details of eddy diffusion but if I were designing a gas turbine combustor, the gross turbulent measurements would give me as much information.

CHIGIER - If you are only interested in knowing the length of the flame, you are right. But many people now want to know the rate of formation and destruciton of pollutants inside the flame. For that information you have to know what is happening locally.

GLASSMAN - Are you sure that you are going to obtain all the detail you need (ρ', v', etc....) with your laser instrumentation?

CHIGIER - Definitely. I am confident that a clear support of this viewpoint will come out of this workshop.

COMBUSTION MODELING
an APS summer study assessment

DONALD R. HARDESTY
Sandia Laboratories

These remarks are intended to summarize the principal objectives and conclusions of an American Physical Society Workshop which was conducted at Princeton University during July 1974.[1] The study was jointly sponsored by the National Science Foundation, the Federal Energy Administration and the Electric Power Research Institute. One aspect of the study involved an exploration of the role of physics in combustion.[2] Seventeen physicists and engineers* examined in some depth the interrelationship between experimental diagnostics and modeling in combustion science, while a parallel effort focused on the prospects for the utilization of emulsified fuels in a variety of combustion applications.

The principal objectives of the study were to identify, for the physics community, areas of research related to combustion science which have promise of impact on practical systems, specifically with regard to efficient and nonpolluting fossil fuel utilization. Further, within selected areas, an assessment was made of the potential for significant contribution by the physics community. Our audience was assumed to be scientists who are not "combustion oriented" but who may have special expertise in particular diagnostic or computational techniques.

My comments here are restricted principally to our general assessment of problems which are related to the formulation of advanced models of combustion systems, where new or unconventional insight from physicists might be helpful. The characterization of the physics of the combustion of emulsified fuels represents a specific example where improved modeling capability is required.

---

*Participants in the combustion study included: R. Barnes (Battelle), F. Bracco (Princeton), J. Chang (LLL), P. Chung (U. of Illinois), J. Dooher (Adelphi U.), F. Dryer (Princeton), H. Dwyer (UC-Davis), R. Gelinas (LLL), F. Gouldin (Cornell), D. Hardesty and D. Hartley (Sandia), P. Hooker (Stanford), M. Lapp (GE), S. Marsden (Stanford), A. Prosperetti (Cal. Tech.), M. Ross (LLL), and S. Self (Stanford).

As suggested schematically in Figure 1, combustion modeling involves the mathematical description of the essential chemical and physical processes and boundary conditions which comprise laboratory-scale devices and practical systems such as gas turbine combustors, furnace flames and internal combustion engines. The coordination of "subscale" models of these processes, through a system of conservation laws which may be approximated for numerical solution in large computers is a task which contains many of the elements of physical experimentation. It is generally required that the end result of such an effort should reliably predict the performance, scaling characteristics, efficiency and pollutant emission characteristics of the system, as well as the sensitivity of the predictions to changes in the major variables and fundamental processes. For laboratory-scale investigations, an accurate prediction of the local properties of the flow field is a minimal requirement.

An axiom of the combustion study is the fundamental importance of experimental diagnostics in the formative and final stages of the modeling effort. Certain other biases are evident in the report. In particular, it was accepted that improvement of practical combustion systems will benefit from an understanding of the details of multiphase, multidimensional, nonadiabatic, turbulent flows, in general and as they occur under unique temporal and geometric constraints imposed by particular devices. As suggested in Figure 2, for the example of an internal combustion engine, there exists an inverse correlation between our ability to predict, through modeling, certain aspects of practical systems and the sensitivity of these aspects to local details of the combustion process. There is good reason for this circumstance. Previously, there was little need for consideration of details since only gross properties of combustion systems were of interest - or, indeed, could be measured. As a consequence, relatively crude models sufficed for estimating these properties. Recently, however, we have observed positive reinforcement for more comprehensive combustion modeling. For example, relatively simple models of the combustion process in liquid and solid propellant rockets and gas turbine combustors have been successful in the prediction of performance and exhaust gas composition. Minor empirical modification in the combustion process and fuel-air charge composition in conventional spark-ignition and diesel engines have produced significant changes in emission characteristics and fuel economy (not all of which have been desirable!). Similarly, pollutant emissions and performance in utility boilers and furnaces have responded to empirical adjustment of the burners. Some of these adjustments have been based on predictions derived from improved combustion models.

Accepting the premise that understanding of the details is important for the management of combustion processes, two questions were posed. First, what is it about the current status of combustion science that gives

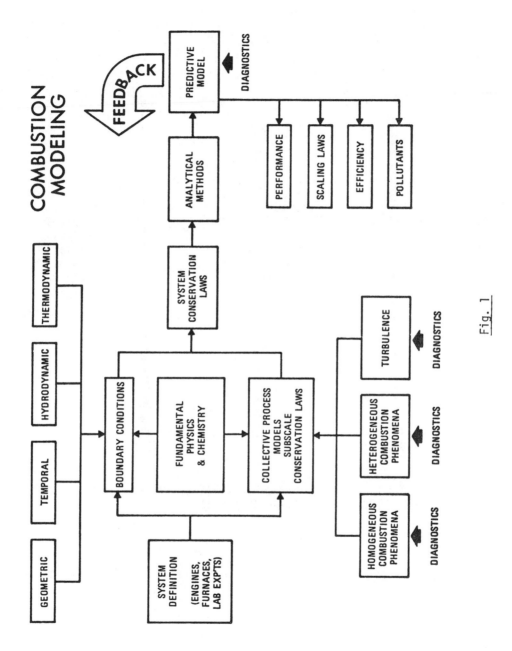

COMBUSTION MODELING

Fig. 1

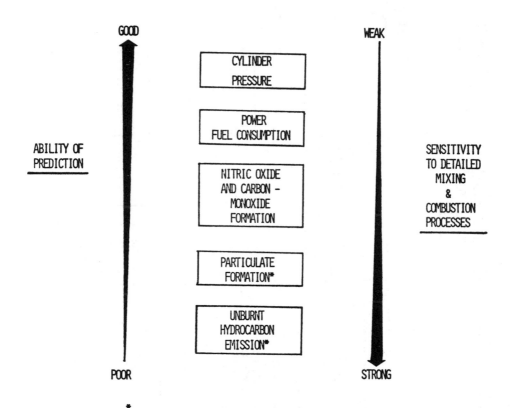

GOOD   WEAK

ABILITY OF
PREDICTION

SENSITIVITY
TO DETAILED
MIXING
&
COMBUSTION
PROCESSES

CYLINDER
PRESSURE

POWER
FUEL CONSUMPTION

NITRIC OXIDE
AND CARBON –
MONOXIDE
FORMATION

PARTICULATE
FORMATION*

UNBURNT
HYDROCARBON
EMISSION*

POOR   STRONG

* UNACCEPTABLE EMISSION LEVELS COULD RESULT FROM
LESS THAN 1 PERCENT OF TOTAL FUEL INPUT.

Fig. 2 Predictive Accuracy vs. Sensitivity to Detailed Description

us an inspired hope for improvement through better modeling? Secondly, which are the key areas where new input and insight (e.g., from physicists) will be most beneficial?

Our response to these questions emphasized the significance of recent and simultaneous developments:

a. In the application of new experimental diagnostic techniques, particularly those based on nonperturbing, laser light scattering phenomena (Doppler velocimetry, Raman, Rayleigh and Mie scattering, and fluorescence), in order to make local, instantaneous measurements of flow field properties.

b. In computational fluid dynamics, that is, in the ability to mathematically describe and efficiently compute complex reacting flows.

c. In the rapidly growing availability of scientific computers which have large storage capacities and, thereby, in the mere capability of processing the equations which describe two - and three - dimensional flows.

The formulation of more detailed mathematical models of laboratory-scale and practical systems requires an understanding of and, at least a semi-quantitative. representation of the subscale processes which compose the "subroutines" for the overall combustion model. Thus, accurate mathematical descriptions are required, e.g., for gas phase and heterogeneous chemical kinetics, for the combustion of fuel droplets in sprays and for the nucleation, agglomeration and oxidation of particulates, such as soot. In addition to the impetus provided by economic and environemental considerations, it is the coincidental development of diagnostic probes to permit the measurement of details and the development of computational schemes and facilities to permit the incorporation of this new information in new models which accounts for the resurgence of interest in detailed combustion modeling.

The answer to the second of our rhetorical questions is not nearly so straightforward. Given our limited ability to formulate precise models of most of the physical elements of combustion systems, it is difficult to accord priority among selected studies or new inputs by physicists. Within the traditional fields of combustion science (which include, for example, the study of processes such as the ignition, propagation and quenching of laminar and turbulent flames, droplet and spray combustion, and the kinetics of the principal energy-releasing and pollutant-forming reactions), a key requirement is for the development of advanced diagnostic and signal processing tools to be applied to laboratory-scale studies and practical system evaluation. In the so-called <u>advanced</u> or <u>novel</u> areas of combustion science, which include very lean and catalytic combustion, fluidized-bed and porous bed combustion, and electromagnetically augmented combustion, there is possibly the greatest likelihood for significant impact by the physicist. This is true because in all of these areas, there is a vacuum created by the nonexistence of unified theories and, to some extent, uncertainty as to the forms of governing conservation equations.

It is in the developing fields of combustion science, where we place such efforts as turbulent flow modeling and the development of numerical techniques to treat reacting compressible flows, that we find the most obvious promise for substantial immediate impact by the physicist through contributions to detailed modeling and diagnostic probe development. It is also in these areas that close interaction among the physicist, the numerical fluid dynamicist and the combustion engineer is most required. In the space that remains, I would like to elaborate on our assessments in these areas.

In view of the fundamental importance of turbulence in all practical combustion systems, a requirement of any overall combustion model is the need to describe quantitatively the interaction between turbulent fluctuations and each subscale process, as well as the influence of turbulence on the system flow field. While the important engineering variables are the time-averaged flow properties, the occurence of turbulent fluctuations fundamentally alters the form of the equations which govern the conservation of mass, momentum, species and energy. In lieu of other information, <u>ad hoc</u> approximations, called closure models, are presently required in order to model terms which appear in the equations, due to turbulence. Relief from this unacceptable situation is hopefully in sight.

We are optimistic that pulsed or chopped laser light sources, which may be tuned in wavelength to avoid spectral interferences, will enable instantaneous and local measurements within turbulent reacting flows. Many of these techniques and their potential for providing the kind of information required by more detailed models were examined in detail[2] and are discussed in these proceedings by Lapp and Hartley[3] and Chigier.[4] Their application to studies of combustion flows should enable measurement of single and joint probability distribution functions of fluctuating properties. It is not unrealistic to envision the direct measurement of one-point averaged correlations (which now must be modeled in reacting flows, with only meager information), such as the Reynolds stress terms, $[\overline{(\rho u_i)' u_j'}]_{,j}$ in the momentum equation; turbulent diffusion terms, $[\overline{(\rho u_i)' \sigma_\alpha'}]_{,i}$ in the species conservation equations; and the turbulent heat conduction terms, $[\overline{(\rho u_i)' h'}]_{,i}$ in the energy equation.*

While many of the new diagnostic techniques remain in a proof of principle stage, as regards their application to combustion flows, clearly the prospect for their rapid development reinforces an expanded effort to mathematically describe and compute unsteady, multidimensional turbulent flows. It was argued that perhaps the most sensible approach is a broad one which would involve heuristic modeling of complex flows, as well as more precise treatment of idealized flows with fewer essential elements requiring separate treatment. By way of example, many of the details of turbulent reacting flows can be examined without recourse to testing in complex practical systems. At the most basic (laboratory) scale, well-designed one- and two-dimensional steady turbulent flows may be examined. Here, measurement of single and joint probability distribution functions may be compared with predictions obtained from numerical solutions to simplified forms of the turbulent Navier-Stokes equations. Results should be systematically compared for incompressible, compressible, and finally, for reacting flows. In all cases, the search should be for universality in the description of the correlation terms. That is, can we model, for example, the distribution functions for various scalars (T, ρ, σ_α) and their correlations, in like fashion for similar flows.

Of particular interest is the possibility of determining the importance of new turbulence production terms in the equations, which arise due to energy release. This question of "flame generated" turbulence has long been the subject of controversy which must be resolved if the governing equations are to be properly posed and solved. It will be important to distinguish between changes which arise due to alteration of the mean flow pattern due to density changes and effects due to some new mechanism of interaction.

In addition to problems of turbulent closure, perhaps the major new difficulties due to physical phenomena, which will appear in more detailed numerical solutions for most combustion flows, will be associated with the use of chemical species equations and strongly temperature-sensitive reaction rates. Detailed treatment of these reactions will considerably increase the total amount of computer time necessary for solution. This

---

*The notation is as follows: subscripts (i,j), spacial coordinates; subscripts (,i), (,j) or (,t), derivative with respect to coordinate (i,j) direction or time, t; ρ, density; u_i, velocity in i'th direction; σ_α, mass fraction of species α; h, stagnation enthalpy; T, temperature; primes, fluctuating components; barred quantities, time-averaged or mean value.

follows from the circumstance that chemical reactions provide the significant energy and species sources and sinks which in turn drive the fluid
dynamic properties of the system. The dynamic properties of the reactive
components, in turn determine the extent of possible kinetic reactions.
The fact that the dynamic and kinetic coupling in combustion systems is
nonlinear gives rise to many time scales which generally are disparate
and variable. The solution of stiff differential equations may be required.
In view of the unhappy realities it will be important to develop compact
kinetic schemes which adequately simulate the detailed chemistry.

A particularly troublesome area is that of describing the interaction
between turbulent fluctuations in the species concentrations and temperature, and the mean and fluctuating components of the source and sink terms
in the species and energy conservation equations. One difficulty is associated with the influence of "unmixedness" on formulation of the time-
averaged reaction (energy release) rate. For reaction to occur, fluctuations of reacting species must essentially be in-phase, whereas the time-
averaged concentrations are insensitive to phase. As an extreme example,
two reacting species in a turbulent diffusion flame may never actually be
simultaneously at a particular location, thus precluding reaction at that
point. However, the time-averaged (mean) concentrations of each of the
two species may be nonzero. Consequently, a local mean energy release
rate simply described in terms of a product of the mean values of the two
species concentrations will also be nonzero when physically, at the point
in question, the rate must be zero.

In fact, the usual procedure of computing a time-averaged reaction
rate term on the basis of time-averages of the instantaneous concentrations,
total density and temperature may also be unacceptable if temperature fluctuations are appreciable or the kinetic rates are very slow. This can be
demonstrated by evaluating the higher order terms in the Taylor expansion
of the temperature-sensitive terms in the reaction rate expression. As
Gouldin[5] has shown, for assumed Arrhenius dependence of the reaction rate,
ω, on the temperature, e.g., according to $T^n \exp(-E/RT)$, (where E is the
activation energy, R is the gas constant and n is a constant) a criterion
for the convergence of the averaged reaction rate term, $\bar{\omega}$, is simply
$\overline{(T')^2}/\bar{T}^2 \ll (E/R\bar{T})^{-2}$, where $\bar{T}$ is the mean temperature and $\overline{(T')^2}$ is the
time average of the square of the temperature fluctuations. It is found
that $\bar{\omega} \sim \bar{T}^n \exp(-E/R\bar{T})$ represents the true time-averaged temperature dependence only for low activation energy, high mean temperature and low
intensity temperature fluctuations.

It is reasonable to assume that numerical computation of idealized
flows will be more readily interpreted than similar analyses for complex
practical geometries. This is important since much of numerical fluid

dynamics remains in an evolutionary state. In fact, Gosman, et al. have
euphemistically termed mathematical models of combustion flows as exer-
cises in the "controlled neglect of reality." It may be fair to say that,
in the near term, detailed numerical treatments of practical systems will
be primarily useful as proving frounds for numerical techniques. This is
an important and perfectly valid justification for the construction of
such models. However, circumspection regarding the results is advised.

On the other hand, for simplified flows, it will be easier to assess
the sensitivity of the results to artifacts of the numerics, to assumptions
about initial and boundary conditions, and to specific models of the physical components of the flow. For example, for the realistic case of finite rate chemical kinetics, the numerical calculation of the equations
governing a steady axisymmetric turbulent diffusion flame can only proceed
after the prescription of suitable initial (boundary) conditions at the
inlet. It is not obvious that this is a trivial exercise, to which the
final results will be insensitive. By way of analogy to the "cold boundary"
difficulty[7] encountered in the calculation of the structure of a laminar
premixed flame, an artificial ignition mechanism must be imposed to start
the steady flame calculation. The sensitivity of the final solution to
the intensity and nature (e.g., high temperature, finite radical concentration, hot gas entrainment) of the numerical pilot or spark must be
determined.

Thus, while combustion modeling will remain, for the foreseeable
future, a mixture of science and art, we may reasonably anticipate a relative increase in the former and an improvement of the latter. Important
inputs from diagnosticians and numerical analysts, physicist or engineer,
is expected.

REFERENCES

1.   American Institute of Physics Proceedings, Volume 25, Efficient
     Energy Use, 1975, American Institute of Physics, New York, 1975.

2.   The Role of Physics in Combustion, Ed. by D. L. Hartley, D. R.
     Hardesty, M. Lapp, J. Dooher and F. Dryer, APS Report PB-242-688, NTIS,
     Springfield, Virginia, 1975.

3.   M. Lapp and D. Hartley, "Scattering Measurement Techniques for
     Combustion Research," this workshop.

4.   N. Chigier, "Laser Velocimetry," this workshop.

5.   F. C. Gouldin, "Role of Turbulent Fluctuations in NO Formation,"
     Comb. Sci. and Tech. 9, 1974.

6.   A. D. Gosman, W. H. Pun, A. K. Runchal, D. B. Spalding and M.
     Wolfshtein, Heat and Mass Transfer in Recirculating Flows, Academic
     Press, London, 1969.

7.   F. A. Williams, Combustion Theory, Addison-Wesley, Palo Alto, p. 109,
     1965.

ELEMENTARY COMBUSTION REACTION KINETICS MEASUREMENTS AT REALISTIC TEMPERATURES

ARTHUR FONTIJN
AeroChem Research Laboratories

I.   INTRODUCTION

For proper understanding, prediction and modelling of combustion kinetics, reliable knowledge of the reaction mechanisms and the rate coefficients of the individual reactions involved are needed. Unfortunately,
the actual flame/combustion environments are usually too complicated to
provide a suitable medium for such measurements. While various techniques
for obtaining some such kinetic rate have been used over the years, these
techniques have only allowed measurements over limited temperature ranges,
which often excluded the actual temperatures of interest in practical combustion systems. Extrapolation and interpolation over various temperature
regimes is often not a reliable procedure since (i) it is becoming increasingly evident that the simple Arrhenius description of the temperature
dependence of rate coefficients, $k(T) \propto \exp(E/RT)$, is unreliable, (ii)
different reactions may dominate at different temperatures, and (iii) systematic errors due to differences in measurement techniques for different
temperature regimes may unduly influence the apparent T-dependences. Thus
techniques that span wide temperature ranges are needed.

II.   THE HTFFR TECHNIQUE

Our recent development [1-3] of high-temperature fast flow reactors
(HTFFR) has opened up the 1000-2000 K regime for study by the same fast-
flow technique used in 100-1000 K experiments. A single technique thus
now can be used to span the 100-2000 K temperature range. A schematic of

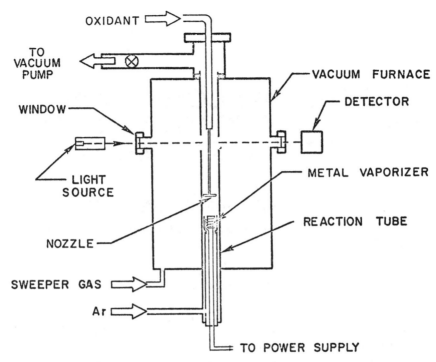

Fig. 1 – Schematic of the AeroChem High–Temperature Fast–Flow Reactor.

the HTFFR, as used in metal atom oxidation studies, is given in Figs. 1
and 2 and details are described in Ref. 1. (Recent improvements are discussed in Ref. 3). The HTFFR consists of a 2.5 cm i.d. alumina reactor
tube contained in a 25 cm i.d., 95 cm long vacuum chamber, Optical observations are made through ports in the reactor and vacuum furnace jacket.
The metal to be studied is vaporized and entrained in a stream of inert
carrier gas. The oxidant (Ox) is introduced into this gas stream at concentrations several orders of magnitude larger than those of the metal
atoms. The reactor is heated using Pt-40% Rh resistance wire; three con-
tiguous, independently controlled, heating zones are used. This has the
advantage that the metal to be vaporized can be placed in the upstream
heating zone which is kept at the vaporization temperature, while the
reaction zone temperature can be independently controlled. For example
in the case of Al, which requires a vaporization temperature of $\sim$ 1700 K,
we have achieved reaction zone temperatures as high as 1900 K and as low
as 300 K. The low temperature is achieved via a recent modification of the
technique by which a 30 cm long HTFFR is used to generate an Al vapor-Ar

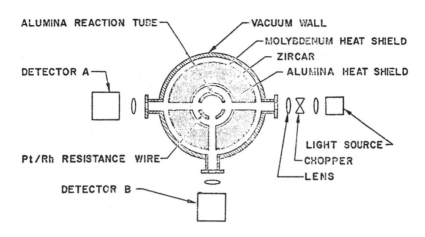

ALUMINA REACTION TUBE

VACUUM WALL

MOLYBDENUM HEAT SHIELD

ZIRCAR

ALUMINA HEAT SHIELD

DETECTOR A

Pt/Rh RESISTANCE WIRE

LIGHT SOURCE

CHOPPER

LENS

DETECTOR B

Fig. 2 – Cross Section of the AeroChem High–Temperature Fast–Flow Reactor
Detector A for metal atom absorption and for chemiluminescence measurements.
Detector B for fluorenscence and for ratio reference measurements.

mixture which then flows through a cooled reactor in which the rate coefficients are measured.

 Rate coefficients, k, are obtained from the observed variation in the
relative metal atom concentration, $[Me]_{rel}$, as a function of oxidizer concentration, [Ox], total pressure, P, and temperature, T, in either of two
ways.[2,3]. The first uses a movable Ox-injection nozzle to vary reaction
distance (time); in this mode, pseudo-first order rate coefficients
([Ox] >> [Me]) are measured from which k is obtained by dividing by [Ox].
Alternatively a time-integrated mode using a fixed Ox-nozzle position and
varying [Ox] may be employed. $[Me]_{rel}$ is measured using a hollow-cathode
resonance line source of the element of interest. In the course of a
series of measurements of a rate coefficient we now employ both absorption
and fluorescence measurements, cf. Fig. 2, of $[Me]_{rel}$, to obtain a large
variation (factor of ~ 100) in the initial (reaction time zero) absolute
metal atom concentration.

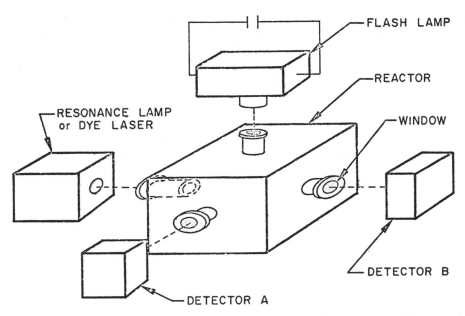

<u>Fig. 3</u> - Schematic of Flash Photolysis Resonance Fluorenscence/Aborption
Apparatus.

Recently the versatility of the technique has been further increased
to allow the study of molecular free radicals in addition to metal atoms,
by the use of laser-induced fluorescence. Specifically, we have studied
the reaction of $A\ell O + O_2 \rightarrow A\ell O_2 + O$ by this technique. [4] The laser provides the much higher intensities needed to measure $[A\ell O]_{rel}$ as compared
to the intensities needed to measure atomic species.

A summary of the rate coefficients obtained thus far with the HTFFR
is given in Table I. These measurements are concerned with the reactions
of refractory metal atom and metal monoxide species for which no suitable
techniques existed at any temperature prior to the HTFFR development. It
is important to extend the technique to measurements of major combustion
species such as hydrocarbons and hydrocarbon free radicals. It would appear that a flow tube reactor is not ideal for such measurements above
1000 K because of the rapidity of pyrolysis and wall reactions, though
HTFFR experiments could certainly be attempted for some reactions. However, another near room-temperature technique for the study of such reactions, the flash photolysis resonance fluorescence/absorption technique
which has been developed in recent years could be adapted to measurements
over the 100-2000 K range using the high-temperature techniques developed
in the HTFFR studies. This photolysis technique makes use of flash photo-

TABLE I. SUMMARY OF HTFFR MEASUREMENTS

Reaction		H kcal mole^{-1}	k ml molecule^{-1} sec^{-1}	Temperature K	Refs.	Remarks
(1) Fe + O$_2$ →	FeO + O	≈ 20	(4 ± 2) x 10^{-13}	1600	2	No discernible T dependence; measurements are now being extended down to ≈ 300°K.
(2) Al + O$_2$ →	AlO + O	-3 ± 1	(3 ± 2) x 10^{-11}	1000-1700	3	
(3) Al + CO$_2$ →	AlO + CO	5 ± 1	(9 ± 3) x 10^{-12}	1500	-	k_3 (T) measurements are to be completed by including ≈ 400 and ≈ 1900°K.
			(1.6 + 0.3) x 10^{-12}	730		
(4) AlO + O$_2$ →	AlO$_2$ + O	≤ 8	(3 ± 2) x 10^{-13}	1400	4	Measurements are now being extended down to ≈300°K.
(5) Sn + N$_2$O →	SnO + N$_2$ + hν	-86 ± 1	≈ 5 x 10^{-13}	1000	5	Reaction studied for its 310-750 nm chemiluminescence which makes it a visible chemical laser pumping candidate; only a limited number of experiments were made to measure k_5.

lytic production of atoms and free radicals in a static or slowly flowing
system, followed by spectroscopic monitoring of their transient behavior
in the presence of added reactants. To date, the reactions of species
such as H, O, OH, Cl, Br, S, N, etc. have been investigated at temperatures from 150 to 500 K. [6-9] A schematic of such an apparatus is shown in
Fig. 3. With some simple modifications this technique can be extended to
measurements of CH_3, CHO, etc. Studies of elementary combustion reactions
at higher temperatures by this technique will be initiated if and when a
funding agency is found to support the work.

ACKNOWLEDGMENTS

The initial development of the HTFFR was made possible by contracts
from Sandia Laboratories. I am grateful to them and to the agencies, quoted in Table I, which have supported the studies of the individual reactions. My principal collaborators in these studies, without whom these
developments would not have been possible, are Dr. W. Felder, Dr. S. C.
Kurzius and J. J. Houghton and more recently also R. Ellison and R. Revo-
linski.

REFERENCES

1.   Fontijn, A., Kurzius, S.C., Houghton, J.J. and Emerson, J.A.
     "Tubular Fast-Flow Reactor for High Temperature Gas Kinetic Studies,"
     Rev. Sci. Instr. 43, 726 (1972).

2.   Fontijn, A., Kurzius, S.C. and Houghton, J.J., "Fast Flow Reactor
     Studies of Metal Atom Oxidation Kinetics," in Fourteenth Symposium
     (International) on Combustion (The Combustion Institute, Pittsburgh,
     1973), p. 167.

3.   Fontijn, A., Felder, W. and Houghton, J.J., "Homogenous and Hetero-
     geneous Kinetics of the Atomic Al/O_2 Reaction in the 1000-1700 K
     Range," in Fifteenth Symposium (International) on Combustion (The
     Combustion Institute, Pittsburgh, 1975), p. 775.

4.   Felder, W. and Fontijn, A., "High-Temperature Fast-Flow Reactor Kinetic
     Studies. The AlO/O_2 Reaction Near 1400 K, J. Chem. Phys. 64, 1977 (1976).

5.   Felder, W. and Fontijn, A., "High-Temperature Fast-Flow Reactor Study
     of Sn/N_2O Chemiluminescence, Chem. Phys. Lett. 34, 398 (1975).

6. Braun, W. and Lenzi, M., "Resonance Fluorescence Method for Kinetics of Atomic Reactions," Disc. Faraday Soc. 44, 752 (1967).

7. Kurylo, M.J., Peterson, N.C., and Braun, W., "Absolute Rates of the Reactions H + C_2H_4 and H + C_2H_3," J. Chem. Phys. 53, 2776 (1970).

8. Kurylo, M.J., Peterson, N.C., and Braun, W., "Temperature and Pressure Effects in the Addition of H Atoms to Propylene," J. Chem. Phys. 54, 4662 (1971).

9. Davis, D.D., Huie, R.E., Herron, J.T., Kurylo, M.J., and Braun, W., "Absolute Rate Constants for the Kinetics of Atomic Oxygen with Ethylene over the Temperature Range 232-500 K," J., Chem. Phys. 56, 4868 (1972).

SOME CONCLUSIONS REGARDING COMBUSTION MEASUREMENTS REACHED AT THE SAI/NSF(RANN) WORKSHOP ON THE NUMERICAL SIMULATION OF COMBUSTION

A. A. BONI
Science Applications, Inc.

Numerical simulation, which involves finite difference solutions of the governing conservation equations in conjunction with appropriate physical models to account for disparate length and time scale phenomena, can be used effectively to provide detailed space-and time-resolved information of the parameters of interest for chemically reacting, multi-dimensional flows. This approach will afford considerably more information related to the operating principles, operating characteristics and limitations of air-breathing propulsion systems than currently available models. While such an approach has been used extensively in fields such as plasma physics, it has just begun to evolve in the field of combustion research. One reason for this state is the lack of well-defined, well-controled and fully-instrumented experiments to elucidate the fundamental mechanisms, and to allow the construction of accurate sub-grid scale models. Thus, while numerical simulation has much potential, there are many uncertainties in the physical models (turbulence, atomization, spray dynamics) and physical/chemical data base (chemical rates and mechanisms) as well as in the computational techniques themselves which must be clarified before any numerical simulation (computer) code can be used to perform meaningful numerical experiments of an entire air-breathing combustion device. Consequently, there exists a need for concurrent experimental programs in combustion research to provide the data required to construct and validate numerical models.

The recent SAI/NSF(RANN) workshop on the numerical simulation of combustion was held with the principal purpose of establishing the current state-of-the-art in the areas related to combustion modeling.

During the workshop, discussion centered around the following issues:

   a.   which experimental and analytical/numerical
        techniques are currently available and appli-
        cable?

   b.   what is the current state of the physical and
        chemical model/data base?

   c.   what further advances are required?

   d.   what are the most productive approaches for
        achieving these advances?

   e.   which combined numerical/theoretical pro-
        grams are required to develop and validate
        numerical simulaion models over the next
        several years?

        General conclusions formulated as a result of the workshop indicated
that currently available finite-difference computer codes in conjunction
with two-and three-equation turbulence models, quasi-global chemical models
and fuel-spray droplet vaporization models could be fruitfully employed to
elucidate many aspects of air-breathing combustion phenomenology, but that
full-scale simulation is still several years away. Programs to accomplish
such work should be funded immediately. However, the accuracy and general
applicability of such an undertaking is currently limited by the assumptions
and validity of the sub-grid scale models, which in many circumstances are
unknown. Therefore, it was concluded that such research efforts should be
undertaken, but not without concurrent and well-coordinated experimental
programs. We indicate in Table I, a partial list of computer simulation
models which have been applied or are applicable to combustion R&D.

        Since the purpose of the present workshop is to focus on the experi-
mental and instrumentation state-of-the-art and research needs in combustion R&D, we present herein the principal conclusions related to those developments in instrumentation and experimentation required to be performed
in support of a numerical simulation program. Numerical simulation will
result in detailed space-and time-resolved predictions of temperature, pressure, species concentrations, turbulent intensity and turbulent length
scale throughout the combustion cavity. Thus, the diagnostic techniques
must be capable of yielding unambiguous space-and time-resolved information
as well. Since non-perturbing measurements in a severe environment are required, optical diagnostic techniques, many based on laser-scattering principles seem to be best suited.

TABLE I.

PARTIAL SUMMARY OF FINITE-DIFFERENCE COMPUTER SIMULATION

MODELS FOR AIR-BREATHING PROPULSION APPLICATIONS

Flow Configuration	Application	Status of Fluid Flow Solutions	Status of Auxiliary Models				Comments
			Turbulence	Chemistry	Radiation	Droplets/Particles	
1-D, time-dependent 1-D variable area, time-dependent	Internal Combustion Engines	Current capability, all speeds	2 & 3 equation models	Global oxidation kinetics, finite rate pollutant kinetics	zonal methods	Discrete size distribution, single particle convective augmentation	Sirignano, Bracco (Princeton (method of quasilinearization) Boni and Chapman (arbitrary Lagrangian-Eulerian)
2-D boundary layer	Jet flames	Current capability	2 & 3 equation models	Equilibrium HC, pollutant kinetics	directional flux approximations	same as above	Patankar and Spalding, Many others
2-D axisymmetric, recirculating flows	Gas turbine Combustors	Advanced Development	2 & 3 equation models, swirl in advanced development	same as above	same as above	same as above	McDonald (UARL) Gosman & Spalding
3-D recirculating flows	Furnaces, Gas turbines	Advanced Development	—	—	—	—	McDonald (UARL) (just operable), Spalding (Imperial College)
2-D, time-dependent compressible	Stratified Charge and Diesel Engines	In development					Boni and Chapman (SAI) (used previously for fireball calculations with simplified chemistry and turbulence). Work at Sandia & Livermore
2-D & 3-D time-dependent compressible	Fireballs, Laser Interactions, internal ballistics, chemical lasers	Operational	2-equation models	—	directional flux approximations, spectral	—	Used extensively by Hirt, et al., at LASL Simulation movies impressive

It was concluded that the only viable approach to obtaining a useful, multi-dimensional simulation code would be for <u>parallel, interactive</u> theoretical and experimental programs to proceed. The <u>experimental approach would yield data</u> that is necessary as input parameters to the computer codes, <u>as well as to verify the ultimate outputs or predictions obtained</u> from the <u>numerical experiments</u> themselves. It was recognized, furthermore, that both the numerical simulations themselves as well as the optical diagnostic techniques must undergo considerable developments before their results can be interpreted unambiguously. Accordingly, we concluded that an <u>evolutionary approach</u>, beginning with rudimentary <u>model experiments</u> should be followed. The model experiments, which may utilize steady state laminar flames, turbulent diffusion flames, rapid compression machines, stratified charge bombs, spray experiments, etc., serve two distinct purposes; viz,

> they serve as means for the development and validation of advanced diagnostic techniques

> they are designed to isolate and study specific processes without the difficulties associated with introducing several unknown and interrelated processes simultaneously.

The importance of continued dialogue with feedback loops between the experimental and analytical communities cannot be overemphasized.

For the purpose of illustration, we indicate one such experiment in Appendix A. In Table II, we list the information required to be obtained from such experiments and the various experimental/diagnostic methodologies which may be employed.

The workshop discussions centered explicitly on the advanced, laser-based diagnostic techniques which appear to have great hope for providing the required in-situ measurements within the next 5 years. However, it was implicitly recognized that more "conventional" diagnostics such as infrared and visible spectroscopy, pressure measurements, ionization probes, sampling probes, high-speed photography, etc., must also play a role. For instance, pressure, flame front location, and general characterization of the flow field can be obtained by use of such "conventional" techniques. For temperature, density, species concentrations, turbulence levels, and turbulence correlations, laser-based techniques appear to be necessary.

Laser velocimetry appears to be in sufficiently advanced state to provide the required turbulence fluctuations. For highly transient phenomena where temporal resolution of 10-100μ sec is necessary, complications arise since multiple measurements are necessary to provide adequate statistics. Trade-offs between sampling times and acceptable accuracy can and should be performed.

TABLE II.
MEASUREMENT NEEDS

PARAMETER	RANGE	APPLICABLE METHODS	COMMENTS
Temperature	$300 \text{ K} \leq T \leq 3000 \text{ K}$	Raman scattering, thermocouples, relative intensity, line reversal	Thermocouples not applicable for automotive applications and at high temperature range. Optical techniques must employ cross-beam or utilize inversion methods.
Pressure	$5 \text{ psi} \leq T \leq 1000 \text{ psi}$	Pressure transducers, inference from T and ρ via equation of state	Pressure response in engine mount must be determined
Density	—	Raman scattering, laser fluorescence, interferometry, emission/absorption	—
Flame front location	—	High speed photography, laser source, incoherent source, self-illumination	
General characterization of the flow field	—	same as above	—
Droplet atomization, droplet size and number density distribution	$1\mu < d < 100\mu$	Holography, laser elastic scattering	Burning and non-burning environments
Species concentration	variable: major species 1-30% minor species 1-3000 ppm	Raman scattering, optical spectroscopy (possibly cross beam) sample probes	HC (differentiation if possible), HCN, CN, N, N_2, NH, NH, NH_2, NO, NO_2, NO_3, SO, SO_2, SO_3, O, O_2, OH, H_2O, C, C_2, CH, CH_2
Flow velocity	subsonic	laser velocimetry, hot wire	—
Turbulent fluctuations	Depends on application	hot wire, laser velocimetry	u_i', ρ', T', y_α', should be available now, signal processing software, and repeatable experiments needed for cyclic systems
Turbulent correlations	Depends on application	laser velocimetry in conjunction with Raman scattering	$\overline{u_i'u_j'}$, $\overline{u_j'y_\alpha}$, $\overline{u_i'T'}$, not yet done, requires resolution of the particle-induced "noise" on the Raman signal problem.

Regarding temperature and concentration measurements in cyclic combustion systems, spontaneous, vibrational Raman scattering should be sufficiently advanced to provide temperature and major species concentrations. Existing lasers indicate that time resolution of about 100 μsec is possible. For instance, ruby lasers operating in the simmer mode appear to be useful. It was indicated, however, that many of the new high-powered, pulsed visible and UV lasers being designed for fusion and isotope separation applications may result in high-energy systems suitable for application to Raman scattering experiments with microsecond resolution and minor species detection capabilities.

In any event, other than vibrational Raman scattering which is presently limited to major species, laser fluorescence may be able to provide minor species detection. However, the absolute quenching cross-sections and pressure broadening parameters for many species of interest are currently unknown. Coherent Anti-Stokes Raman scattering has the potential for better detectability, but is complicated by uncertainties due to nonlinearities associated with beam focusing and turbulent interactions.

In addition to the need for the evolution of the LDV and Raman scattering techniques individually, there also exists a serious need for combined Raman/LDV systems to obtain the turbulence correlations associated with mass and energy transport. Raman interferences from scatter-induced fluorescence required by the LDV system need to be circumvented. Additionally, holographic techniques appear to be well suited to studies of particle and droplet dynamics and jet atomization processes.

With the difficulties associated with the Raman schemes, it appears that the use of line-of-sight emission/absorption techniques, extended to two-beam, cross-correlation type systems may be worth pursuing. Also, such line-of-sight techniques used in conjunction with inversion procedures could as well be useful.

RECOMMENDATIONS REGARDING INSTRUMENTATION

As a result of the workshop and subsequent discussions with several of the participants, we recommend the following diagnostic developments and/or applications for combustion R&D. We have not listed the recommendations according to priorities at the present time.

1. Inelastic and Elastic Scattering from Combustion
    Gases

        experimental studies of pre-flame, post-flame
        zone regions

experimental studies of species within boundary layers

techniques to improve sensitivity

search for techniques to eliminate temperature
effects, characterization of temperature effects
and cross-sections

techniques to eliminate broad-band fluorescence
in Raman experiments

theoretical analysis to aid interpretation of
Raman spectra, particularly free radical parameters

"continuous", repetitively pulsed measurements -
particularly for application to transient phenomena

investigations with high-energy, short pulse
duration lasers

studies to determine applicability of coherent
Anti-Stokes Raman scattering to turbulent
media

2.   Laser-induced Fluorescence

application to combustion systems

determination of quenching and pressure-
broadening parameters

3.   Line-of-sight Techniques, either Source or Receiver
with Spectral Resolution

inversion techniques

cross-beam application

4.   Laser-velocimetry

application to combustion systems

improved techniques for transient system
applications

5. Combined Raman/laser-velocimetry Developments
   and Applications

   techniques to eliminate signal interferences

   gating problems

6. Holography

   application to combustion systems

   improved data reduction procedures

7. Application of "conventional" techniques, such as
   hot-wire, ionization gages, sampling probes, etc.
   in concert with advanced techniques for development
   and validation

Finally, it was the general consensus of the workshop participants that there is a serious and urgent need that a well-coordinated federal program be initiated to support the basic and development studies such as those outlined above. It should be recognized from the outset that a minimum commitment of approximately 5 years of intensive research, with adequate funding will be required. Without such a commitment, we will not be in a position within the next 5 years to impact design modifications or new technologies necessary to be implemented approximately 10 years from now.

APPENDIX A - PILOT MODEL EXPERIMENT

During the workshop session, an attempt to outline a representative pilot experiment was made. Since cyclic combustion was of particular interest, a single-stroke, rapid-compression machine was selected as the test bed. It is recognized that a current experimental/numerical program constructed along similar, but not identical lines is in progress at the University of California, Berkeley under the direction of Professors R.F. Sawyer and A.K. Oppenheim.

The test bed is a single-stroke rapid-compression machine of rectangular cross-section. The walls of the channel are constructed of optical quality windows to permit almost unlimited diagnostic access.

The goal of the experiment is to measure the temporal and spatial distribution of pressure, density, velocity, turbulence level, temperature and species concentrations for a simple fuel/oxidizer combination without the complications imposed by two-phase mixtures. As such, we are particularly interested in the fluid mechanical and turbulence aspects of the cyclic combustion problem. Numerical simulation of such an experiment is considered to be possible within the same period of time required to develop the diagnostics and de-bug the experimental intricacies. Accordingly, both numerical and physical experiments could proceed concurrently.

Because a validated global model for a complex hydrocarbon fuel does not exist yet, a more simplified fuel must be selected. Of the two choices, CH_4/Air and $CO/O_2 + H_2$), the latter mixture is preferred because it represents the "actual" mixture better than methane. Also the kinetic rates and mechanisms are well characterized. For the flow field visualization, high-speed photography and/or holography was considered.

For the measurement of the pressure, pressure gauges in the wall and piston head were deemed to be sufficient.

The entire event has a time duration of 20-40 msec and the time resolution of 1 msec was considered adequate.

For the measurement of v and v' with a spatial resolution of 1 cm, the LDV system was selected. Difficulties associated with adequate sample collection from a statistical point of view could be a problem with the single stroke device. Multiple-stroke operation is to be preferred, although cycle to cycle variations may pose a problem.

For the measurement of the temperature and species concentration, Raman scattering, coherent Raman anti-Stoke scattering, fluorescence and emission/absorption spectroscopy were considered.

In a subsequent sensitivity analysis, the following tentative conclusions were drawn:

The high power required in Raman scattering leads to the application of a 100 mJ ruby laser with a pulse length of about 100 μsec. However, the ruby laser has a repetition rate of only $\simeq$ 1 pps, thus only one data point per event (single-stroke) can be recorded. Operation in the simmer mode may be a possible approach to obtaining multiple data points for a single shot.

Another question is the interference by the UV and visible radiation from the hot chemical species present.

This particular experiment represents one possible approach to obtaining an improved understanding of the fluid dynamics, turbulence and interplay with reaction kinetics as experienced in cyclic combustion systems. The performance of such experiments, while not full-scale simulations, wtill represents an advancement in the state-of-the-art in combustion R&D. Other approaches should also be tried. After this experiment, switching to an alternate fuel with more complex kinetics could be tried to validate chemical models with fluid mechanical effects.

Combustor Design and Phenomenology

PHENOMENOLOGY AND DESIGN OF JET ENGINE COMBUSTORS

RICHARD ROBERTS

Pratt & Whitney Division, United Technologies Corporation

This presentation addresses the design of gas turbine engine main combustors, with emphasis on how the measurement techniques which are the subject of this workshop relate to combustor design problems. The title of this presentation refers to our attempts to understand the physical and chemical processes within the combustor, to express these phenomena by means of mathematical models, and to thereby relate combustor performance to the various design variables. The focus of the presentation will be an analytical combustor model which has been used as a design tool. The model was developed as part of the Air Force Low Power Turbopropulsion Combustor Exhaust Emissions Contract, F33615-71-C-1870, now complete, and is described in greater detail in Reference 1. It is my intention in describing this modeling work to illustrate both the inadequacies in our understanding of turbulent combustion which might be remedied by the existence of appropriate measurements, and to suggest ways in which an analytical combustor model might be used in concert with detailed internal measurements to better identify the controlling processes.

A distinction is necessary between modeling for design purposes and modeling in the perhaps purer sense, where it is desirable to reduce the analysis to the most basic level. First and foremost, a successful design system must come up with a prediction. As is summarized in Figure 1, a successful design tool must relate predicted performance to hardware parameters, such as liner contour, air hole sizes, swirl strength, fuel injection parameters, etc. It is additionally desirable that the computation be performed in a relatively short time (order of a few minutes). Given the requirement to provide a prediction, and the fact that we have yet to formulate a truly adequate turbulent combustion model, premixed or otherwise, the formulation of a design system necessarily involves a number of assumptions. It is my intention here to illustrate the logic that has gone into this model rather than dwell on the virtues of the model itself. In particular, the critical assumptions will be identified. When compared to previously existing design procedures, the

MATHEMATICAL EMISSIONS MODELING

OBJECTIVES:

- IDENTIFY EMISSIONS FORMATION MECHANISMS
- ANALYTICALLY INVESTIGATE EMISSIONS REDUCTION TECHNIQUES
- ASSIST IN HARDWARE DESIGN AND DEVELOPMENT

INPUT VARIABLES

- BURNER GEOMETRY
- INLET PRESSURE, TEMPERATURE
- AIRFLOW DISTRIBUTION
- FUEL FLOW, PROPERTIES
- FUEL INJECTOR CHARACTERISTICS

OUTPUT VARIABLES

- GAS TEMPERATURE, PRESSURE, VELOCITY
- FRACTION FUEL BURNED
- CHEMICAL COMPOSITION

VS TIME AND POSITION

Fig. 1

BUILDING BLOCK APPROACH TO COMBUSTOR MODELING

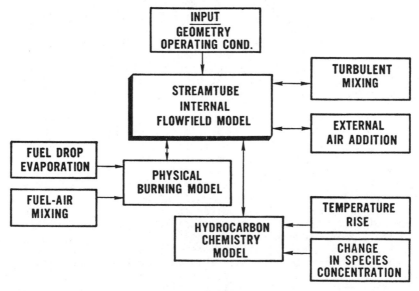

Fig. 2

primary improvement is that empiricism of a gross type has been replaced with empiricism at a more fundamental level.

The number and description of submodels which comprise the combustor model is illustrated in Figure 2. Implicit in this arrangement is the assumption that these submodels indeed represent the controlling phenomena, and that the various submodels can be uncoupled. This approach to model formulation allows a particular submodel to be removed, modified and reinserted without major revision to the model as a whole. The streamtube internal flowfield submodel is the heart of the formulation, defining the physical time-space continuum upon which the physical combustion and chemical reaction models are mounted.

The classic swirl-stabilized combustor internal flowfield is shown in Figure 3 (Ref. 2). Features of this flowfield are reversed flow in the primary zone and introduction of discrete dilution air jets. Essentially all combustors employ some variation on this theme. It is this flowfield which must be modeled in sufficient detail to predict the fundamentally different nature of the flow in various parts of the combustor. A simple one-dimensional plug flow reactor will not provide sufficient detail. An ensemble of stirred reactors is probably valid from a phenomenological point of view, but suffers the major weakness (from our point of view) of being difficult to relate to the geometric combustor configuration. The streamtube arrangement utilized in this model for can and annular combustors is shown in Figure 4. The recirculation zone is positioned at the combustor headplate in all cases. In this manner, an arrangement of one-dimensional, axisymmetric streamtubes can be made to represent a two-dimensional flowfield. Other combutor configurations can be treated by superposition of these two flowfield models.

Recirculation zone contour, and implicily exchange rate with the adjacent streamtube, is determined from the empirical information summarized in Figure 5. A similar set of relationships is used to define non-swirl stabilized recirculation. By the manner in which the model is formulated, the recirculation zone serves as a heat source to initiate combustion in the outer streamtubes. It is currently assumed, for non-premixed systems, that the recirculating mixture is at stoichiometric equivalence ratio and that chemical species transferred out of the zone are at equilibrium concentration (including NO).

The steady-state streamtube conservation equations are listed in Figure 6. Angular momentum and radial equilibrium equations are written only for can-type combustors. A swirl velocity component about the combustor centerline is not treated for annular combustors. In the calculation procedure, the energy equation is replaced by a set of equations governing conservation of reacted fuel and chemical species of interest. These equations are written for each streamtube and solved via matrix inversion. Auxiliary calculations for streamtube transfer, fuel droplet evaporation, dilution air jet entrainment, hydrocarbon thermochemistry, etc., are performed external to the matrix. The streamtube calculation procedure is an initial value, Runga-Kutta type integration down the length of the combustor.

PRIMARY ZONE FLOW PATTERN OBSERVED
IN A CAN-TYPE COMBUSTION CHAMBER

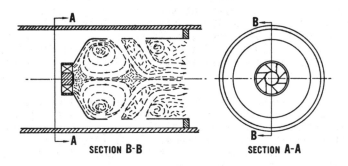

SECTION B-B   SECTION A-A

Fig. 3

STREAMTUBE SCHEMATIC

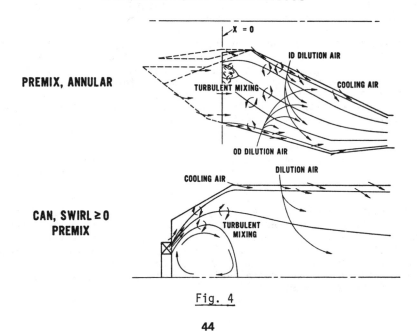

PREMIX, ANNULAR

CAN, SWIRL ≥ 0
PREMIX

Fig. 4

SWIRL-STABILIZED RECIRCULATION ZONE MODEL

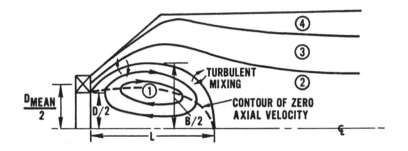

COLD FLOW
RECIRCULATION ZONE DIMENSIONS

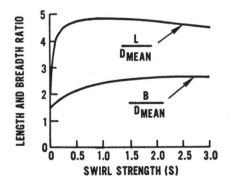

TEMPERATURE CORRECTION FACTORS

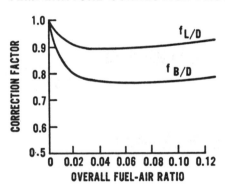

Fig. 5

45

STREAMTUBE CONSERVATION EQUATIONS

MASS

$$\frac{d}{dx}\left(\rho_j\, U_j\, A_j\right) = \frac{dw_c}{dx}\bigg|_j$$

LINEAR MOMENTUM

$$\frac{d}{dx}\left(\rho_j\, U_j^2\, A_j\right) = -\frac{d}{dx}\left(P_j A_j\right) + 2\pi\, P_{j+1}\, R_j\, \frac{dR_j}{dx} - 2\pi\, P_{j-1} R_{j-1}\, \frac{dR_{j-1}}{dx}$$

$$+ \frac{dw_t}{dx}\bigg|_{j-1,j}\left(U_{j-1} - U_j\right) + \frac{dw_t}{dx}\bigg|_{j,j+1}\left(U_{j+1} - U_j\right)$$

ENERGY

$$\frac{d}{dx}\left(\rho_j\, U_j\, A_j\, C_{p_j}\, T_{o_j}\right) = \rho_j\, U_j\, A_j\, \dot{Q}_j + \frac{dw_t}{dx}\bigg|_{j-1,j}\left(C_{p_{j-1}}\, T_{o_{j-1}} - C_{p_j}\, T_{o_j}\right)$$

$$+ \frac{dw_t}{dx}\bigg|_{j,j+1}\left(C_{p_{j+1}}\, T_{o_{j+1}} - C_{p_j}\, T_{o_j}\right)$$

ANGULAR MOMENTUM

$$\frac{d}{dx}\, \rho_j\, U_j\, A_j\, V_j\left[\frac{(R_j + R_{j-1})}{2}\right] = \frac{dw_t}{dx}\bigg|_{j-1,j}\left(V_{j-1}\, \frac{(R_{j-1} + R_{j-2})}{2} - V_j\, \frac{(R_j + R_{j-1})}{2}\right)$$

$$+ \frac{dw_t}{dx}\bigg|_{j,j+1}\left(V_{j+1}\, \frac{(R_{j+1} + R_j)}{2} - V_j\, \frac{(R_j + R_{j-1})}{2}\right)$$

RADIAL EQUILIBRIUM

$$\frac{d}{dx}\left(P_j - P_{j-1}\right) = \frac{d}{dx}\left[\rho_j\, V_j^2\left(\frac{R_j^2}{R_j^2 - R_{j-1}^2}\, \ln\left(\frac{R_j}{R_{j-1}}\right) - \frac{1}{2}\right)\right]$$

Fig. 6

46

The turbulent mixing model, described in Figure 7, is a very simple expression for mass exchange between adjacent streamtubes based on centerline velocity difference. Implicit is the assumption that the streamtube flow is well ordered and that exchange occurs in a uniform manner, at least on a scale of the combustor characteristic dimension. Recent measurements (Ref. 3) indicate that the turbulent flowfield exhibits a periodic nature, which is obviously not reflected in this model. Experimental confirmation of this fact will, it is hoped, include a determination of whether this behavior can be treated as a sequence of quasi-steady states or whether all phenomena are coupled in a time varying manner. The indicated value of eddy viscosity, including the dependence on combustor inlet temperature, is based on an empirical fit to a body of P&WA combustor emissions data.

The transverse dilution air jet mixing model includes empirical submodels for jet penetration, based on hole size, combustor liner pressure drop, etc., and for rate of mixing of jet air into the combustor internal flow. The expression relating dilution air addition rate to jet parameters is given in Figure 8. As a consequence of the one-dimensional nature of each streamtube, azimuthal variation in dilution rate is not accounted for.

The key assumptions of the physical combustion model are presented in Figure 9. For turbulent mass diffusion flames, where liquid fuel enters the combustor without substantial premixing, it is assumed that vaporization rate controls the rate at which fuel is made available for combustion, and that the turbulent diffusion rate acts to ensure that combustion occurs at stoichiometric mixture strength. The so-called stoichiometric assumption is relaxed when the deficient component, either vapor fuel or air, is exhausted within a particular streamtube. Subsequent reaction occurs at bulk streamtube conditions. For premixed systems, it is recognized that turbulent diffusion (temperature, species) controls the reaction rate rather than hydrocarbon kinetics. An empirical reaction rate expression is employed until the parent fuel species is depleted.

The evaporation rate expression presented in Figure 10 is used to predict the availability of vapor fuel (Ref. 4).

Kinetic hydrocarbon oxidation models of the type shown in Figure 11 have been developed by a number of investigators. This prediction model reflects P&WA thinking, with acknowledgement to the work of Edelman and Fortune (Ref. 5). Three principal events are postulated to occur, separated by the dashed lines:

1. ignition delay;
2. rate and magnitude of heat release;
3. final equilibration of CO.

While simultaneous solution of this reaction system is practical for simulation of shock tube and plug-flow reactor processes, it is not applicable to a model of this type for computational reasons. The approach taken in the formulation of this model has been to compute reaction profiles using the above kinetic system for a plug flow reactor over a range of inlet pressure, tempera-

TURBULENT MIXING MODEL

$$\tau = \text{REYNOLDS SHEAR STRESS} = \frac{1}{A} \, (U_j - U_{j-1}) \, W_t$$

$$\tau = \mu_t \, \frac{\partial U}{\partial R} \sim \mu_t \, \frac{(U_j - U_{j-1})}{(r_j - r_{j-1})}$$

FROM EXPERIMENTAL TURBULENT JET MIXING DATA:

$$\mu_t = \rho K b \, |U_{max} - U_{min}|$$

$$\frac{dW_t}{dx} \bigg|_{j-1,j} = 2\pi R_{j-1} \ \text{EDMU} \ | \ \rho_j \, (U_j^2 + V_j^2)^{\frac{1}{2}} - \rho_{j-1} \, (U_{j-1}^2 + V_{j-1}^2)^{\frac{1}{2}}$$

$$\text{EDMU} = 0.00684e^{-0.0032432 \, T_{t4}}$$

Fig. 7

TRANSVERSE JET MIXING MODEL

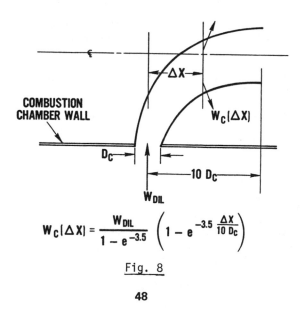

$$W_c(\Delta X) = \frac{W_{DIL}}{1 - e^{-3.5}} \left(1 - e^{-3.5 \frac{\Delta X}{10 \, D_c}} \right)$$

Fig. 8

48

KEY ASSUMPTIONS OF COMBUSTION MODEL

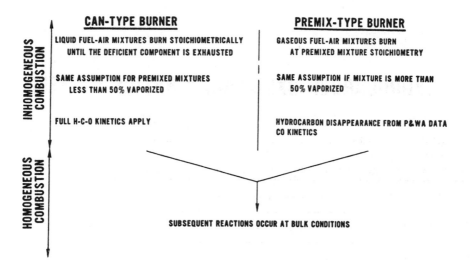

CAN-TYPE BURNER

LIQUID FUEL-AIR MIXTURES BURN STOICHIOMETRICALLY
UNTIL THE DEFICIENT COMPONENT IS EXHAUSTED

SAME ASSUMPTION FOR PREMIXED MIXTURES
LESS THAN 50% VAPORIZED

FULL H-C-O KINETICS APPLY

PREMIX-TYPE BURNER

GASEOUS FUEL-AIR MIXTURES BURN
AT PREMIXED MIXTURE STOICHIOMETRY

SAME ASSUMPTION IF MIXTURE IS MORE THAN
50% VAPORIZED

HYDROCARBON DISAPPEARANCE FROM P&WA DATA
CO KINETICS

INHOMOGENEOUS COMBUSTION

HOMOGENEOUS COMBUSTION

SUBSEQUENT REACTIONS OCCUR AT BULK CONDITIONS

Fig. 9

PHYSICAL COMBUSTION MODEL

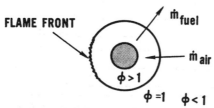

FLAME FRONT

$\dot{m}_{fuel}$

$\dot{m}_{air}$

$\phi > 1$

$\phi = 1$ $\phi < 1$

FROM STATIONARY DROPLET BURNING DATA:

$$\dot{m}_f = \frac{2\pi D_d \lambda g}{C_{Pg}} \ln \left[\frac{1 + QY_{02}}{Li} - \frac{C_{Pg}}{L} \left(T_l - T_g \right) \right]$$

MULTIPLIED BY:

$$\left[1.0 + 0.276 \, Re^{1/2} \, Pr^{1/3} \right]$$

TO ACCOUNT FOR CONVECTION

Fig. 10

49

HYDROCARBON OXIDATION KINETICS SYSTEM

Reaction	Rate Constant

1) $C_8H_{16} + O_2 = 2C_4H_8O$ $\quad k_1 = 7.5 \times 10^6 \ T^{1.5} \ e^{-7900/T}$

2) $C_4H_8O + O_2 = HO_2 + CO + CH_3 + C_2H_4$ $\quad k_2 = 10^{11} \ T^{1.5} \ e^{-10000/T}$

3) $C_8H_{16} + OH = H_2CO + CH_3 + 3C_2H_4$ $\quad k_3 = 3 \times 10^{10} \ T \ e^{-4500/T}$

4) $CH_3 + O = H_2CO + H$ $\quad k_4 = 2 \times 10^{13}$

5) $CH_3 + O_2 = H_2CO + OH$ $\quad k_5 = 10^{12}$

6) $H_2CO + OH = H_2O + CO + H$ $\quad k_6 = 10^{14} \ e^{-4000/T}$

7) $C_2H_4 + O_2 = 2H_2CO$ $\quad k_7 = 3 \times 10^2 \ T^{2.5}$

8) $C_2H_4 + OH = CH_3 + H_2CO$ $\quad k_8 = 5 \times 10^{13} \ e^{-3000/T}$

9) $CH_3 + H_2 = CH_4 + H$ $\quad k_9 = 6 \times 10^{11} \ e^{-5500/T}$

10) $C_2H_4 = C_2H_2 + H_2$ $\quad k_{10} = 7 \times 10^8 \ e^{-23250/T}$

11) $C_2H_2 + OH = CH_3 + CO$ $\quad k_{11} = 10^{13} \ e^{-3500/T}$

12) $2H + M = H_2 + M$ $\quad k_{12} = 2 \times 10^{18} \ T^{-1}$

13) $2O + M = O_2 + M$ $\quad k_{13} = 10^{17} \ T^{-1}$

14) $OH + H + M = H_2O + M$ $\quad k_{14} = 7 \times 10^{19} \ T^{-1}$

15) $H + O_2 = OH + O$ $\quad k_{15} = 2.24 \times 10^{14} \ e^{-8400/T}$

16) $O + H_2 = OH + H$ $\quad k_{16} = 1.74 \times 10^{13} \ e^{-4730/T}$

17) $H + H_2O = H_2 + OH$ $\quad k_{17} = 8.41 \times 10^{13} \ e^{-10050/T}$

18) $O + H_2O = 2OH$ $\quad k_{18} = 5.75 \times 10^{13} \ e^{-9000/T}$

19) $HO_2 + M = H + O_2 + M$ $\quad k_{19} = 2.4 \times 10^{15} \ e^{-22950/T}$

20) $HO_2 + H = 2OH$ $\quad k_{20} = 6 \times 10^{13}$

21) $CO + OH = CO_2 + H$ $\quad k_{21} = 5.6 \times 10^{11} \ e^{-540/T}$

Fig. 11

ture and equivalence ratio. A reduced hydrocarbon oxidation system is then
fitted to the predicted reaction profiles to duplicate the behavior of interest. A representative reaction profile for the plug flow reactor is shown on
Figure 12. For the purposes of our model, the ignition behavior and final
equilibration aspects are of most importance. The latter is particularly
important in the presence of CO and hydrocarbon "quenching" due to over-rapid
dilution air addition. The rapid temperature rise stage occurs on a time scale
which is generally an order of magnitude less than the reaction zone residence
time.

   The reduced hydrocarbon oxidation system is described in Figure 13. It
is based on the definition of a partial equilibrium state (wherein CO_2 is
excluded) and the assumption that the CO + OH reaction solely determines the
local CO concentration. Reactions R1 and R2 are taken directly from the preceding set. Reaction R3 represents a fit to the results of the series of plug
flow reactor calculations. Reaction R4 allows for the oxidation of relatively
simple hydrocarbons, not limited to C_2H_2. A composite rate was determined
from the literature. Local species concentrations and temperature, falling
between limits set by the "partial" and full equilibrium states, are defined
by the ratio of $CO_2/CO_{2_{eq}}$.

   An empirical fuel consumption rate for turbulent premixed systems was
derived by correlating experimental turbulent reactor concentration data using
the parcel burning model of Howe and Shipman (Ref. 6). The resulting expression is presented in Figure 14. An NO kinetic system (Ref. 7) based on the
Zeldovich reactions is employed, as shown in Figure 15.

   Predicted combustor internal flow and pollutant concentration profiles
are presented in Figures 16-19 for a JT8D can-type combustor operated at nominal engine idle power setting. The temperature profiles behave in the expected manner, but the predicted axial velocity profiles exhibit considerable
change in character with position down the combustor. The temperature and
velocity variations can result in substantial differences (up to 30%) in residence time between adjacent streamtubes. We currently have no way of knowing
whether this is indeed occurring. Velocity and local temperature data within
the combustor would confirm or refute the validity of the internal flowfield
and mixing model. The corresponding concentration profiles for CO, UHC and NO
reflect combustion activity in the forward or primary zone portion of the combustor. There are similar profiles of local equivalence ratio, percent fuel
droplet evaporation, etc. The double peak in the UHC profile for streamtube 2
is a consequence of passage from overall fuel-lean to overall fuel-rich equivalence ratio. The initial lean to rich transition is due to the competing
droplet evaporation and fuel vapor combustion reactions, while the return to
lean condition is due to the addition of dilution air. With some verification
of the internal flowfield model, the existence of experimental species concentration profiles would permit assessment of the combined physical/chemical
combustion model. Specific information on percent liquid fuel, percent vapor
fuel and percent reacted fuel, accompanied by measurements of chemical species
such as CO, OH and NO, would permit direct assessment of the physical combustion model and would perhaps allow us to infer the role of turbulent mixing.

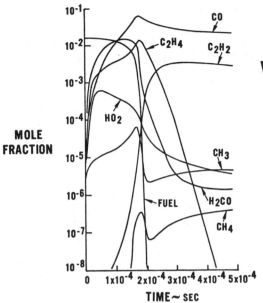

VARIATION IN SPECIES
CONCENTRATION
WITH TIME
AT $\phi = 1.0$,
P = 2 ATM,
Ti = 1000°K

Fig. 12

REDUCED HYDROCARBON OXIDATION SYSTEM

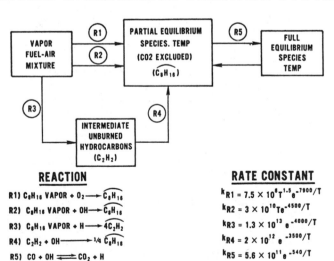

REACTION

R1) C_8H_{16} VAPOR + O_2 ⟶ $\widehat{C_8H_{16}}$

R2) C_8H_{16} VAPOR + OH ⟶ $\widehat{C_8H_{16}}$

R3) C_8H_{16} VAPOR + H ⟶ $4\widehat{C_2H_2}$

R4) C_2H_2 + OH ⟶ ¼ $\widehat{C_8H_{16}}$

R5) CO + OH ⇌ CO_2 + H

RATE CONSTANT

$k_{R1} = 7.5 \times 10^6 T^{1.5} e^{-7900/T}$

$k_{R2} = 3 \times 10^{10} Te^{-4500/T}$

$k_{R3} = 1.3 \times 10^{13} e^{-4000/T}$

$k_{R4} = 2 \times 10^{12} e^{-3500/T}$

$k_{R5} = 5.6 \times 10^{11} e^{-540/T}$

WHERE $\widehat{C_8H_{16}}$ = PARTIAL EQUILIBRIUM PRODUCTS OF COMBUSTION (CO_2 EXCLUDED)

Fig. 13

TURBULENT COMBUSTION RATE

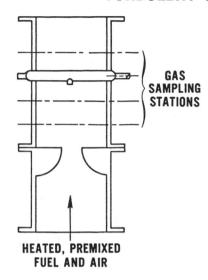

GAS SAMPLING STATIONS

HEATED, PREMIXED FUEL AND AIR

BURNING RATE DATA CORRELATED WITH PARCEL BURNING MODEL:

$$Q_{FUEL} = -68 \rho \, S_L \left(X_{F_0} - X_F\right) 2^{0.31 \times 10^4 \tau}$$

S_L = LAMINAR FLAME SPEED

$\left(X_{F_0} - X_F\right)$ = MASS FRACTION OF UNBURNED FUEL

τ = RESIDENCE TIME

Fig. 14

NITRIC OXIDE FORMATION KINETICS RELATIONSHIPS

REACTION		FORWARD RATE CONSTANT,
NUMBER	EQUATION	cm³/mole-sec
1	$N_2 + O = NO + N$	$R_1 = k_{11} = 1.35 \times 10^{14} e^{-37,500/T}$
2	$N + O_2 = NO + O$	$R_2 = k_{12} = 6.4 \times 10^9 \, T e^{-3,125/T}$
3	$N + OH = NO + H$	$R_3 = K_{13} = 7 \times 10^{11}$
4	$H + N_2O = N_2 + OH$	$R_4 = k_{14} = 4 \times 10^{13} e^{-6,000/T}$
5	$O + N_2O = O_2 + N_2$	$R_5 = k_{15} = 5 \times 10^{13} e^{-14,000/T}$
6	$O + N_2O = NO + NO$	$R_6 = k_{16} = 2.5 \times 10^{13} e^{-13,450/T}$

$$\frac{d[NO]}{dt} = 2 (1 - \alpha^2) \left[\frac{R_1}{1 + \alpha K_1} + \frac{R_6}{1 + K_2} \right]$$

WHERE

$$\alpha = [NO]/[NO]_e \qquad K_1 = \frac{R_1}{R_2 + R_3} \qquad K_2 = \frac{R_6}{R_4 + R_5}$$

R_i IS THE ONE-WAY RATE OF THE ith REACTION: FOR INSTANCE:
$R_1 = k_{11} [N]_e [NO]_e$

Fig. 15

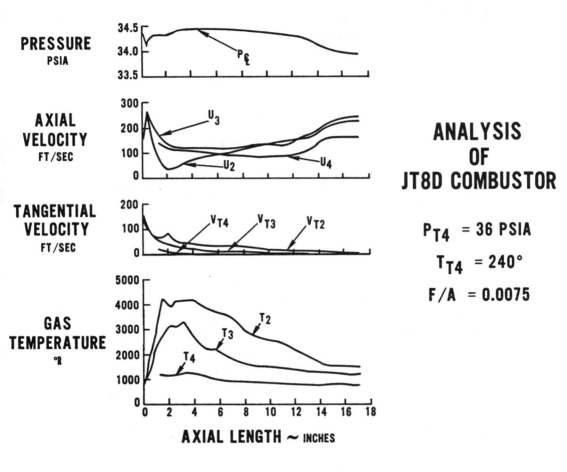

PRESSURE PSIA

AXIAL VELOCITY FT/SEC

TANGENTIAL VELOCITY FT/SEC

GAS TEMPERATURE °R

AXIAL LENGTH ~ INCHES

ANALYSIS OF JT8D COMBUSTOR

P_{T4} = 36 PSIA

T_{T4} = 240°

F/A = 0.0075

Fig. 16

54

EMISSION PROFILES OF JT8D COMBUSTOR
P_{T4} = 36 PSIA T_{T4} = 240° F/A = 0.0075

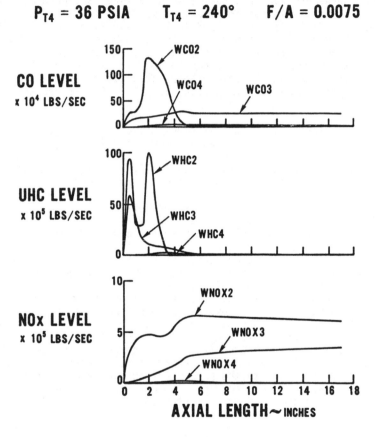

Fig. 17

In this instance, the model could assist in the interpretation of experimental data which might otherwise be ambiguous.

Figures 18 and 19 present a comparison of predicted concentration profiles with measurements taken in the front end of a JT8D combustor at idle. A cooled gas sample probe was used at three axial locations in the primary zone. Generally good agreement was obtained between measured and predicted local fuel-air ratio. This would suggest that the aerodynamic flow field model is valid insofar as radial distribution of fuel and air is concerned. It was not possible to experimentally distinguish gas phase unburned hydrocarbons from raw vapor or liquid fuel. Agreement between measured and predicted CO concentration profiles is somewhat poorer, as can be seen in Figure 19. Although the correct trends are obtained, it should be pointed out that the comparison is made on a logarithmic scale and that relatively few streamtubes were employed in the prediction.

Figure 20 compares predicted average exit plane pollutant concentrations with measured data for variation in engine power setting. Agreement is generally good over the engine operating range. These are the predictions of most interest to the combustor designer, who typically will vary combustor design parameters (model input) in a search for minimum emissions.

As a combustor modeler, there is great temptation to view the model as an end in itself, and to orient my specification of needed combustor measurement techniques to those which will either verify the model or permit me to reduce the arbitrariness of the assumptions. Certainly from the viewpoint of gaining a better understanding of turbulent combustion, this is not an unreasonable position to adopt. However, the justification for a mathematical model such as this must be in terms of its usefulness as a design tool. To date, the model appears to be suitable for this purpose, although there are shortcomings when applied to combustors far removed from so-called "conventional" designs. Combustors of this category tend to be encountered in the search for very low emissions, and tend to result from an idealized control strategy. It usually happens that emission data measured from such a combustor substantially exceeds the predicted values. The existence of appropriate internal measurements would permit detailed comparison with the predicted profiles and would perhaps allow inference of the actual combustion mechanism and, incidentally, the actual potential of the various emission control strategies.

REFERENCES

1. R. J. Mador and R. Roberts, "A Pollutant Emissions Prediction Model for Gas Turbine Combustors," AIAA Paper 74-1113, San Diego, October 1974.

2. J. S. Clarke, "The Relation of Specific Heat Release to Pressure Drop in Aero-gas-turbine Combustion Chambers," Proceedings of the 1955 IME-ASME Conference on Combustion, pp. 354-361, 1955.

COMPARISON OF PREDICTED AND MEASURED FUEL-AIR RATIOS
JT8D BURNER (RIG)

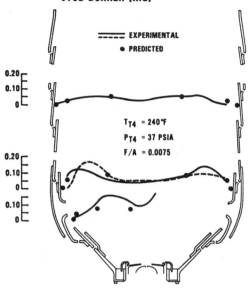

----- EXPERIMENTAL
• PREDICTED

T_{T4} = 240°F

P_{T4} = 37 PSIA

F/A = 0.0075

Fig. 18

COMPARISON OF PREDICTED AND MEASURED
CARBON MONOXIDE CONCENTRATIONS
JT8D BURNER (Rig)

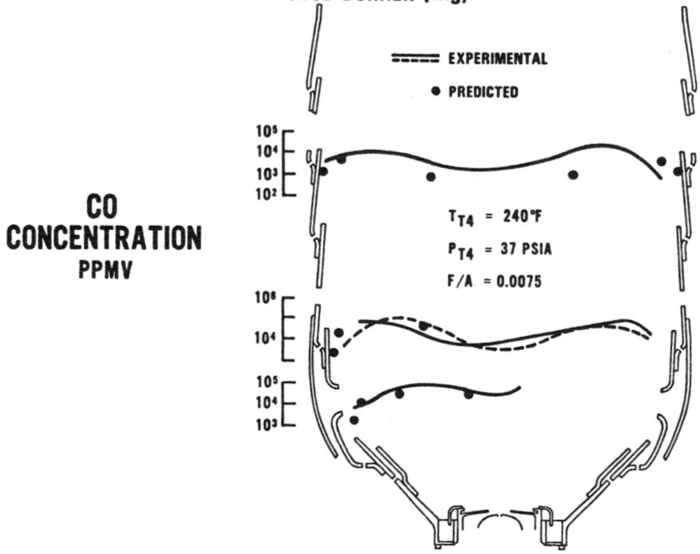

----- EXPERIMENTAL
• PREDICTED

**CO
CONCENTRATION
PPMV**

T_{T4} = 240°F

P_{T4} = 37 PSIA

F/A = 0.0075

Fig. 19

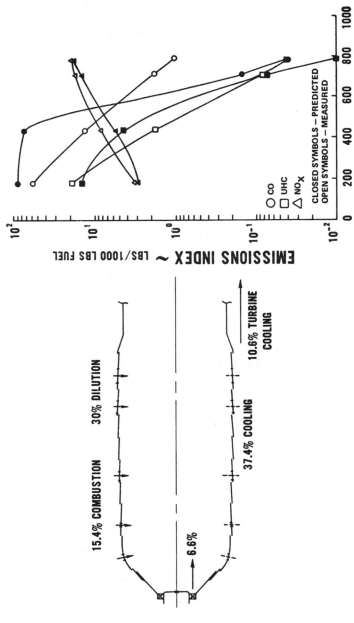

COMPARISON OF MEASURED AND PREDICTED EMISSIONS

P&WA JT8D-15 COMBUSTOR

Fig. 20

58

3. R. R. Dils, "Dynamic Gas Temperature Measurements in a Gas Turbine Trans- ition Duct Exit," ASME Paper No. 73-GT-7, Washington, April 1973.

4. B. J. Wood, W. A. Rosser and H. Wise, "Combustion of Fuel Droplets," AIAA Journal, Vol. 1, pp. 1076-1081, 1963.

5. R. B. Edelman and O. F. Fortune, "A Quasi-Global Chemical Kinetic Model for the Finite Rate Combustion of Hydrocarbon Fuels with Application to Turbulent Burning and Mixing in Hypersonic Engines and Nozzles," AIAA Paper 69-86, New York, 1969.

6. C. W. Shipman and N. M. Howe, Jr., "A Tentative Model for Rates of Combus- tion in Confined Turbulent Flames," Tenth Symposium (International) on Combustion, The Combustion Institute, pp. 1139-1149, 1965.

7. G. A. Lavoie, J. B. Heywood and J. C. Keck, "Experimental and Theoretical Study of Nitric Oxide Formation in Internal Combustion Engines," Combustion Science and Technology, Vol. 1, pp. 313-325, 1970.

DISCUSSION

BLOOM, PINY - How does this relate to fuel economy?

ROBERTS - Carbon monoxide and incompletely burnt hydrocarbons are the products of reactions where all the energy available from the fuel-air mix- ture has not been extracted. Therefore one can state that substantial pollu- tion levels correspond to less than high combustion efficiency and vice versa.

Specifically the fuel consumption in an aircraft gas turbine engine is de- termined primarily by the cycle efficiency, i.e. by burner, compressor and turbine efficiencies. Cruise efficiency, as far as we know, is very close to 100% for current engines. At takeoff power, CO and hydrocarbon numbers are low enough to indicate a very high efficiency. The idle efficiency of a typical engine is, say, 92 to 96% or something in that range, but then relatively little fuel is consumed at idle.

Cruise is the mode of operation where most of the fuel is consumed. Therefore the fuel economy of an engine is only affected when the burner ef- ficiency falls a lot below 100% at steady state high power operation. The very low integrated CO and hydrocarbon levels required by the current EPA aircraft emission standards, imply a correspondingly high combustion effi- ciency. The model is thus used to identify those combustor designs which exhibit high combustion efficiency, particularly at the cruise point. With the achievement of high combustion efficiency, fuel economy is essentially independent of emission levels.

BILGER, Univ. of Sidney - Aren't there secondary effects? You may require too much pressure drop to do what you want to do, or you may end up with an awful temperature profile; both of these will affect engine efficiency. Do you get these in your model?

ROBERTS - The pressure drop is a design input. You are correct in stating that the pressure drop of a combustor is something that affects engine cycle efficiency and we often find ourselves at the limit. Current day engines are running about 2-1/2% across the liner and perhaps 5% in the combustion system including all losses. We find that those numbers are pretty much determined by other people and we have to live with them. Pressure drop affects the combustion process by affecting jet penetration and perhaps the rate of turbulent mixing, although the relationship is not well defined. The scale of turbulence in a combustor is related to the pressure drop and a lot of other factors.

The model is able also to predict combustor exit temperature profile, since a temperature is calculated for each streamtube. Goals are defined for average radial temperature profile and "pattern factor", as we call it, which are considered acceptable levels. If we come up with a prediction which clearly violates those levels, the particular design is discarded.

PENNER, UCSD - How do you reconcile your predicted and observed CO profiles? Do you go back and try to find out how you would have to modify your input data or your computational techniques to accommodate your results?

ROBERTS - What we do is called sensitivity analysis by some people, and fudging by others. Certain of the hydrocarbon rates, the CO reaction rates, the turbulent eddy diffusivity, etc...., are really only known within a certain band of accuracy. We have gone back into the model and varied some of these rates to see what happens. Intellectual honesty, I hope, prevented us from going well beyond the allowable limits on these things. We found that the predicted emission trends were determined by things other than chemical rates. I think it has to do with the rate of heat release, with the turbulent combustion model.

PENNER - That is not clear, though, because your combstion model omits completely the coupling between nitrogen chemistry and H-C-O kinetics. It would be conceivable that if you put this coupling in your model, the CO would agree within the limits of error.

ROBERTS - It would be difficult for us to do that without putting in a full kinetic mechanism. However, my personal opinion is that it is not the reason. I may be wrong.

MELLOR, Purdue Univ. - Have you used your model to predict smoke production?

ROBERTS - We have only very briefly worked on smoke prediction. There are some analyses available. Appleton at MIT, for instance is probably closest to coming up with a real phenomenological model for smoke formation. We have tried to put in very crude smoke formation and consumption models based on local stream tube equivalence ratios. In something as simple as laminar flames, we find that smoke forms above an equivalence ratio of 3 and is consumed when the temperature is above 2000° and that sort of thing. We have found that we are grossly unable to predict smoke. We can fudge things around and get the correct range of smoke levels but not get the right trend, i.e. one which would fit our observation, that we often find maximum smoke at less than full power, say at something like climb power or 80% power. We just don't get that trend at all with this kind of a simple model but we really have not worked on it very much.

WOLFSON, AFOSR - I would like to suggest that noise might be also an important factor in the combustion process. Combustion-driven noise - as you well know - is of great interest to EPA and of course to NASA: this was one of the things that led towards the elimination of the SST.

Conversely, noise may be a contributor towards triggering combustion instability, whether it be low frequency or high frequency combustion instability in augmentors and also in main burner systems especially in military aircraft, with the very high energy density and power distributions of their augmentation systems. Work is currently underway in this area (Summerfield, Strahle, etc...), and possibly more research is needed.

PENNER - If I may ask Dr. Roberts again, I am puzzled about his disposition of the CO question. A logical procedure has to be the kind that he prescribed, where we begin with some kind of model, make measurements and then improve the model in order to get better understanding. I think it is essential that when there is a discrepancy of the sort that came up with CO, that it be really tracked down and that the model be changed until the drastic differences in behavior between observed and predicted results are eliminated. What better encouragement for amending your model can you hope for, than to get totally different slopes for the measured and observed data?

ROBERTS - The principal mechanism responsible for CO being in the exhaust of the engine, is really quenching in the dilution zone, especially in the case where you have low hydrocarbons in the exhaust. I think it is not a case of CO forming in the latter part of the burner but that of CO not being eliminated by the addition of air at the streamtube temperature. To say that one particular part of kinetics is the reason why this slope is different is not realistic. There are so many other aspects.....

PENNER - I didn't suggest that it was. But if you are not changing your model to try to account for the results, what is the point of making more measurements?

ROBERTS - I feel that such changes would not be meaningful. What I really need is some better information on the rate of mixing of cold jets for example, and on other aspects of the model, rather than to concentrate strictly on changes to the model which may appear to improve it but not reflect the reality of jet engine combustors.

GOULDIN, Cornell Univ. - This CO burning problem is also one area where turbulence is going to have a marked effect on the chemistry because of the mixing of cold jets with hot jets. As Hardesty pointed out earlier, the effect of temperature fluctuations on temperature-dependent reaction parameters is usually accounted for by Taylor series expansions and by their time averaging.. What you find is that convergence is possible only if the time-averaged temperature fluctuations are less than a certain function of the activation energy and temperature (see p. 20). This of course depends on the activation energy of the particular reaction of interest. In general though, my impression is that there is no place in the gas turbine combustor where this condition is ever met. Certainly when you are injecting a cold stream and you quench the hot stream with it, these fluctuations are going to be very large. Hence, I don't see how you can ignore the fact of turbulence. This particular quenching case looks like a fairly simple problem which you could resolve with some basic turbulence research.

MELLOR - Also, it could be that the CO which comes out of the exhaust results from three-dimensional effects, whereas this is an axisymmetrical program.

ROBERTS - In addition, there is some evidence to indicate that besides the so called uniform turbulence (as predicted by mixing length theory, for instance), there is also a periodic component. There have been some measurements made at United Technology Research Center which indicate a 150 Hz variation in the recirculation zone position. In other work at Pratt & Whittney where we measured temperature at the exit of a burner, we noticed that a lot of power is generated in the 100 to 500 Hz range. This raises the question of whether the primary effect of turbulence is accounted for in this model or whether we should be using a quasi steady state equation having a sequence of alternating phases. I really don't know the answer to that.

McGREGOR, ARO - How were those measurements made? Could it be that the problem with agreement is a measurement problem rather than a modeling one?

ROBERTS - Very briefly, the exhaust plane measurements were taken with a stainless steel steam cooled probe. The sample was ducted by a teflon line heated above 450°F to the more or less conventional, on-line instrumentation. These included a flame ionization detector for hydrocarbons, nondispersive infrared detectors for CO, CO_2 and NO, and nondispersive ultraviolet detector for NO_2. It is pretty much the type of instrumentation that is specified by the EPA aircraft emission standards.

The internal probing measurements were taken with a copper tipped, steam cooled stainless steel probe. The sample was ducted into the same type of instrumentation in one case. We also captured samples in evacuated stainless steel bottles and analyzed them. I think your point is well taken, however. The potential for error in these measurements is probably at least of the same order of magnitude as the uncertainties of the model.

RHODES, ARO - On the basis of our work of the past few years, I would concur that we need help in deciding how fluctuations influence the chemistry and vice-versa. In many of the good experiments I have seen the flow is locally in equilibrium. What we need very badly are experiments where the flow is definitely out of equilibrium so that the kinetics - turbulence interaction can be evaluated. For instance ignition delay is very sensitive to initial conditions. A small change in the concentration of free radicals may change the solution by an order of magnitude. Hence you need to have not only an experiment where the flow is out of equilibrium, but you also need one where the initial conditions are very well defined.

On the other hand, I would agree with Glassman on a couple of points. One of them is that over the years, very little help in terms of solving practical problems has come from fundamental studies on turbulence. The other has to do with a very extensive set of calculations we ran, using some of the experimental work that Dr. Bilger did with Dr. Kent in Australia. There, if you are interested in things like flame length or concentration along the centerline, or velocity fields, the fluctuations did not seem to make any great amount of difference.

But then, in terms of overall combustion efficiency, fluctuations do make a fair amount of difference because unmixed fuel-rich gas may exist even though the average flow is lean.

ROQUEMORE, AFAPL - There are some non-technical problems in applying in situ optical techniques to practical combustion problems which I believe should be recognized. To make optical measurements in practical combustion environments requires: the combustion facilities, the optical systems, spectroscopists and combustion scientists. This combination of talent and equipment does not exist in many organizations. In my opinion, the problem in bringing these ingredients together are impeding combustion measurement technology more than the technical problems associated with actually making the optical measurements. I don't know the solution to this problem; but, whatever it is, it will require that diagnostic and combustion people work closely together. In doing this, it must be realized that the two groups have very different backgrounds and speak different technical languages. For example, many of the people well versed in advanced laser measurement techniques know very little about combustion. Because of this, they don't know what experiments should be performed and what to expect in terms of measurement environment. I believe that modelers can help with this prob-lem by using the models to design good experiments which would identify the combustion source, parameters to be measured, spatial and time characteristics of measurements, accuracy of the parameters needed to check models and describe the predicted combustion environment. In this way the diagnostic people can assess what optical techniques might be effectively used and in some cases may be able to perform the experiment.

LENNERT, ARO - There is another point that should be brought out here, regarding the need to bring the diagnostics people and the combustion people together. I think it is of paramount importance that the workers who are developing the instrument should appreciate the problems of the engine de-velopment people as well. It is a three-fold interaction as opposed to a two-fold interaction.

WOLFSON - Standards and specifications insofar as improvement of instru-mentation are also an area of concern. In many instances identical instru-ments which detect and evaluate certain effluent constituents from air breathing systems, produce very different results. Yet these systems have apparently the approval of EPA and of their own manufacturers. The coordi-nation between the user and the manufacturer of instrumentation techniques should be improved.

PART II
Optical Measurement Techniques

SESSION 4
Laser Velocimetry
Chairman: NORMAN A. CHIGIER

LASER VELOCIMETRY
FOR COMBUSTION MEASUREMENTS
IN JET PROPULSION SYSTEMS

NORMAN A. CHIGIER
University of Sheffield

I.   INTRODUCTION

In the first session which we are holding on the subject of optical
measurement techniques, I wish to very specifically examine the subject
of laser velocimetry for jet propulsion systems. We wish to examine the
extent to which developments in instrumentation have succeeded in providing an instrument which is both suitable and capable of making measurements
in a proto-type gas turbine combustion chamber. Laser velocimeters, also
known as laser anemometers are sold commercially by a number of companies
and are in a sufficiently developed state that they may be purchased and
used without the need for the user to undertake any major developments in
the instrumentation. Before considering the use of a laser velocimeter,
it is, however, necessary to design an optical system and make decisions
concerning the laser power and signal diagnostic system. It is also necessary for the user to have an understanding of the general principles of
laser velocimetry, optics and signal diagnostics in order that a measurement can be made. The problems associated with making measurements in combustion systems are common to the various types of combustor and measurements are in the process of being made in gas turbine combustion chambers,
internal combustion engines and industrial furnaces.

I shall concentrate my remarks on the special problems which arise
when the laser anemometer is used in combustion systems. These problems
are associated with making measurements in a high temperature radiating

67

environment, which can be laden with solid or liquid particles. There
is also the special problem of viewing windows in the high temperature
and high pressure combustors, since measurements with laser anemometers
are not possible unless an optically clear path is provided to the region
in which measurements are to be made. I shall also address the problem
of the need for making velocity measurements and the use that can be made
from such measured data.

II. VELOCITY MEASUREMENTS

Attempts to measure velocity in combustion systems have been made
over a long period of time. These were mainly carried out with Pitot
and other pressure impact probes which required to be water cooled to
withstand the high temperature conditions in the flame. Probes have also
been constructed of platinum which required no cooling and of high temperature steels which required partial cooling, but allowed the probe end
to be heated. These heated probes were found to be particularly successful
in making measurements in flames laden with liquid oil or solid carbon
particles which caused blockage in water cooled probes. In order to determine velocity from such probes, it was necessary to measure the temperature and preferably to measure the local concentration in order that
the local density could be determined. The measurements of temperature
and concentration were usually carried out at different times to the pressure measurements and it was only possible to determine time average velocities.

The hot wire anemometer became the principal velocity measuring device in non-combusting systems and because of its high frequency response,
measurements of turbulence intensity and stress could be made. Many attempts were made to adapt the hot wire anemometer for combustion and it
was found that iridium wires could be used in non-oxidizing atmospheres
provided that temperature levels were not excessive and there were no
particles in the system. Even though attempts are continuing to develop
this instrument for more general flame use, the experiments to date suggest that it will not be a general purpose velocity measuring device in
combustion systems. Again, it is necessary to have separate information
on the temperature, since the hot wire is sensitive both to velocity and
temperature change. The hot wire anemometer has been used in internal
combustion engines under non-firing conditions, but the problems of interpretation of the measurements are complex and the overall accuracy of
the measurements is uncertain.

It is in the light of the lack of any suitable alternative that we must consider the great interest and potential of the laser anemometer. The laser anemometer measures velocity directly without the requirement of making a separate measurement of temperature and species concentration. No calibration is necessary with a laser anemometer since the fundamental principles of the instrument are based upon the measurement of frequency. The measurement of frequency is one of the most accurate and reliable that can be made in the whole field of instrumentation. Many questions have been raised in the literature concerning the accuracy of the laser anemometer. In my opinion there is no question about the accuracy of measurement of velocity of an individual particle passing through the control volume. There is, however, some doubt about the accuracy of the averaging procedure which is used in determining the time- or space-average velocity. There is also little doubt in my mind that both the potential and actual accuracy of a laser anemometer is far in excess of that of the water cooled Pitot tube.

In principle, there does not appear to be any limitation to the range of velocities that can be measured by a laser anemometer. Measurements have been made and reported in the literature in creeping flow systems with very low velocities in the order of mm/s. The velocity of fluid being entrained into jets has been measured and it has been demonstrated that the theoretical arguments concerning entrainement are valid. These extremely low entrainment velocities could not be previously measured by other measuring devices. At the other end of the scale, measurements have been made in supersonic wind tunnels and except for the special problem of seeding there seems to be no reason why measurements of velocity cannot be made in hypersonic wind tunnels.

III. SPECIAL PROBLEMS ASSOCIATED WITH COMBUSTION SYSTEMS

Since no physical probe is required, the high temperatures in flames present no special problems for measurements with a laser anemometer. In all systems it is essential that there is an optically clear path for the laser beams to penetrate and cross at the control volume. The very high intensity of radiation of a laser within an extremely narrow band of wavelength is usually so much in excess of the radiation intensity from the flame, within the laser wavelength band, that radiation from the flame does not seriously affect the measurement. Filters are however, generally used in order to prevent radiation from other wavelengths entering into the photomultiplier detector. Provided that a particle is sufficiently small to follow the local gas movement and that sufficient light is scattered, measurements can and have been made in a wide range of combustion systems.

Windows are required to be optically transparent and must be able to
withstand the high temperatures. Quartz and sapphire windows have been
used and where there is a likelihood of deposition or fogging on the windows, provision of an air screen may be required. When the forward scatter
mode is used windows have to be provided both for the beams entering on one
side of the system and the beams leaving, on the other side of the system.

When there are particles of liquid fuel or soot in the system it may
not be possible to make measurements in regions where the quantity of particles is so large that the system is too dense to permit penetration of
sufficient light to the control volume or that insufficient light will be
scattered from the control volume. Such a situation can arise in the close
proximity to atomizer nozzles, but it does not generally apply throughout
the remainder of the combustion system. In particle laden systems, a special problem arises in that the anemometer will record velocities of all
sizes or particles passing through the control volume and large particles
will have considerable differences in velocity to that of the local gas.

When the temperature and/or density variations are sufficiently large
to cause significant differences in refractive index, the possibility arises of the laser beams being deflected. The extent of this deflection is
dependent upon the total size of the system as well as the temperature gradients. If the refractive index of the gases through which the two laser
beams entering the system are not the same, the possibility arises that
the laser beams will not cross, and under these conditions a measurement
is not possible. Alternatively, in turbulent systems the temperature
fluctuations can cause the laser beams to jitter and this could also affect the measurements. In practice, it has been found that when measurements have been made in open flames where temperature gradients are large,
that the beams have crossed and that there was no significant deflection
of the beams. On the other hand, attempts to use the laser anemometer in
large industrial furnaces have been only partially successful and the difficulties have been ascribed to deflection of beams caused by changes in
refractive index. It should be noted that even though systems are at high
temperatures, the temperature gradients are not always large, particularly
in turbulence systems with high rates of mixing.

The high frequency response of a laser anemometer allows us to make
measurements of turbulence characteristics in flames. In the process of
taking averages from individual velocity measurements, errors can easily
arise. These errors are associated with the non-uniformity of seeding and
with taking averages of particles which may have a wide spectrum of velocities. In jet systems, if particles are only added to the jet and there
are no particles in the surroundings, flow measurements of average velocity will become increasingly in error as the edge of the jet is approached.
This is due to the fact that only velocity of fluid which originally ema-

nated from the nozzle is being measured, and the velocity of fluid from
the surroundings is not being measured, simply because it is unseeded.
In the extreme, zero velocity is recorded in the surroundings simply due
to the lack of seeding. In many practical systems there is sufficient
recirculation to ensure that some particles do enter into the gases surrounding the jet. It is therefore necessary to make some assessment of the
extent of seeding in the various regions of the combustion chamber.

The overall frequency response of the laser anemometer as a whole, is
usually restricted by the detection and recording system. When punched
tape recorders are used, the response will be as low as one or two measurements per second. When a digital counter is used and interfaced with
a computer, overall frequency response for individual velocity measurements of the order of 20 kHz has already been achieved. It appears to be
quite feasible that this response can be extended further. For practical
purposes such a system measures the instantaneous velocity of particles.
In order to have an indication of the continuous variation of velocity
with time at any one point, heavy seeding is required so as to provide an
almost continuous flow of particles entering and leaving the control volume.
The possibility exists, for the first time, of making detailed studies of
the turbulence characteristics and structure of flames by using the high
frequency response laser anemometer.

We shall have reported at this meeting the extent to which the LDV
system has been used in large scale industrial practice. The work currently being carried out at General Electric in Cincinnati[1] on the measurements of three velocity components in the exhaust of a large gas turbine combustor is one example of how far developments have reached in the
making of measurements in a relatively difficult industrial environment.
At the NASA-Ames Research Centre[2], a large scale robust anemometer has
been in use for several years for making measurements in large wind tunnels.
This instrument is transported on a fork-lift truck and is so solidly constructed that the optical system is quite insensitive to rough treatment.
These are only two examples of the fact that the LDV is not only a laboratory instrument, but is fully capable of being used in industrial environments.

IV.   EXPERIMENTS AT SHEFFIELD UNIVERSITY

The experiments at Sheffield University have been carried out on open
flames in a size range intermediate between small bunsen burner flames and
the large flames used in industrial practice. Measurements have been made
of the three velocity components: axial, radial and swirl in both forward
and reverse flow regions of the flame. Turbulence intensity levels are of
the order of several hundreds of percent and it is essential when making

measurements under such conditions that the anemometer be able to distinguish the velocity sense. On the basis of the measurements which have
been made with a laser anemometer it has become clear that many of the previously reported measurements of velocity in flames, using water cooled
Pitot tubes were incorrect in regions where turbulence intensity exceeded
30% and in regions of reverse flow. When turbulence intensities are of
the order of 100%, the errors arising from use of Pitot tubes are extremely
large. Comparisons which have been made between hot wire anemometers and
laser anemometers in non-reacting air jet flows show that the hot wire anemometer points are less accurate than those of the laser anemometer, particularly in the regions of intermittent flow near the outer edge of the jet.
Most of the attempts to calibrate a laser anemometer, simply demonstrate
that the laser anemometer is the most accurate velocity measuring instrument.

A rotating disc can be introduced into the optical path of the laser
beams and the speed of rotation of the disc can be compared with the velocity recorded by the anemometer as a mark on the disc passes through the
control volume. These tests again show that the laser anemometer measurement of velocity is more accurate than the determination of speed of rotation of the disc. When particles are dropped through the control volume
and the speed of the droplets is measured by a stroboscope, close agreement
is found between the measurements of velocity of the particle.

Fig. 1 shows the laser anemometer system which was used for measurements
of velocity components in a natural gas flame with swirl. The results of
this experiment were first reported at the Combustion Symposium in Tokyo[3].
Magnesium oxide particles were added to the air supply. A question has
been raised as to the necessity of adding seeded particles when making
laser anemometer measurements. In order that the measurements can be made,
sufficient light must be scattered by the particle passing through the control volume so that a clear signal will be detected by the signal diagnostic
system. The intensity of scattered light is a direct function of the laser
power and the ability to detect the signal is largely dependent upon the
sensitivity of the photomultiplier. It has been demonstrated that measurements can be made in the atmosphere in which no artificial seeding is added and the measurement is dependent upon the presence of dust normally
found in the atmosphere. In order to make measurements under these conditions high laser powers and photomultipliers with high sensitivity are
required. For the measurements which we have made, we find that it is
possible to obtain some signals without seeding, but the signal to noise
ratio is greatly improved by the introduction of artificial seeding.

We have used a laser with a nominal power of 500mW. The power of the
beams entering the control volume is of the order of 100mW. Our measurements were made under conditions of forward scatter. A number of investigators have attempted to make measurements in flames using a 5MW helium-
neon laser and they have found great difficulty in obtaining signals of

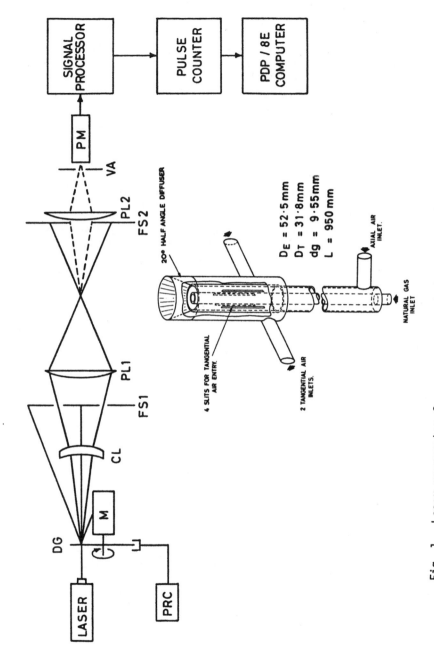

Fig. 1 – Laser anemometer for measurement of three velocity components and reverse flow in open swirling flame.

73

sufficient intensity.

The size of the magnesium oxide particles was between 1 - 3μm. The
flames in which we made our measurements were sufficiently luminous that
measurements can be made without seeding by using lasers with powers of
2 - 4W.

Many of the systems developed for laboratory use have been based upon
forward-scatter. For large scale industrial application it is preferable
to have all the measuring equipment on one side of the apparatus and thus
back-scatter is preferable. In many of the laboratory experiments when
traverses are made, the flame is moved while the anemometer is maintained
stationary. In industrial practice, it is usually essential for the instrument to be used when traverses are made. When using back-scatter it
is essential that both the laser power and sensitivity of the photomultiplier be sufficiently large to obtain detectable signals. The measurements carried out at the NASA-Ames Research Center[2] have shown a clear
ability to make measurements in large wind tunnels and the measurements
that are being made at General Electric, Cincinnati[1] on gas turbine exhaust gases again show that provided laser powers of the order of 4W are
used, measurements can be made.

V.   SOME RESULTS IN TURBULENT FLAMES MEASUREMENTS

The size of the control volume is generally larger than the effective control volume for a hot wire anemometer so that the spatial resolution of the laser anemometer is considerably less than that of the hot
wire. The size of the control volume is largely dependent upon the width
of the laser beams and the angle of crossing of the beams. For studies in
which there is a special requirement to reduce the size of the control
volume, focussing of the laser beams and increasing the beam separation
can lead to some reductions in the size of the control volume.

In order to obtain measurements in reverse flows and high turbulence
intensity we have used a rotating disc diffraction grating. This has
proved satisfactory for the velocity range, less than 50m/s, which we
have used. In order to make the axial measurements we initially maintain
the laser beams in the same plane as the axis of the burner. For radial
components the plane of the beams is rotated 90° and measurements made
along a radial line of the flame. The flame can also be traversed along
another radial line from which measurements of the tangential velocity
components can be measured. By this means, the three velocity components
are measured in sequence and the assumption is made that the system is

steady. It is possible to make simultaneous measurements of the velocity components, but this requires either double or treble the signal diagnostic system. With such a multiple diagnostic system, space and time correlations of velocity fluctuations can be made.

Velocity measurements in flames are of particular interest for development and testing of turbulence models in numerical prediction methods. Fig. 2 shows a set of steamlines determined by integration of axial components of velocity measured by the laser anemometer in the flame. These measurements show the region of reverse flow and the amount of flow entrained from the surroundings. A major problem in determining stream functions is that the density is now known. Large scale density variations occur in the flow due to variations, mainly in temperature, and, to a lesser extent, in specie concentration. In order to determine local stream functions we require to measure simultaneously velocity, temperature and specie concentration. It is possible at this time to measure temperature and temperature fluctuations with fine wire thermocouples, but measurement of specie concentration is still largely dependent upon removing samples by probe measurements. The streamlines shown in Fig. 2 have been computed, without taking into account the density. term.

The subjects of flame generated turbulence and isotropy of turbulence in flames have been discussed for many years. The laser anemometer provides us with an instrument in order to test the assumptions and suggestions which have previously been made. Fig. 3 shows comparisons of mean velocity and velocity fluctuations as measured in a flame and in an air jet without combustion. The mean velocity measurements show increases in both the forward and reverse flow directions as a direct consequence of combustion. The magnitude of the velocity fluctuations are also increased as a consequence of combustion. The ratio of the fluctuating components to the time average components (local turbulence intensity) shows a reduction in the flame, compared to the non-reacting system. These results show the necessity of separate examination of the influence of combustion on mean velocity and fluctuating components. The conclusion that some authors have come to in the past that turbulence intensities in flames are lower than in corresponding air jets, is true, but the conclusion that there is no influence of combustion on turbulence is not valid. It can still be argued that the results shown in Fig. 3 are not conclusive evidence of flame generated turbulence and that the increases in turbulence velocity fluctuations may simply be a consequence of acceleration of the time average flow.

Fig. 4 shows that combustion affects the velocity fluctuations. The magnitude of axial components is increased to a greater extent than that of the radial and tangential velocity components, whereas in the central region of a jet without combustion the turbulence is nearly isotropic.

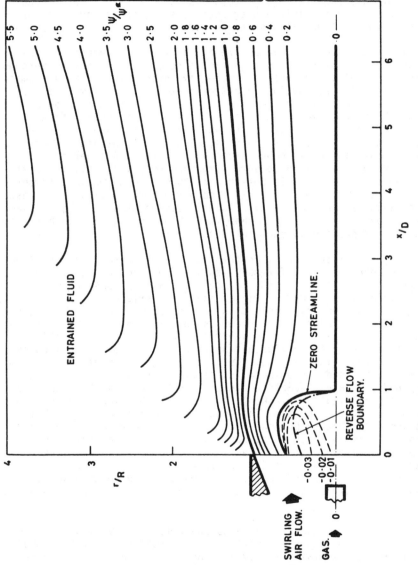

Fig. 2 - Stokes streamlines in swirling flame calculated from axial velocity components measured by laser anemometer.

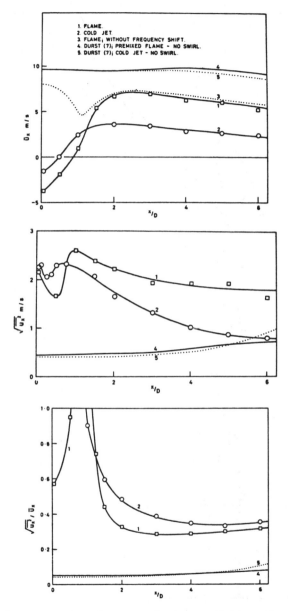

<u>Fig. 3</u> - Axial components of mean velocity, velocity fluctuations and local turbulence intensity on axis of swirling flame.

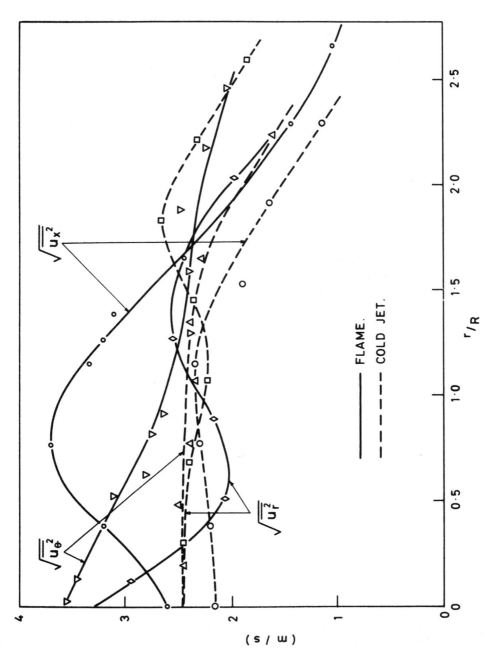

Fig. 4 - Comparison between axial, radial and tangential fluctuating velocity components in flame and cold jet at x/D - 1.0.

Combustion results in anisotropy of the turbulence. Fig. 5 shows the changes in kinetic energy of turbulence as a consequence of combustion.

On the basis of the measurements which we have made as well as other measurements which are being undertaken we feel that the laser anemometer can be recommended as an instrument capable of making accurate measurements of velocity in three dimensional industrial flame systems.

When measurements are made in flames laden with particles, there is a special problem in the use of the laser anemometer. For measurements in liquid spray flames or pulverized solid fuel, particles of the order of 100 μm and larger are present. These large particles do not follow the streamlines of the gas flow and we are interested in simultaneously measuring the size and velocity of particles. In the laser anemometer we only utilize the information on frequency in order to determine the velocity. The intensity of the scattered light can easily be measured using the same optical and diagnostic system. We have introduced the laser anemometer into a spray flame of the type shown in Fig. 6. In attempting to make measurements of velocity in the system the need was recognized for obtaining additional information concerning the size of particles.

A series of calibration studies have been carried out in which large particles of known size and velocity are passed through the control volume and measurements are made of signal intensity. Fig. 7 shows the system in which a stream of monosized droplets were passed through the centre of the control volume. Fig. 8 shows that when a particle of 810μm is passed through a control volume with fringe spacing of 5μm, an excellent signal is obtained. Velocity measured by the anemometer is in close agreement with velocity measured by a stroboscope. It was thus demonstrated that even when a droplet occupies almost the entire control volume, it is still possible to obtain accurate velocity measurements. Fig. 9 shows the signal duration and amplitude of droplets as measured by the oscilloscope.

The factors influencing the intensity of scattered light are: intensity of laser light, angle of beam crossing, scattering angle, absorptivity of the partcile and location of the particle in the control volume. If it is possible to control the intensity of laser light and the angles of beam crossing and light scattering, it may be possible to obtain an indication of particle size by measurement of signal intensity. The number of fringes crossed by the particle may serve as an indication of the location of the particle within the control volume. Considerably more work is required before the laser anemometer will be ready for making measurements of both size and velocity.

Fig. 10 shows velocity profiles measured in a kerosene spray flame[4]. The points on the graph are time averaged and include particles of both

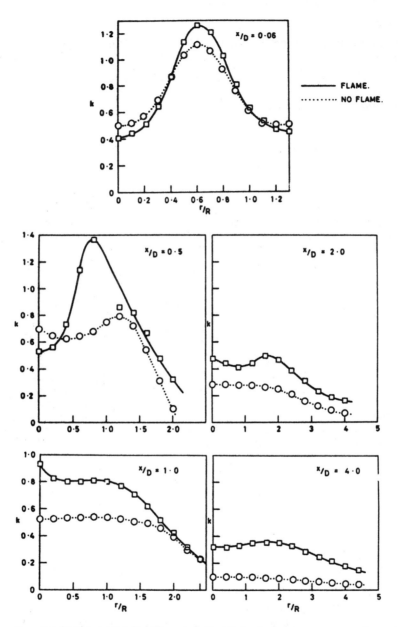

<u>Fig. 5</u> - Variation of kinetic energy of turbulence - comparison between flame and jet without combustion.

Fig. 6 - Kerosene spray flame - twin fluid atomizer.

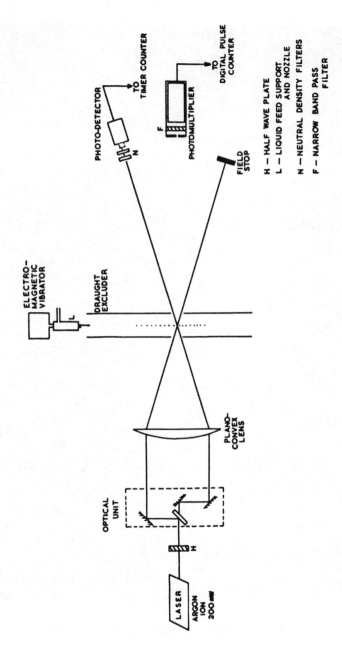

Fig. 7 - Stream of monosize droplets passing through control volume of laser anemometer.

82

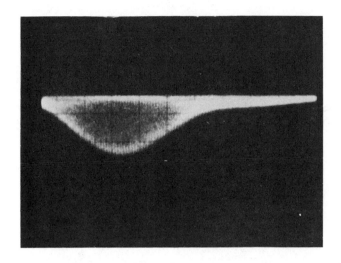

TIME BASE
50 μsecs/div

AMPLIFIER
0·05 volts/div

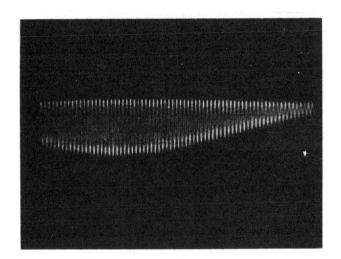

TIME BASE
20 μsecs/div

AMPLIFIER
0·05 volts/div

Fig. 8 - Oscilloscope trace of laser anemometer signal for 810μm droplet
passing through control volume with 5μm fringe spacing.

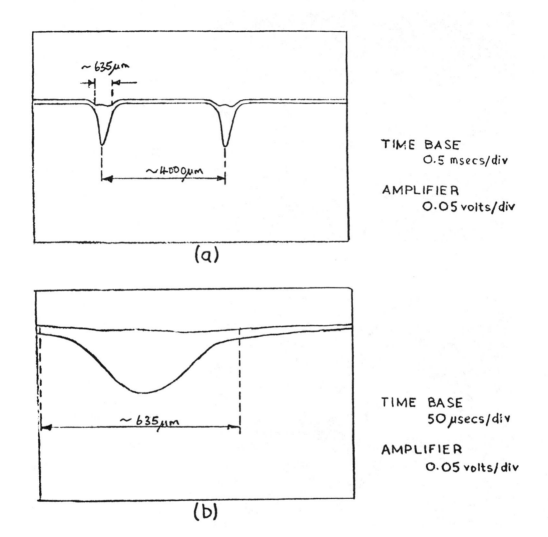

TIME BASE
0.5 msecs/div

AMPLIFIER
0.05 volts/div

(a)

TIME BASE
50 μsecs/div

AMPLIFIER
0.05 volts/div

(b)

Fig. 9 - Signal duration and amplitude from oscilloscope trace for stream of
particles passing through the control volume.

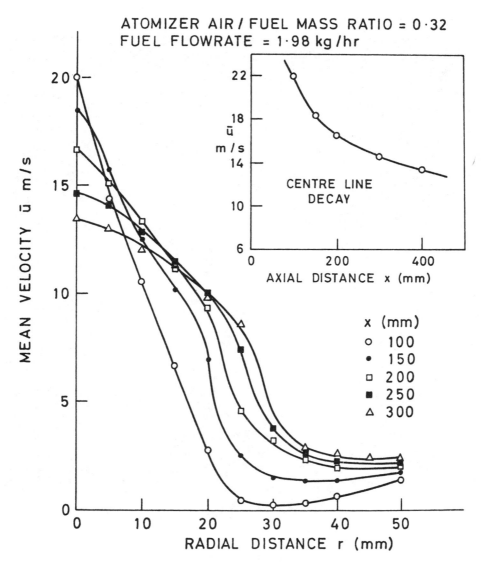

Fig. 10 - Velocity profiles measured in kerosene spray flame by a laser anemometer.

Fig. 11 - Photograph of a turbulent methane diffusion flame impinging on
a flat plate, showing the cellular structure of turbulent eddies.
(The editor and publishers are grateful for Dr. Chigier's permission
to use this photograph on the book jacket.)

large and small size. Such measurements do not provide the essential information, but the laser anemometer has the potential of being able to separately measure the velocity of particles of different size.

During our discussion at this meeting we have been largely concerned
with time average measurements. In many combustion systems there is clear
evidence of the presence of large eddy structures. Fig. 11 shows an example of the cellular structure of a turbulent impinging diffusion flame[5].
This photograph shows that eddies of fuel pass through the air and that
burning only takes place at the interface of the eddies where mixture ratios are within the limits of flamability. If we wish to obtain information in such flow systems, it is necessary to make measurements with instruments having a high frequency response. The laser anemometer provides
us with a tool for making velocity measurements with a high frequency response and thereby opens up the possibility of determining the structure
and movement of large eddies in flame systems.

REFERENCES

1.   General Electric Co., Cincinnati. Private Communication.

2.   Grant, G. R. and Orloff, K. L. Two-Color Dual-Beam Backscatter Laser
     Doppler Velocimeter, Appl. Opt., 12, 2913 (1973).

3.   Chigier, N. A. and Dvorak, K., Laser Anemometer Measurements in Flames
     with Swirl, 15th Symposium (International) on Combustion, the Combus-
     tion Institute, 573-585, (1975).

4.   Chigier, N. A. and Styles, A. C., Laser Anemometer Measurements in
     Spray Flames, Deuxieme Symposium European sur la Combustion, The Com-
     bustion Institute, 563-568 (1975).

5.   Chigier, N. A., Measurements in Turbulent Flows with Chemical Reactions.
     Analytical and Numerical Methods for Investigation of Flow Fields with
     Chemical Reactions, Especially Related to Combustion, AGARD-CP - 164,
     Section IV - 4 (1975)

DISCUSSION

HARDESTY - What can you do near a wall?

CHIGIER - These problems are being examined and are being solved. We now have biomedical people making LDV measurements in veins, arteries, and capillaries. I do not know that this problem has been solved for combustion systems but I believe that we can use the experience of others.

Basically, the laser beams must penetrate the system. Thus in all practical systems, it is essential to have an optically clear window. Near the wall, truncation of the control volume gives rise to problems, although systems have been already developed which allow the control volume to touch the wall itself. Of course you have also the question of how do you get the particles into the flow. Those problems are special ones, but I think we are approaching a measurement capability, which will allow internal combustion engine researchers to make measurements in quench layers. Many of the problems associated with present systems, are due to the large size of the control volumes.

PENNER - How about the length scale?

CHIGIER - In turbulence theory, there have been many statements about length scale, all of which were deduced by time-averaging at one or several points. I believe that we have now the capability for simultaneous two-point measurements and we can get a real measure of length scale by using correlations between these two points, using discriminate averaging. It will be necessary to double or triplicate the optical and diagnostic system to accomplish this.

PENNER - Spatial resolution?

CHIGIER - It is governed first of all, by the diameter of the laser beam, i.e. the "waist" diameter. A laser beam does not have a uniform intensity, rather it has a Gaussian distribution. Beam diameter is calculated from theory but I think we have to carry out an experiment equivalent to that which we use in spray photography. That is, one has to take a particle and slowly pass it through the control volume and decide what is truly the effective control-volume. We will have on the outer edges the equivalent of a fuzzy section because the intensity is decreasing towards the outer edges. The calculations which have been made in the past need to be examined more closely.

In practice, you can vary the size of the control volume by focusing, by varying the angle between the beams and also by varying the beam separation. At the present time, we are dealing with heights of a control volume of the order of about 0.6 mm and with lengths in the order of 2 or

3 mm. In principle, it is possible to reduce these somewhat, but I think that obtaining very small volumes will be a tough problem. It will depend upon the ability to focus and reduce the diameter of the laser beams.

SELF - I feel that Dr. Chigier's comments on the ease of LDV application to the study of combusting flows should be taken with caution. Dr. Chigier's remarks do not seem to include realistic burner configurations and conditions. Suffice it to say that there is all the difference in the world between measurements in open laboratory flames and real world combustors.

BILGER - I'd like to get your comments on several points. You haven't talked about things like fine particulates of unknown cross section which act as seeds. Also, since seeding is correlated with velocity, don't you have a counting bias for the regions of large velocities? Finally, when you discussed the refraction index changes, you just talked about the beams not crossing altogether. But in turbulent flow, the fringe pattern is also going to be bouncing around due to the refraction index changes, and aren't you measuring then the particle velocity plus the fringe velocity?

CHIGIER - Indeed, there are many things which are worrying us, but I would like to say that all these points, I believe, are points of detail which we will refine as we go along. The accuracy in making an individual LDV measurement is, I believe, extremely high. Is the time average measurement representative of the flow at that point? This is the question.

Now if you have particles of high velocity and low velocity and there is partial seeding, you can run into problems, as I mentioned earlier. As to the questions of moving fringes, and of particles which are nonspherical, they can cause problems, which need to be watched. But I believe that these problems do not prevent us from making the measurements in combustion systems.

WOLFSON - Don't you believe that to introduce additives will present serious problems of interaction with the system?

CHIGIER - Yes, but let's just say generally on that point that if you think that we have problems, then it should be recognized that the people who have succeeded in making measurements in supersonic windtunnels, at NASA - Ames for instance, had much bigger ones. They are very reluctant to introduce any foreign particles inside their tunnels. The answer to this lies in the signal processing system. The earlier systems used spectral analyzers and frequency trackers. But now we have a whole range of signal diagnostic systems. We have single particle counters which we are using at Sheffield. There are filter banks and photon correlators.....Many of these instruments have been specifically designed to deal with the problem of minimizing the amount of seeding. We are trying to push down to an absolute minimum the quantity of particles that have to be introduced and also their size. But, if you want to get a continuous measurement of velocity as a function of time, then ideally you want to have one particle going through the control volume at all times; i.e. as soon as one has left, you want another one to come in.

So, in that type of measurement, you have to have a sufficient number of particles. On the other hand, the people who work in the atmosphere use a filter bank and every time a particle happens to come by, they measure it, and then take a long time averaging. In general, I believe that it is not essential to have a large amount of seeding which may disturb the flow.

BERSHADER, Stanford Univ. - The use of higher power laser to overcome other limitations is an interesting concept. I don't think though, that you can go very far without producing an electrical breakdown in the control volume.

LENNERT, ARO - We have had several units in operation at the ARO center since 1967, and we are not introducing any artificial seeding. Currently, we are installing a two-component LV system, in a carefully designed housing, into the plenum chamber of a transonic wind tunnel, where it shakes, rattles, and rolls all over the place. We have been successful in making multi-component measurements of velocity in this and other operating wind tunnels.

In regard to power density, we're talking roughly about several watts of continuous laser power for LV measurement. This is far below the range of megawatt per cm^2 which corresponds to the self focusing regime which leads to ionization breakdown. If you were using a _pulsed_ ruby laser in the Joule/pulse range, you might encounter this problem.

We are also making holograms and multi-component velocity measurements in the cylinder of modified diesel engines for the Army Tank and Automotive Command. The experiment is set up so that measurements are made as a function of crank-shaft angle. We just cycle the engine a number of times to get a sufficient amount of data at a particular point in the cylinder and then scan step-wise through the entire volume. There are problems in making these measurements but as Dr. Chigier pointed out they are details and are being resolved.

As to probing into the boundary layers, one can get the probe volume rather close to the region of the wall. However, if you wish to enter the boundary layer, an off-axis receiver optics system is required because you are then making measurements in only a portion of the probe volume as opposed to the whole focal region.

Finally, we have made two component back-scatter measurements from a distance of 30 meters. We measured falling raindrop velocities, wind velocities, trailing vortices stemming from a flying aircraft, etc...These were not laboratory situations but actual field work. There are a host of applications dependent upon your imagination, resources and ingenuity.

LASER VELOCIMETER MEASUREMENTS
OF A CONFINED TURBULENT
DIFFUSION FLAME BURNER

F. K. OWEN
United Technologies Research Center

SUMMARY

Laser velocimeter measurements have been made in the initial mixing region of a confined turbulent diffusion flame burner. Measurements of the axial velocity profiles and the RMS and probability density distributions of the velocity fluctuations show that there are significant variations in the time dependent flow field.

INTRODUCTION

It is now generally accepted that changes in operating conditions which alter the flow patterns in combustion devices can have significant effects on combustion efficiency and pollutant emissions. In particular, the size, shape, recirculated mass flow and local turbulence levels associated with recirculation zones are critical to flame stability, intensity, and overall performance.

Unfortunately experimental examination of the interaction between fluid dynamic and chemical processes inside combustors is difficult since the mean flow fields and turbulence properties of combusting flows with recirculation are difficult to document with any degree of reliability using conventional instrumentation. For instance, flows with severe adverse pressure gradients, which normally give rise to separation and recirculation, are difficult to document as they are extremely sensitive to local geometry and probe interference. In addition, streamline curvature and associated static pressure variations make conventional mean flow instrumentation techniques unreliable. Problems associated with turbulent structure measurements are even more acute because linearized hot wire data interpretations are not

91

accurate in these highly turbulent flows. Fortunately, with the advent of
the laser velocimeter, linear nonperturbing fluid mechanical measurements
of complex three-dimensional flow fields are now possible provided light
scattering particles can be relied upon to follow the local fluid velocity.

This present study of recirculating flows is part of a combustion research
program directed towards developing an understanding of the chemical and
physical processes occurring in combustion devices and their effects on
pollutant formation. Specifically, this study was initiated to obtain
reliable mean velocity and turbulence data in recirculating combusting flow
fields and to determine the effects of swirl and ambient pressure on these
parameters. To achieve this aim, a laser velocimeter which could determine
the direction as well as the magnitude of the instantaneous velocity was
required. Such a system has been developed and used to obtain detailed mean
and turbulence measurements in combusting flow fields.

EXPERIMENTAL DETAILS

Measurements have been made in the initial mixing region of confined
coaxial streams of methane and preheated (850°F) air. A schematic of the
burner geometry employed is shown in Fig. 1. In the present work an outer
(air) to inner (methane) jet exit velocity ratio of 22:1 was maintained
throughout the experiments. This high velocity ratio ensured that a recirculation zone was present above the center jet in all test cases. This aero-
dynamically produced flameholding capability provides an opportunity to
contrast swirling and nonswirling flow fields without the disruptive presence
of a physical flameholder.

As mentioned previously experimental documentation of recirculating
flow fields is a difficult task. Even conventional laser velocimeters are
subject to directional ambiguity which results in data interpretation errors
in highly turbulent and/or recirculating flows. This problem is illustrated
in Fig. 2 where Gaussian probability density distributions of the instantaneous velocities corresponding to local turbulent intensities of 20 and 70
percent are presented. It can be seen that, with directional ambiguity, the
negative velocities are assigned their equivalent positive values which
leads to errors in the calculated mean value and standard deviation. For a
Gaussian probability distribution these errors rise sharply for turbulence
intensities above 40 percent. For example, errors are approximately 5 and
10 percent, respectively, in the mean and RMS fluctuating velocities for a
local turbulence level of 70 percent.

To overcome this problem the mean flow velocity and turbulence measurements were made with a dual beam laser Doppler velocimeter utilizing a liquid
Bragg cell which acted as a beam splitter and frequency shifted the first
deflected beam so that zero velocity frequency offset was achieved. The two
beams were combined at the detection volume where they generated moving
fringes so that a stationary particle produced a Doppler frequency (f_0).

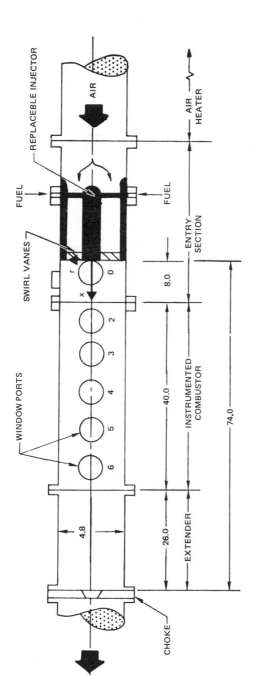

INJECTOR AND SWIRL VANE GEOMETRIES

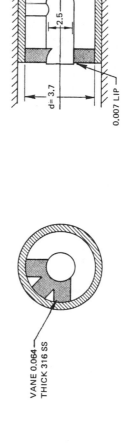

Fig. 1 - Schematic Diagram of Axisymetric Combustion System (All Dimensions in Inches)

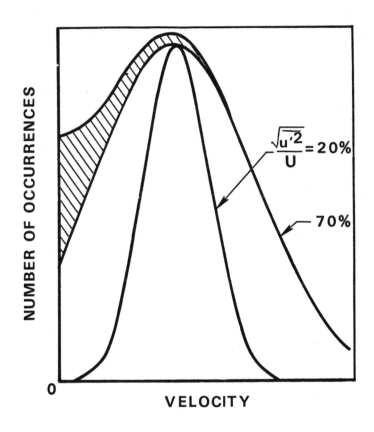

Fig. 2 - Velocity Probability Distributions For Moderate
and High Turbulence Intensities

Thus, in the flow field, particles generated Doppler frequencies of $f_0 \pm f$ depending on their velocities normal to the moving fringes. A detailed description of the optics and signal processing instrumentation was presented in Ref. 1.

SAMPLE RESULTS

Results of an extensive study of the combustor mean and fluctuating velocity fields over a range of ambient pressure and swirl will be presented at a later date. To illustrate some of these data the measured mean and turbulent intensity profiles of the axial velocity component at a swirl number of 0.3 are presented in Fig. 3. The mean velocity profile shows the lateral extent of and reverse velocities within the time averaged recirculation zone 0.75 inches downstream of the injector. There is also evidence of a second recirculation zone caused by flow behind the backward facing step at the nozzle exit plane.

In Fig. 3 it can also be seen that the turbulent fluctuation levels are extremely high in the initial mixing region and reach a maximum near the combustor centerline. It will be shown in the paper that there is an extensive region of highly time dependent flow in the region of the time averaged recirculation zone and that intense turbulent fluctuation levels such as those shown near the centerline in Fig. 3, are due to large scale recirculation unsteadiness about its mean location.

Quantitative insight into the large scale turbulent (unsteady) nature of the recirculation zone may be obtained from the velocity probability density distributions (see Fig. 4 for example). These measurements, which can only be obtained with a zero velocity frequency offset velocimeter, show the unsteadiness of the flow field in the initial mixing region. Within the time averaged recirculation zone there are significant numbers of positive velocity occurrences (approximately 30 percent) which are the result of bubble breakdown and/or convection downstream. This large scale movement is also evident in the photographs taken from a high speed movie film recently produced by Dr. C. T. Borman, UTRC. Successive frames show the extensive movement of the instantaneous flame front. These results emphasize the time dependent nature of the combusting flow field.

REFERENCES

1.   F. K. Owen, AIAA Paper No. 75-120, January 1975.

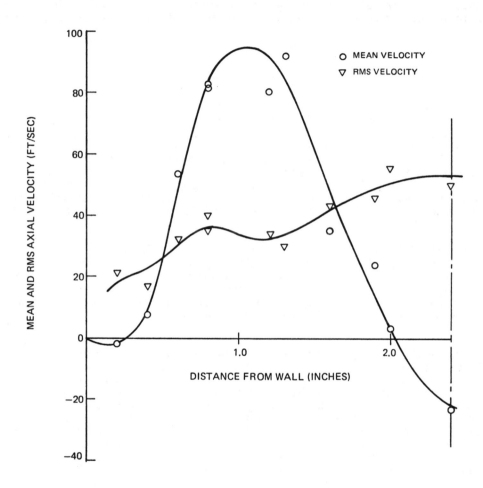

Fig. 3 - Combustor Axial Velocity Distributions
At X = 0.075 Inches

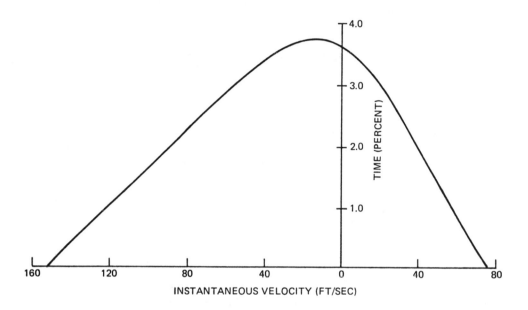

Fig. 4 - Velocity Probability Density Distribution
On Centerline of Combustor At X = 0.75 inches
(J = -24 ft/sec; u'/Ū = -208%)

PROBLEMS OF LASER VELOCIMETER APPLICATION TO COMBUSTION SYSTEMS

JEFFREY A. ASHER
General Electric Company

Professor Chigier indicated earlier that recent developments in laser velocimetry (LV) have made this technique quite viable in a wide range of combustion applications. In the last five years General Electric has used the LV technique in applications which include gas flows in speed ranges from low subsonic to hypersonic at temperatures of -100° to 3000°F.

Unfortunately the LV techniques still requires more than the understanding of an instrument instruction book. It is to this point that I would like to focus these informal remarks.

As a rule the LV still requires that the user have a fundamental understanding of optics, electronic processing and particle dynamics. To begin with, Dr. Owen of P&W mentioned the use of the velocity probability distribution or histogram in representing LV data output. The histogram (Fig. 1) is the velocity (x axis) versus the number of particles at that velocity (y axis). If a sufficient number of velocity data are used, this distribution becomes time independent.   "Sufficient" here depends on the turbulence level (first standard deviation or RMS) of the velocity. For low turbulence (Fig. 1, x/d=3) the distribution can be obtained rapidly owing to the small fluctuations in velocity. As shown in Fig. 1, x/d = 10, high turbulence requires significantly longer data acquisition times since a much wider fluctuating velocity range is being encountered.

For high turbulence flows, which tend to characterize combustion systems, the problems encountered with applying the LV technique appear in the following areas:

1. Optics Set Up - Turbulent eddies or thermal gradients in a flow can
    provide refraction effects. These effects may either alter the
    position of or destroy the scattering volume where the incident
    laser beams cross and from which measurements are made. The
    altered position of the scattering volume results primarily

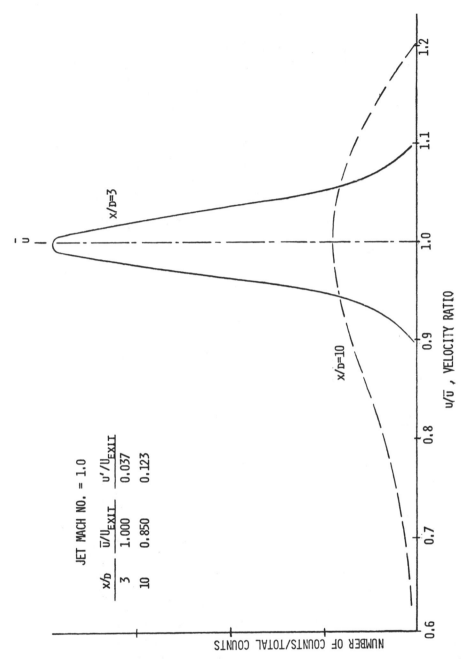

Fig. 1. Laser Velocimeter Velocity Probability Distribution on the Center-line of an Axisymmetric Air Jet.

99

in the uncertainty of the measurement location (in backscatter)
or the ceasing of data acquisition entirely (for non-back-
scattered angles). In some cases a steady or unsteady lapse
in data rate may occur if the two beams do not cross due to
refraction effects (i. e., no scattering volume exists).

2. LV Processor - Due to the use of a finite Doppler frequency
    range selection for both a counter and tracker types of process-
    or, significant bias can occur if the frequency occurs outside
    this range. Unfortunately for high turbulence (i. e., a wide
    velocity range), the associated Doppler frequency range usually
    prevents data acquisition within one band selection. Special
    care and more test time is required to laboriously merge the
    two histograms together without adding a velocity bias. This
    is especially important in deducing the turbulence level.

3. Particle Dynamics - The LV technique utilizes the particle
    velocity from which the flow velocity is deduced directly.
    Particles are no asset to the experimentalists. They must be
    put into the flow in sufficient quantity to obtain relatively
    rapid test times and additionally be small enough so that the
    particles will follow the fluid. (In combustion systems the
    problem of particle phase change must be considered.) Our
    philosophy at GE has been to add particles of a known size and
    constituency where these parameters can be defined à priori to
    meet the requirements imposed by the flow and the objectives of
    the test.

    In connection with turbulence measurements, significant errors
    can occur if the fluid turbulence spectra contains high frequency
    components to which larger particles will not respond (Fig. 2).
    In highly decelerating (or accelerating flows) such as with a
    normal shock, pronounced changes between true and measured mean
    velocities are shown to occur for particle sizes larger than 0.5
    microns (Fig. 3).

    Lastly, the problem of concentration and the resultant effect on
    velocity measurements can be important. An example of this effect
    is in the high shear region of a jet flow ($r/d \geq 0.5$). Here mixing
    occurs between the fluid molecules from the higher speed jet
    efflux and the lower speed entrained air. For the case where
    only the jet efflux is seeded with particles, the mean velocity
    may be high by 5 or 6% and the data acquisition time increased
    many fold. In the case of an axisymmetric jet, the amount of seed
    concentration in the entrained and main flows has still not been
    quantitatively ascertained and confirmed by experiments.

In summary, the use of the LV system in a combustion system for the
measurements of velocity has a long way to go before it can be used as easily
as a pitot - static probe. With care, though, this technique has already and
will continue to harvest data in situations which have defied measurement
until now.

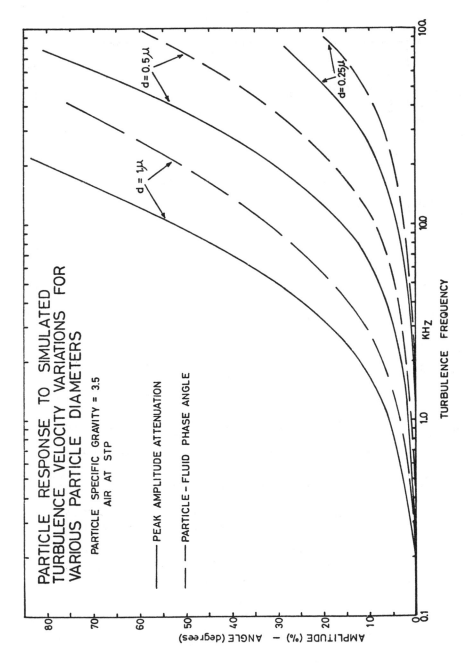

Fig. 2

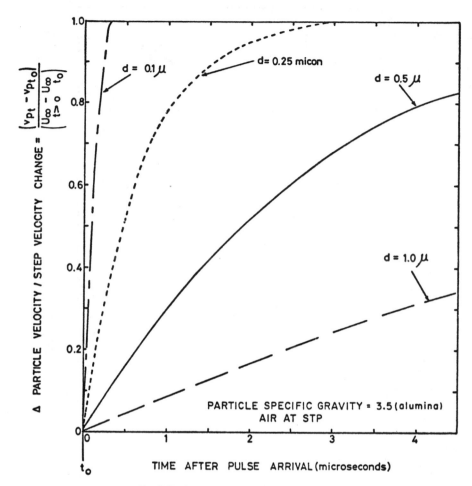

PARTICLE MOVEMENT WITH STEP
CHANGE IN VELOCITY FOR VARIOUS
PARTICLE DIAMETERS

PARTICLE SPECIFIC GRAVITY = 3.5 (alumina)
AIR AT STP

Fig. 3

102

LASER ANEMOMETRY IN
HIGH VELOCITY, HIGH TEMPERATURE
BOUNDARY LAYERS

S. A. SELF
Stanford University

A laser anemometer has been developed and used to measure profiles of average velocity and turbulence intensity in the electrode wall boundary layer of the Stanford 2 MW and 8 MW MHD systems. The instrument design is described in some detail in Ref. 1. Recent measurements under subsonic and supersonic conditions are reported and compared with computations in Ref. 2.

The basic design of the anemometer was primarily determined by the constraints imposed by the goemetry of the MHD rig, as well as the conditions of the measurements required. The latter called for measurements of high spatial resolution (0.1 mm) close to the electrode wall in hot, turbuent, high velocity gases in which there were negligible scattering centers naturally present. The geometry of the magnet and channel, together with the desire to avoid access ports which might perturb the flow in the measurement region, led to the scattering geometry shown in Fig. 1.

A single beam coaxial backscatter system is employed, with optical access from the downstream sidewall at a small angle ($\theta = 26°$) to the flow axis. With this arrangement the Doppler frequency shift $\delta\nu = 2$ u $\cos\theta/\lambda$ is too large for heterodyne detection and is determined optically, before detection, by a scanning Fabry-Perot interferometer.

The lens and mirror arrangement shown is used to separate the coaxial incident and scattered beams. The Ar^+ laser beam (100 mW single mode at 514.5 nm) is expanded to 25 mm diameter by telescope T_1, focused by lens L_1 (f = 60 cm), deflected by the elliptic-section plane mirror M_1 and then deflected again by the scan mirror M_2 into the port. The latter is purged with dry N_2 to prevent water and seed condensate on the window,

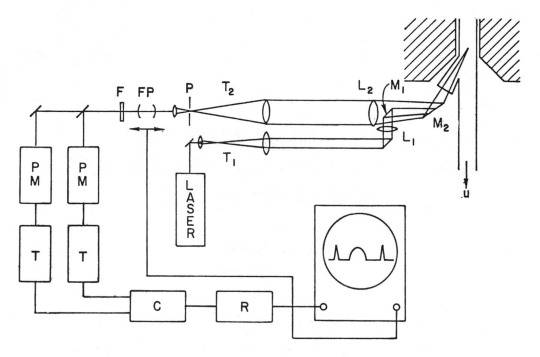

Fig. 1 - Laser Anemometer Design

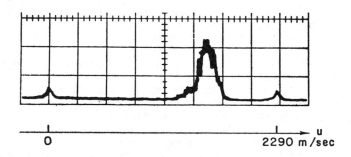

Fig. 2 - Velocity Probability Density Function
P(u) - Signal Trace

104

which is protected by a shutter when not in use. The incident beam is
focused to a waist (< 50 μm diameter) opposite the electrode, which it
intercepts near grazing incidence, and the focus can be traversed through
the boundary layer between y = 0 and 30 mm by the micrometer controlled
scan mirror.

Light scattered from the focus is collected and collimated by lens
L_2 (f = 60 cm) and compressed to 1 mm diameter by telescope T_2. The
latter includes a spatial filter (pinhole 25 μm chain) to reject radiation
from the plasma and walls, and particularly to reject laser light scattered
from the walls. After compression the collected beam enters the confocal
spherical Fabry-Perot interferometer, whose mirror separation is scanned
piezoelectrically, over one free spectral range, in synchronism with the
oscilloscope timebase at ~ 1 Hz. Alternate sets of mirrors, with free
spectral ranges Δf = 2 GHz and 8 GHz were used for subsonic and supersonic
measurements respectively. The Fabry-Perot finesse is > 50, corresponding
to a velocity resolution < 2% of $u_{max} = \lambda\Delta f/2\cos\theta$, corresponding to the
free spectral range.

Scattered light pulses due to the passage of individual particles
through the focus, which typically contains ~ 100 photons, then pass through
a spectral filter (width 2 nm) and are divided between two photomultiplier
detectors. This allows the use of coincidence techniques to discriminate
against dark current pulses and shot noise pulses due to laser light scattered from the walls. The rate of signal pulses in coincidence is determined by the ratemeter, and the signal rate is displayed on the storage
oscilloscope.

Subject to the requirement that the particle flux is uniform, this
signal rate represents the velocity probability density function P(u),
from which the average velocity and turbulence intensity may be measured.
A typical display (Fig. 2) shows P(u) and reference signals at u = 0 and
u = u_{max}, which arise from laser light, unshifted in frequency, scattered
from the channel wall.

The flow was seeded with ZrO_2 particles of ~ 1 μm diameter at a rate
of ~ 1 g/sec, to yield displays of P(u) at a given y-position in a single
sweep (~ 1 sec). The particles were dispersed into the plenum chamber
preceeding the channel, via a subsidiary flow of N_2, from a specially
designed powder feeder, which was activated only when a velocity measurement was required.

Results of measurements of the boundary layer profiles of average
velocity and turbulence intensity in a test under supersonic conditions,
with a free stream velocity of 1600 m/sec, are shown in Fig. 3. Plotted on
logarithmic scales the profile u(y) approximates a straight line corresponding to a dependence $(u/u_\infty) = (y/\delta)^{1/n}$ with n ~ 5. The relatively
high level of free-stream turbulence is thought to consist primarily of low
frequency (< 1 kHz) fluctuations associated with the combustor. A comparison of these measurements with calculations is made in Ref. 2, where
further velocity measurements under subsonic conditions are also detailed.

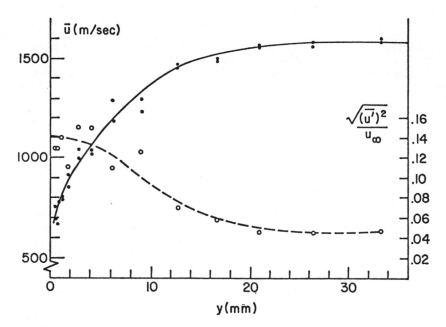

Fig. 3a - Mean Velocity and Turbulence Intensity Profiles

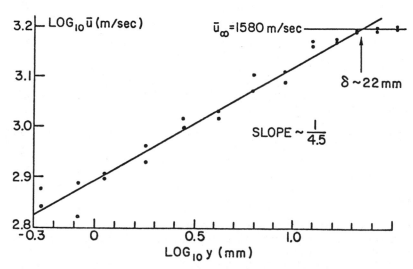

Fig. 3b - Logarithmic Fit of the Mean Velocity
Profile (n = 4.5)

The principal limitations of this instrument are:

(i) Interference from laser light-scattered from the channel walls, especially for small y. This results in unduly large reference signals which preclude measurements for y $\lesssim$ 0.5 mm.

(ii) The need to seed the flow with relatively large fluxes of particles to obtain a satisfactory rate ($\sim 10^5$/sec) of detected signals. This arises from the very small scattering volume combined with the use of a single-channel spectrum analyzer, which makes use of only $\sim$ 2% of the available signals. Furthermore fluctuations of particle flux due to imperfect mixing in the plenum chamber can give rise to sperious modulation of the observed P (u).

Nevertheless, valuable measurements of considerable gasdynamic significance for MHD can now be made. Future plans include a study of changes of velocity profile as a result of MHD interaction.

REFERENCES

1. S. A. Self, "Boundary Layer Measurements in High Velocity High Temperature MHD Channel Flows," Proceedings of Second International Workship on Laser Velocimetry, Vol. II, edited by H. D. Thompson and W. H. Stevenson, Purdue University, March 1974.

2. J. W. Daily, C. H. Kruger, S. A. Self and R. H. Eustis, "Boundary Layer Profile Measurements in a Combustion Driven MHD Generator." Sixth International Conference on MHD Electrical Power Generation, 1975.

Absorption-Emission and
Resonance Techniques
Chairman: W. K. McGREGOR

ABSORPTION-EMISSION MEASUREMENTS
IN JET ENGINE FLOWS

W. K. McGREGOR
ARO, Inc.

Absorption and Emission Techniques are the oldest forms of optical measurement, going back at least a century, to the times of Beer. Thus they are well understood and somewhat mundane in the context of concentration measurements in homogeneous samples for instance (Ref. 1). Temperature is routinely obtained by measuring the concentrations of two excited states in the same band and by using the Boltzmann equilibrium relationship (Ref. 2). Their clear advantage over scattering techniques, in such a case, lies in their much larger cross section.

There exists combustion situations of near-uniform properties across the flow. In such a case the absorption-emission techniques are straightforward (Refs. 3, 4). However, many important flows - flames, engine exhausts, combustors... - are definitely non-homogeneous and considerable difficulty arises in the interpretation of the data. A mathematical inversion technique must be used to extract the concentration distribution along the path of observation. It can be based on either spatial or frequency scanning.

Spatial scanning is a well established absorption-emission technique for axisymmetrical geometries. It was developed in closed form by Abel (Ref. 3). In most practical problems, the "onion peeling" technique (Fig. 1) enables the observer to make a direct measurement (a) of the absorption or emission of the outer zone (1) first, then to move one step closer to the center (b) where the acquired knowledge on zone 1 leaves the properties of zone 2 as the only unknown, and so on... This method works rather well in a number of applications (Refs. 4, 5) but some cumulative loss of accuracy takes place as we approach the center zone especially if turbulence is present. Furthermore, in the very practical case where the flow of interest is not axisymmetrical, the method loses its validity.

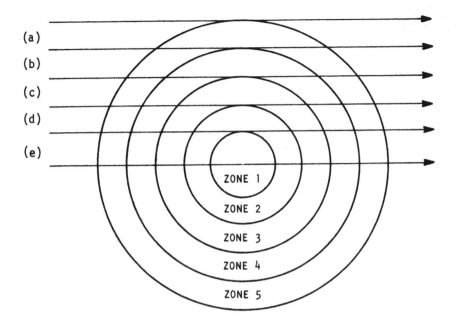

Fig. 1 - The Onion-Peeling Technique (Ref. 4).

Frequency scanning consists in measuring emission at different frequencies on a given path. The choice of frequencies should be such that the strongly absorbing frequencies would reflect mostly the contribution near the emerging point of the beam (self absorption), whereas the cumulative emission of the whole beam would emerge at less absorbing frequencies (optically thinner cases). An inversion technique of these measurements yields a temperature or concentration profile. This technique has been used in some flame applications (Refs. 6, 7). Its advantage is that it does not depend on any assumption of axisymmetry; each line of sight is resolved separately.

As you can see, non-homogeneous flows present a serious problem to line-of-sight optical techniques. However, most combustion products have been measured successfully for concentration by this technique in near uniform conditions. Water vapor is measured on a regular basis, mostly at 2.5μ where there is little atmospheric interference. Carbon oxide (CO_2) is measured with even better accuracy around 4.5μ. The hydroxyl radical (OH) is also in good shape, through measurements of the $^2\Sigma$-$^2\pi$ electronic transition around 3090 Å (Ref. 6). Nitric oxide (NO) is also measured with good accuracy by using a resonance lamp in the γ-band interval (2200-2270 Å) (Ref. 8). Carbon monoxide (CO) electronic resonance lines lie too far in the UV for practical measurements. Infrared measurements, such as those performed near 2200 cm^{-1} by gas filter correlation methods have shown good accuracy (Ref. 9).

Hydrocarbons (C_xH_y) are difficult to separate from each other, since we have not been able to find tunable lasers in the 3μ range, where the fine structure of the CH bonds could be scrutinized. For temperature, a lot of work has been done on atomic lines, such as Na and K, by reversal or absorption techniques.

REFERENCES

1. R. W. Ladenburg et al, ed. "Physical Measurements in Gas Dynamics and Combustion," High Speed Aerodynamics and Jet Propulsion Series, Vol. IX, Princeton University Press, 1954.

2. R. H. Tourin, "Spectroscopic Gas Temperature Measurements - Pyrometry of Hot Gases and Plasmas," Elsevier Publishing Company, New York, 1966.

3. H. R. Griem, "Plasma Spectroscopy," McGraw Hill, New York, 1964.

4. M. Griggs and F. C. Harshbarger, "Measurements of Temperature Profiles at the Exit of Small Rockets," Applied Optics, Vol. 5, No. 2, February 1966.

5. W. Herget, "Temperature and Concentration Measurements in Model Exhaust Plumes Using Inversion Techniques," from the Proceedings of the Specialist Conference on Molecular Radiation, October 5-6, 1967, NASA TMX-53711.

6. M. E. Neer, "Numerical Calculations of UV Emission and Absorption Spectra of OH," AFSC-ARL, TR-74-0109, August 1974.

7. C. M. Chao and R. Goulard, "Nonlinear Inversion Techniques in Flame Temperature Measurements," from "Heat Transfer in Flames" (N. H. Afgan and J. M. Beer, eds.), Scripta Book Company, Washington, DC, 1974.

8. M. G. Davis, W. K. McGregor and J. D. Few, "Spectral Simulation of Resonance Band Transmission Profiles for Species Concentration Measurements: NO_γ-bands as an example," AEDC TR-74-124, January 1975.

9. D. E. Burch and D. A. Gryvnak, "Infrared Gas Filter Correlation Instrument for In-Situ Measurement of Gaseous Pollutants," A Philco-Ford Corporation Report EPA-650/2-74-094, December 1974 (See also AIAA Paper 76-110).

ABSORPTION-EMISSION SPECTROSCOPY APPLIED TO THE STUDY OF POLLUTANT KINETICS AND REACTION INTERMEDIATES IN HIGH-INTENSITY CONTINUOUS COMBUSTION

PHILIP C. MALTE
STEPHEN C. SCHMIDT
DAVID T. PRATT
Washington State University

An area currently receiving major attention in the Combustion and Thermal Fluids Research Group at Washington State University is the application of optical diagnostic techniques to the characterization of high-intensity continuous combustion. Path-integrated absorption and emission spectroscopy is being used in conjunction with a laboratory jet-stirred reactor (a modified Longwell reactor, Ref. 1) to measure gas temperatures and molecular and atomic free-radical concentrations in situ. Application of the research is to the understanding and definition of combustion and pollutant kinetic mechanisms relevant to high-intensity continuous combustors, such as gas turbine combustors. Particular attention has been directed to NO_x formation chemistry.

The jet-stirred reactor is an attractive device for laboratory studies of chemical kinetics. The nearly spatially uniform (with respect to time-mean composition and temperature), highly nonequilibrium combustion volume of a reactor operated near chemically rate-limited blowout may be sampled by both physical probes and path-integrated optical techniques and also may be theoretically modeled as "perfectly stirred reactor" (PSR) (Ref. 2). Within the reactor, the nonequilibrium chemistry is not confined to narrow one-dimensional flame zones (as in "flat flame" laboratory burners) but is distributed nearly uniformly throughout the combustion volume. Hence, the spatial problems inherent in experimental characterization of chemical processes within one-dimensional flames due to steep temperature and concentration gradients have been circumvented. Physically, the reactor is designed with very small inlet jets ($\sim$1 mm) and operated at high inlet jet velocities ($\sim$sonic) in order to minimize mixing times. Analytically, the chemical kinetic system within the reactor is described to good approximation by a matrix of algebraic reaction rate equations, rather than by a system of differentail reaction rate equations. Elementary chemical reactions occur simultaneously (zero-dimensional) rather than sequentially as in one-dimensional flames.

One of the jet-stirred reactors fitted with optical access ports which is currently being used at Washington State University is shown in Fig. 1. The reactor, which is cast from calcia-stabilized zirconia with an internal volume of 50 cm^3 and a single-pass optical path length of 5 cm, is designed as a conical segment of a Longwell reactor (Ref. 1) with optimization for micromixing based on the work of Hottel et al, (Ref. 3) (see Ref. 4 for a further description of the reactor). Reactor fuels have been mainly methane and carbon monoxide (with 20 to 2000 ppm inlet H_2O).

Spectroscopic path-integrated emission and absorption measurements are performed using scanning monochromators (e.g., Spex 1-meter) and appropriate optical transfer systems (e.g., multiple-pass). A Xe lamp is used as a light source. Data interpretation employs conventional spectroscopic techniques which have been established in one-dimensional combustion experiments over the past 30 years (see, for example, Gaydon, Ref. 5; Penner, Ref. 6; Kaskan, Refs. 7 & 8; and Bowman, Ref. 9).

Specifically, measurements of concentrations of the key reaction inter-mediates O, OH and CH in the high-intensity CH_4/air and CO/air combustion are being used to study the effects that the nonequilibrium processes of energy release and oxidative pyrolysis exert on the kinetics of pollutant formation and destruction. Major emphasis has been on N_2 fixation to NO and NO_2. CO combustion efficiency has received some attention, "fuel NO_x" studies have been initiated, and an investigation of the SO_3 formation mechanism has been proposed.

A finding of NO_x experiments completed to date (using PSR chemical kinetic modeling for data interpretation) is that, for combustion temperatures below approximately 1800°K, the super-equilibrium O-atom concentrations appear to directly affect NO_x formation through a kinetic mechanism which involves nitrous oxide (N_2O) as an NO_x precursor, (Ref. 4) and not solely through the well known Zeldovich mechanism, as previously assumed.

Prior to discussing specific spectroscopic measurements, a comment regarding probe-sampling techniques is felt to be in order. In principle as well as in practice, it is important that concentrations of stable species also be measured in situ, rather than relying completely on gas-sampling techniques. In gas sampling from highly nonequilibrium combustion regions, gas-phase and gas-probe wall reactions involving free radicals and stable species are quite likely. For example, it is well established that the NO-NO_2 proportionality is affected by probe sampling (Ref. 10)(see footnote below*) and that the CO-CO_2 proportionality also may be affected, (Ref. 11) albeit

---

*Our experience, particularly for fuel-lean CO/air combustion giving super-equilibrium O-atom concentrations of 1000 to 6000 ppm, indicates that NO concentrations of order 30 ppm are oxidized completely to NO_2 in a water-quenched quartz sampling probe. In this process the following reactions are believed important: $O + O_2 \rightarrow O_3$, $O_3 + NO \rightarrow NO_2 + O_2$, $NO + O \rightarrow NO_2$, $NO_2 + O \rightarrow NO + O_2$.

to a lesser extent. A most promising course of action for determining trace
concentrations of important stable species such as SO_3, HCN, N_2O, etc. in situ
is to utilize a tunable dye laser (Refs. 12 & 13) to conduct high-resolution
spectroscopic measurements which can isolate the individual vibration-rotation
lines of these species.

O-ATOM MEASUREMENTS

Quantitative measurements of O-atom concentration are conducted by
utilizing a scanning monochromator to monitor the (apparent) continuum
chemiluminescence in the near ultraviolet (we have worked at 3500 Å) due to
the CO + O recombination (Ref. 14):

$$CO + O \rightleftharpoons CO_2^* \ (^3B_2)$$

$$CO_2^* \ (^3B_2) + M \rightleftharpoons CO_2^* \ (^1B_2) + M$$

$$CO_2^* \ (^1B_2) \rightarrow CO_2 \ (v) + h\nu$$

Since the excited electronic state from which the transition originates is
bent, transitions to highly excited vibration states (large v) are likely and
a vibrational propulation inversion may be feasible. (Ref. 15). The absolute
O-atom response of the reactor/optical system is calibrated by operating the
reactor at fuel-lean mixtures, moderate residence times ($\gtrsim 10$ msec), and
sufficiently high temperatures ($\gtrsim 1500°$K). At these conditions, based on
experiment and analysis, nearly complete conversion of fuel-carbon to CO and
CO_2 exists and partial equilibrium for the $O/H/OH/H_2/H_2O$ subsystem and the
reaction CO + OH $\rightleftharpoons$ CO_2 + H is suspected to occur (Ref. 4). Measurements of
T and [CO] then give a measurement of [O]. The collisional deactivation of
the excited CO_2 is included in data interpretation, and the activation energy
of the overall process CO + O $\rightarrow$ CO_2 + hν is taken as 2.5 kcal/mole. We have
measured O-atom concentrations as low as 30 ppm with this method.

OH-RADICAL MEASUREMENTS

Electronic ground-state OH-radical concentrations are determined from
the continuum radiation absorbed in the first order by the $A^2\Sigma^+ - X^2\pi(0,0)$
transition. The measured equivalent widths of isolated spectral lines are
interpreted using curve-of-growth methods and published line widths to determine the concentration in a specific vibration-rotation level. Total ground-
state populations then are inferred using a Boltzmann distribution based on
a kinetic temperature. Electronically excited OH will receive attention later.

Since it was desired to begin spectroscopic OH measurements with a simple
system, preliminary measurements have been conducted for H_2/air combustion.
It must be noted therefore that because of the rapid H_2 burning rate (i.e.,
short chemical reaction time) the reactor is more susceptible to mixing
influence, and the energy-releasing chemistry more readily approaches
equilibrium.

CH AND C_2 EMISSION MEASUREMENTS

Our spectroscopic analyses for hydrocarbon fragments to date have considered only the prominent emission bands due to CH* ($^2\Delta$) and C_2^* ($^3\pi$). It is instructive to point out for methane combustion that the integrated (with respect to wavelength) emission for the CH $^2\Delta$ - $^2\pi$ electronic transition may be interpreted in light of the following postulated kinetic mechanism (Refs. 16 & 17):

$$CH + OH + H \rightarrow CH^* (^2\Delta) + H_2O$$

$$C_2 + OH \rightarrow CH^* (^2\Delta) + CO$$

$$CH^* (^2\Delta) \rightarrow CH + h\nu$$

The first reaction leads to a linear proportionality $h\nu \sim [CH]$, while the second reaction causes the emission intensity to depend to second order on the concentration of hydrocarbon fragments. That is, for methane oxidative pyrolysis, C_2 apparently results from the reaction of two hydrocarbon fragments; e.g., $CH + CH \rightarrow C_2^* + H_2$ followed by a C_2^* deactivation. Measurements (Ref. 16) conducted for methane traces added to an $H_2/O_2/N_2$ flame indicate a nearly linear relationship between $CH^* (^2\Delta)$ emission intensity and hydrocarbon concentration. The occurrence of this behavior during high-temperature oxidative pyrolysis of methane in jet-stirred reactors would yield an inferred measurement of the product [CH][OH][H] from the observed emission intensity. This steady-state measurement, in conjunction with NO_x measurements and with the O-atom and OH-radical measurements that define the state of the hydrogen/oxygen subsystem, would contribute to an understanding of the pyrolysis fragment/prompt NO_x link. (Ref. 18).

ADDITIONAL PATH-INTEGRATED MEASUREMENTS

Finally, the following in situ measurements and objectives are envisioned as a logical continuation of the above experiments:

1. CH and C_2 in absorption: Prompt NO_x.

2. CH, NH, NH_2, HCN: Prompt NO_x and fuel NO_x.

3. NO and NO_2: Combustion-formed vs. probe-formed NO_2.

4. Oxidation of SO_2 to SO_3 under fuel-lean, low-temperature, non-equilibrium combustion conditions.

INFLUENCE OF FLUID MECHANICS

The measurements described on the preceding page attempt to achieve a chemically rate-controlled environment, i.e., conditions such that the effects of fluid mechanics are small or negligible. In many instances for slow reaction chemistry it is possible to verify the time-mean spatial homogeneity of the reactor's scalar properties using physical probes. However, a question remains concerning the effect of turbulent fluctuations of temperature and concentration on the path-integrated spectroscopic measurements.

It is possible to estimate the magnitude of turbulent concentration fluctuations and therefore to argue that their effect on the path-integrated measurements should be small for a properly designed, chemically rate-controlled stirred reactor. In order to relate the turbulent fluctuations to the chemical process, the time-averaged, Reynolds-decomposed conservation-of-species equation is utilized (neglecting molecular transport). With K_i equal to the mass fraction of species i, and with w_i equal to the corresponding chemical production rate (sec^{-1}), the equation is

$$\frac{\partial \overline{K}_i}{\partial t} + \overline{u}_j \frac{\partial \overline{K}_i}{\partial x_j} + \overline{u'_j \frac{\partial K'_i}{\partial x_j}} = \overline{w}_i$$

where ( )' and $\overline{(\ )}$, respectively, denote at a point a turbulent fluctuation and a time-mean property. For a steady-state stirred reactor with (spatial) homogeneity in time-mean composition, the first two terms of the equation are zero. That is, the mean chemical production rate $\overline{w}_i$ (which is the same everywhere in a homogeneous reactor) must equal the rate at which local fluctuations in concentration are diffused.

The magnitude analysis is conducted by utilizing the following substitutions for the two remaining terms:

$$\overline{u'_j \ \partial K'_i \ / \ \partial x_j} \sim \sqrt{\overline{u'^2}} \ \sqrt{\overline{K_i'^2}} \ \Big/ \ x = \sqrt{\overline{K_i'^2}} \ \Big/ \ \tau_t$$

where x and τ_t are, respectively, appropriate turbulent length and time scales, and

$$\overline{w}_i = (\overline{K}_i - \overline{K}_i^*) \ \Big/ \ \tau$$

where $\overline{K}_i^*$ = reactor inlet mass fraction, $\overline{K}_i$ = reactor mass fraction, and τ = reactor mean residence time; this substitution applies to a chemically rate-limited stirred reactor. Therefore,

$$\sqrt{\overline{K_i'^2}} \ \Big/ \ (\overline{K}_i - \overline{K}_i^*) \sim \frac{\tau_t}{\tau}$$

With the rms velocity fluctuation estimated as about one-quarter of the feed-jet velocity, and with x taken as the inlet jet size (i.e., its maximum value) then for reactor operation with (1mm-dia.) sonic jets $\tau_t \sim 10^{-5}$ sec. Since $10^{-3} \lesssim \tau \lesssim 10^{-2}$ sec, it follows that

$$\sqrt{\overline{K_i'^2}} \Big/ (\overline{K_i} - \overline{K_i^*}) \lesssim 10^{-2}$$

With reduced jet velocities and increased turbulent length scales, that is, for a mixing-influenced combustor (even with reactants premixed), temperature and composition fluctuations within the reactor volume may become large. Spatial variations in the time-mean scalar properties also occur, particularly as aircraft (and industrial burner) conditions are approached. Hence, it becomes extremely interesting and important to study experimentally the actual spatial and temporal species concentration and temperature variations and the interaction of the time-and-space-varying fluid mechanics and chemistry. Laser-Raman scattering is the candidate technique for such measurements and is consequently the subject of a proposed study by the authors. In addition to providing important statistical data on localized spatial ($\sim$1mm) and possibly temporal variations of the reacting species and temperature to support modeling studies of the combined fluid mechanical/chemical system, Raman scattering measurements also can be used to delineate more clearly the effect of inhomogeneities on path-integrated absorption-emission spectroscopy.

Research supported by NSF Grant ENG73 20136-A01 and Gen. Motors Corp.

REFERENCES

1. J. P. Longwell and M. A. Weiss (1955). "High Temperature Reaction Rates in Hydrocarbon Combustion," Ind. and Eng. Chem., 47, p. 1634.

2. H. Jones and A. Prothero (1968). "The Solution of the Steady-State Equations for an Adiabatic Stirred Reactor," Comb. and Flame, 12, p.457.

3. H. C. Hottel, G. C. Williams, and G. A. Miles (1967). "Mixedness in the Well-Stirred Reactor," Proceedings of the 11th Symposium (International) on Combustion, The Combustion Institute, Pittsburgh, PA, p. 771.

4. P. C. Malte and D. T. Pratt (1975). "Measurement of Atomic Oxygen and Nitrogen Oxides in Jet-Stirred Combustion," Proceedings of 15th Symposium (International) on Combustion, The Combustion Institute, Pittsburgh, PA.

5. A. G. Gaydon (1974). The Spectroscopy of Flames, John Wiley & Sons, Inc., New York, New York.

6. S. Penner (1959). Quantitative Molecular Spectroscopy and Gas Emissivities, Addison Wesley, Redding, Mass.

7. W. E. Kaskan (1958). "Hydroxyl Concentrations in Rich Hydrogen-Air Flames Held on Porous Burners," Combustion and Flame, 2, p. 229.

8. W. E. Kaskan and D. E. Hughes (1973). "Mechanism of Decay of Ammonia in Flame Gases from $NH_3/O_2/N_2$ Flames," Combustion and Flame, 20, p. 381.

9. C. T. Bowman (1974). "Non-Equilibrium Radical Concentrations in Shock-Initiated Methane Oxidation," paper presented at 15th Symposium (International) on Combustion, Tokyo, Japan.

10. J. D. Allen (1975). "Probe Sampling of Oxides of Nitrogen from Flames," Combustion and Flame, 24, p. 133.

11. W. E. Kaskan (1959). "Excess Radical Concentrations and the Disappearance of Carbon Monoxide in Flame Gases from Some Lean Flames," Combustion and Flame, 3, p. 49.

12. K. W. Nill (1974). "Tunable Infrared Lasers: Prospects for Instrument Applications," Proceedings of Society of Photo-Optical Instrumentation Engineers, 49.

13. K. G. R. Sulzmann, J. E. Lowder, and S. S. Penner (1973). "Estimates of Possible Detection Limits for Combustion Intermediates and Products with Line-Center Absorption and Derivative Spectroscopy Using Tunable Lasers," Combustion and Flame, 20, p. 177.

14. M. A. A. Clyne and B. A. Thrush (1962). "Mechanisms of Chemiluminescent Combination Reactions Involving Oxygen Atoms," Proc. of Royal Society of London, Vol. 269A, p. 404.

15. A. K. Levine (1968). Lasers, Vol. II, Marcle Dekker, Inc., New York.

16. J. Peeters, J. F. Lambert, P. Hertoghe, and A. Van Tiggelen (1973). "Mechanisms of C_2* and CH* Formation in a Hydrogen-Oxygen Flame Containing Hydrocarbon Traces," Proceedings of 13th Symposium (International) on Combustion, The Combustion Institute, Pittsburgh, PA, p. 321.

17. R. P. Porter, A. H. Clark, W. E. Kaskan, and W. E. Browne (1967). "A Study of Hydrocarbon Flames," Proceedings of 11th Symposium (International) on Combustion, The Combustion Institute, Pittsburgh, PA, p. 907.

18. C. P. Fenimore (1971). "Formation of Nitric Oxide in Premixed Hydrocarbon Flames," Proceedings of 13th Symposium (International) on Combustion, The Combustion Institute, Pittsburgh, PA, p. 907.

BOUNDARY LAYER MEASUREMENTS OF TEMPERATURE AND ELECTRON NUMBER DENSITY PROFILES IN A COMBUSTION MHD GENERATOR

JOHN W. DAILY
C. H. KRUGER
Stanford University

Temperature and electron density measurements have been made in an MHD generator by spectroscopic means. Spatial and spectral resolution are obtained with a rotating mirror optical system and a scanning monochromator. The optical system is shown in Fig. 1.

A tungsten lamp is imaged into the plasma for use with the line reversal temperature measurement. Light emitted from the plasma is collimated and sent to a pivot-mounted mirror that also serves as the aperture stop. From the mirror the light is then passed through a second lens and imaged on the entrance slit of the monochromator. Because the mirror is at the focal point of the field lens, only that light which is parallel to the lens axis and whose solid angles and image size are defined by the mirror stop and the entrance slit respectively will be admitted. Scanning is accomplished by rotating the mirror. Resolution in the y-direction is approximately 0.3 mm. One unusual aspect of the system is that cylindrical lenses are used, allowing a larger aperture in the x-z plane. Thus, for a given resolution in the y direction, the total solid angle can be maximized and the emission signal used in the electron density measurements is increased.

Optical access to the MHD channel is provided by two slots on opposite sides of the channel. The slots are 0.5 cm wide and 4.0 cm in height. These slots are capped with quartz flats which are film cooled with a nitrogen purge gas. In addition, pneumatically-operated shutters are provided which protect the flats when measurements are not being made.

John W. Daily's present address is Department of Mechanical Engineering, University of California, Berkeley, Calif. 94720.

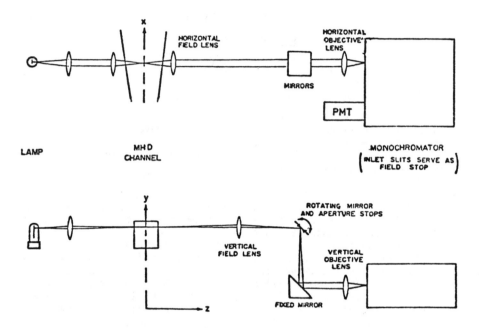

Fig. 1 - Optical Scanning System

120

The gas temperature was measured using the sodium D line wing reversal method which has been described by Brederlow, et. al. (Ref. 1). This is essentially a comparative measurement; in the present case the standard source was a calibrated tungsten strip lamp whose intensity was varied by means of current control. The reversal condition occurs at a given wavelength when the lamp current is such that the intensity from the lamp alone is just equal to the intensity from the lamp plus the plasma. For a uniform plasma it is easily shown that at reversal the gas temperature is equal to the equivalent black body temperature of the lamp. For a non-uniform plasma, with relatively cool sidewall boundary layers, the reversal temperature is reduced below the centerline gas temperature. A correction for this effect was obtained from a numerical solution of the equation of radiative transfer using a calculated sidewall temperature distribution along the line of sight. Fig. 1 shows the resulting temperature correction as a function of wavelength for two sidewall boundary layer thicknesses. In the wings of the line where the optical depth is less than unity, the correction is a minimum, almost constant, and nearly linear in δ. Present measurements were generally made at a wavelength of 5887 Å.

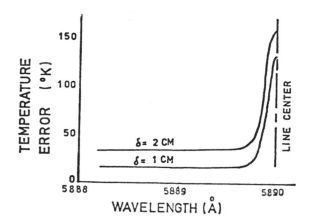

Fig. 2 - Line Reversal Temperature Error as a Function of Wavelength for Two Sidewall Boundary Layer Thicknesses.

Fig. 3 shows the correction in temperature for a fixed boundary layer thickness at λ = 5887 Å as a function of reversal temperature for a sidewall temperature of 1750°K. Again, over the range of conditions considered here, the effect is almost linear. Such numerical corrections to the reversal temperature were applied at each point in the electrode boundary layer profile. The correction is fairly insensitive to the assumed sidewall profile shape.

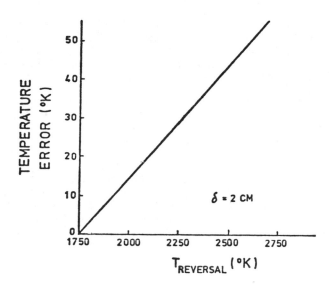

<u>Fig. 3</u> - Line Reversal Temperature Error as a Function of Reversal Temperature

To perform the reversal measurement the intensity was first recorded as the image of the lamp was scanned in the y direction away from the electrode surface with the plasma absent. Then for each MHD operating condition, the procedure was repeated, including an emission only profile at zero lamp current. These results were later cross-plotted to obtain the profile of reversal temperature. This profile was then corrected for the effect of side-wall boundary layer as described above.

Results for subsonic and supersonic MHD experiments are shown in Figs. 4 and 5. The curves marked "theory" are based on the calculations described by Daily, et. al. (Ref. 2).

For comparison with the line reversal measurement, we have obtained temperature profiles from the relative intensity of emission at 5887 Å. Such measurements are based on the hypothesis that the observed intensity at any scan position y in the electrode boundary layer is simply proportional to the density of emitting (excited-level) Na atoms at the channel centerline. (This corresponds to the procedure used by Hohnstreiter, et. al (Ref. 3), except that here only a narrow wavelength interval is used rather than the emission intensity integrated over the line.) The relative-intensity method requires an auxiliary absolute measurement - taken here as the reversal temperature in the core of the flow - but otherwise is simpler, since it

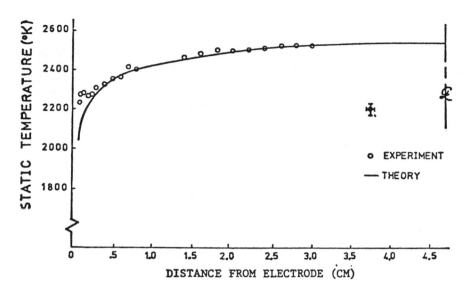

FIG. 4 - Static Temperature - Subsonic Flow

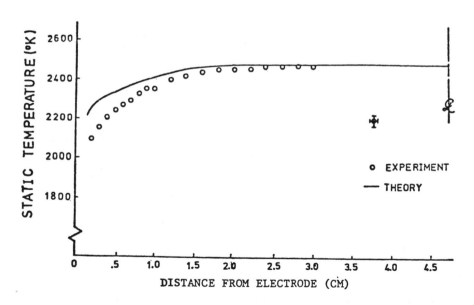

Fig. 5 - Static Temperature - Supersonic Flow, B=2.1T

123

involves only a single scan for a complete profile and does not require use of
the tungsten strip lamp. In addition, calculations were performed in which
these relative intensity measurements were corrected for the presence of the
sidewall boundary layer, using the numerical solution of the equation of
radiative transfer discussed previously. Temperature profiles for the subsonic experiment are shown in Fig. 6 for these three methods: line reversal,
relative intensity, and corrected relative intensity. It can be seen that the
relative intensity profile compares well with the line reversal values and
that the relative intensity correction is small.

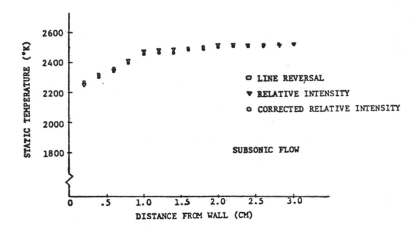

Fig. 6 - Comparison of Temperature Profile Methods

The electron density is measured by use of absolute and relative intensities of non-resonant potassium lines originating from high lying electronic
levels which are in collisional equilibrium with the free electrons. The
populations N_k of these levels are in Saha equilibrium with the free electrons
and hence proportional to the square of the electron density. Emission
measurements were used to obtain profiles of N_k; the tungsten strip lamp
was used as a comparative source to infer absolute intensities. Again there
is a correction resulting from the presence of the sidewall boundary layers,
the measured intensity being proportional to the average value of N_k along
the line of sight. A numerical correction to obtain the centerline value is
obtained from calculated sidewall profiles of N_k. The potassium 6D (5360 Å)
and 7D (5112 Å) lines were used in the present study.

This method was originally devised for relative profile measurements;
however, satisfactory results have also been obtained for the absolute
electron density. For a number of measurements made under equilibrium conditions, the measured electron density agreed with the equilibrium value at
the measured temperature with a mean difference of 10% and a standard
deviation of 16%.

Results corresponding to the subsonic and supersonic conditions of Figs. 4 and 5 are shown in Figs. 7 and 8.

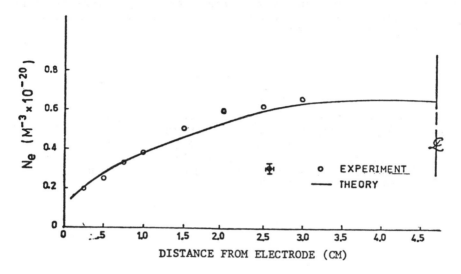

Fig. 7 - Electron Number Density - Subsonic Flow

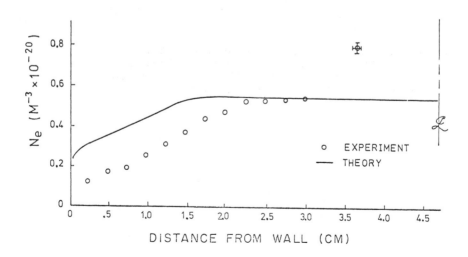

Fig. 8 - Electron Number Density - Supersonic Flow

The theory curve in Fig. 7 is again taken from the calculations of Daily,
et. al (Ref. 2). A supersonic rapid expansion experiment has demonstrated the
suitability of the electron density measurement for nonequilibrium conditions.
For the MHD conditions reported here, however, Fig. 12 demonstrates that no
electron density nonequilibrium was observed.

ACKNOWLEDGEMENTS

    The work reported in this paper was supported by the National Science
Foundation, Grant ENG 73-04116-A01 and Grant NSF AER-72-03487, by the
Energy Research and Development Agency, Contract DI-14-32-0001-1227, and by the
AeroPropulsion Laboratory, Wright-Patterson Air Force Base under Contract
AFAPL F-33615-72-C-1088. The authors acknowledge with gratitude the assis-
tance of Mr. Frank Levy and Mr. Frank Reigal in conducting the experimental
work.

REFERENCES

1. G. Brederlow, W. Riedmüller, M. Slavat, "On the Applicability of the
    Line Reversal Method for Measuring the Electron Temperature in a Weakly
    Ionized Alkali Seeded Rare Gas Plasma," Institute Für Plasmaphysik,
    Report IPP 3/69, Garching Bei Müchen, Germany, 1968.

2. J. W. Daily, C. H. Kruger, S. A. Self and R. H. Eustis, "Boundary
    Layer Profile Measurements in a Combustion Driven MHD Generator."
    Sixth International Conference on MHD Electrical Power Generation, 1975.

3. G. F. Hohnstreiter, C. H. Kruger, R. M. Evans, and M. Mitchner, "The
    Influence of Boundary Layers on Spectroscopic Temperature Measurements
    in MHD Channels," Electricity from MHD, Proc. International Symposium
    on Magnetohydrodynamic Electrical Power Generation, Warsaw, Paper
    SM-107/18, July 1968.

CONCENTRATION MEASUREMENTS BY FLUORESCENCE

ARTHUR FONTIJN

AeroChem Research Laboratories

As a brief introduction to the potential of fluorescence techniques in combustor measurements, I would like to mention the excellent recent report (Ref. 1) by K. Schofield of ChemData Research on the feasibility of using fluorescence to monitor minor stratospheric species. In this report, the various factors influencing the sensitivity of the method are discussed and the available rate coefficient (cross section) data affecting this sensitivity are compiled. For an important species such as NO_2 using a 2-Watt 488 nm line of a cw Ar^+ laser, the theoretical detection limit is 10^{-3} parts per billion (ppb), independent (in these relative units) of pressure from 10 to 760 Torr. Measured sensitivities using a 100 mW Ar^+ laser line at 760 Torr are on the order of 1 ppb. Quenching rate coefficients in air are on the order of a few times 10^{-11} ml molecule^{-1} sec^{-1} or quenching upon every 10 collisions. If we assume that in the jet engine environment quenching could conceivably occur upon every collision the limit-of-sensitivity should be $\approx$<10 ppb. The principal difference between the cold atmospheric environment and the hot jet engine environments is the gas luminosity and the presence of particles. If a proper background subtraction method is used (e.g., an angularly oscillating interference filter and a lock-in amplifier--similar to techniques used in day airglow studies) the gas luminosity should not seriously affect this limit. The particle effect is harder to estimate; however, since the lower concentration of practical interest in a jet engine is $\approx$ 1000 ppb, it strikes me as highly probable that fluorescence can indeed be used to measure NO_2 in the jet engine environment.

The accuracy of fluorescence measurements should also be considered since it presents a special problem. At the pressures of interest the signal intensity is inversely proportional to $\sum_i k_{Q,M_i} \times (M_i)$, where (M_i) represents the concentrations of the different species M_i which constitute the bath (fuel, air and combustion products) in which the species of interest is to be measured.

In general, the quenching rate coefficient k_{Q,M_i} differs for different M_i and will depend somewhat on temperature. This is not a problem if the calibration and the actual measurements are performed in a bath of the same composition and temperature. Where this cannot be done, the magnitude of this effect need be considered. Diatomic species (O_2, N_2) are in general the least efficient quenchers. A triatomic species such as NO_2 is 2 to 3 times more efficient as an NO_2 quencher than these species, CO_2 and H_2O could conceivably be even more efficient.* Hence the ultimate accuracy of the method in jet engine measurements is dependent on the knowledge of the bath composition at the point of measurement, i.e., on the determination of the major species composition which should be an inherent part of the total system characterization. Research preliminary to jet engine measurements of a given species via fluorescence should concentrate on measurement of (relative) quenching cross sections and their temperature dependence.

A final point about fluorescence which I should mention here is that it can be used to measure the concentration of vibrationally excited species just as readily as ground state species, provided the concentration is high enough. The same holds for Raman, but again, in fluorescence much lower concentrations can be measured.

REFERENCES

1. K. Schofield, "Molecular Fluorescence as a Monitor of Minor Stratospheric Constituents," NASA CR-2513, February 1975.

---

*Compare, for instance, the OH quenching data compilation by K. M. Becker and D. Haaks, Z. Naturforschung 28a, 249 (1973).

LOCAL SPECIES CONCENTRATION MEASUREMENT BY RESONANCE SCATTERING TECHNIQUE

C. P. WANG
The Aerospace Corporation

ABSTRACT

An estimate of the possible detection limits for combustion inter-
mediates and products with the resonance scattering technique using tunable
lasers is given. Its advantages and limitations are compared with various
other laser diagnostic techniques. Also proposed here is a two-photon
fluorescence scheme for the local species concentration measurement.

SUMMARY

Resonance scattering or resonance fluorescence can occur only when
the frequency of the incident radiation coincides with an absorption line
or band of the molecules. Transition to a state of higher energy may then
occur, followed by re-emission of light at the same frequency. Since the
absorption line or band is characteristics of the molecule, the scattered
resonance radiation may then be used to identify uniquely the molecules
responsible for scattering.

Because of resonance effects, resonance scattering cross-sections,
which depend on the oscillator strength of the transition and the line
shape, may be as much as a factor of 10^{16} higher than the Rayleigh

This work reflects research supported by SAMSO Contract F04701-74-C-0075.
Part of the results will be included in an invited article entitled "Laser
Applications to Turbulent Reactive Flows: Resonance Absorption and Resonance
Scattering Techniques" to appear in Combustion Science and Technology.

scattering cross-sections. However, when the lifetime of the excited
state is much larger than the mean collision time between molecules,
quenching of the fluorescence becomes important and reduces the effective
scattering cross-section substantially. To estimate the effective resonance
scattering cross-section, first, we calculated the resonance absorption
cross-section σ_{abs} by the relation

$$\sigma_{abs} = \frac{\pi e^2}{mc} \times \frac{f_{12}}{\Delta\nu}$$

where f_{12} is the oscillator strength, $\Delta\nu$ is the effective line width, e is
the electronic charge, m is the electron mass, and c is the speed of light.
For atomic species, f_{12} can be obtained from tabulated atomic transition
probabilities. For molecules, the rotational oscillator strength can be
expressed as

$$f_{J'J''} = \frac{H}{2J' + 1} \quad (2J'' + 1) \quad X_{J''} \quad f_{v'v''}$$

$$f_{v'v''} = \frac{8\pi^2 mc}{3he^2} \quad \frac{\Sigma Re^2(\overline{r}_{v'v''})}{\lambda_{v'v''} \quad g_{e'}} \quad q_{v'v''}$$

where H is the Hönl-London factor, $X_{J''}$ is the Boltzmann factor, J' and J''
are rotational quantum numbers, v' and v'' are vibrational quantum numbers,
Re is the electronic transition moment, $\overline{r}_{v'v''}$ is the internuclear separation
of the atomic nuclei, $g_{e'}$ is the electronic degeneracy and $q_{v'v''}$ is the
Frank-Condon factor.

The effective resonance scattering cross-section can be expressed
as

$$\left(\frac{d\sigma}{d\Omega}\right)_R = \frac{\sigma abs}{4\pi} \quad FQ$$

where F is the fraction of monitored fluorescence and Q is the Stern-Volmer
quenching factor. Based on these equations the resonance scattering cross-
sections for selected atoms and molecules, can be calculated and the minimum
detectable number density can be estimated using certain instrument factor.
The results are listed in Table I. The results indicate the potential of
laser resonance scattering for the measurement of local species concentration
by high resolution detection of trace metals and molecular species. The
substantially higher return signal as compared with Raman scattering, makes
this scheme more attractive for certain species. However, there is a
potential problem due to overlap between individual bands and wings. Care
must be taken in selecting the appropriate line in a given gas mixture and
in interpreting the measurement.

TABLE I. Minimum Detectable Number Densities for Resonance
Scattering for Selected Atoms and Molecules

Species	Wavelength λ [nm]	Resonance-Scattering Cross-Section $(\frac{d\sigma}{d\Omega})_R$ [cm^2 str^{-1}]	Minimum Detectable Number Density* $n\ell_{min}$ [cm^{-2}]
N_a	589.6	2.6×10^{-16}	1.3×10^7
Hg	253.7	2.1×10^{-18}	1.6×10^9
OH	309.4	2.8×10^{-22}	1.2×10^{13}
NO_2	400.0	2.2×10^{-24}	1.5×10^{15}
C_6H_6	250.0	1.7×10^{-24}	2.0×10^{15}
NO	5300.0	1.6×10^{-24}	6.7×10^{20}

*(Assuming laser power 10^{-3}W, circuit bandwidth 10^3Hz and statistical
uncertainty 10%.)

Finally, if the collisional quenching and scattering from background are
still a problem, a new technique using two-photon fluorescence is proposed.
First, let two laser beams intersect at a small volume to be measured. Second,
tune the laser frequencies of these two beams such that the sum of these two
laser frequencies coincide an absorption line or band of the molecules to be
measured. Then the subsequent fluorescence due to the two-photon absorption
is proportional to the species concentration to be measured. Since only
these species in the two-beam intersection region can be excited and the
wavelength of the detected fluorescence radiation is far from the wavelengths
of the incident laser beams, this scheme is more specific. However, the
disadvantage of this scheme is that a high-power tunable laser is required.

DISCUSSION
session on absorption-emission and resonance techniques

PENNER, UCSD - I'd like to make a comment on hydrocarbon diagnostics using tunable lasers. There are tunable lasers available for that application, such as the one developed for C_2H_4 by Hinkley at Lincoln labs, for instance. There are tunable lasers for the study of automotive exhausts

You could also do a precision analysis of methane with the helium-neon laser which happens to have a coincidence with the methane band around 3.3 microns. This will allow a concentration measurement within a few parts in 10^9. I'm not disagreeing with your statement that it hasn't been done. But I am suggesting that you could buy equipment to do it.

MCGREGOR - How do you propose to handle the diagnostics of a complex mixture of hydrocarbons?

PENNER - I think the only hope for that is to go to a tunable laser. There's just no hope in doing it any other way. For ethylene, the measurement technique is already developed in the nine-micron region.

WOLFSON, AFOSR - I'd like to point out that the frequency scanning method has been successfully carried out by Krakow[1] on a flame a number of years ago. His company - Warner and Swaysey - developed the instrument and sold it for $25,000, with the technique.

MCGREGOR - True, but this was done in a laboratory situation and not a real one.

GOULARD, Purdue Univ. - Also you must remember that the solution of a frequency inversion gives the temperature profile only if you know the concentration profile, or the concentration profile only if you know the temperature profile. For instance, the satellite measurements of atmospheric temperature profiles (by scanning the emission of the 15 μCO_2 band) rely on the fact that

---

[1] M. T. Chahine "A General Relaxation Method for Inverse Solution of the Full Radiative Transfer Equations" J. Atmospheric Sciences, Vol. 29, May 1972, pp. 741-747.

the CO_2 concentration is known and uniform across the atmosphere. Krakow[2] and later Simmons were looking through flames of composite burners made of sections i (3 and 7 respectively) of different unknown properties (C_i, T_i) but known dimensions and spacing. Such knowledge does not exist generally in a flame.

MCGREGOR - Still, in many cases you do know one of the two profiles.

LAPP, GE - What about those experimental configurations which are not axi-symmetrical, or where there are small turbulent fluctuations, such as they cannot be resolved into thin concentrical anuli? Is there progress in that direction?

MCGREGOR - I reported on where I believe we are now. We've got to try to handle non-symmetrical situations.* We have a whole series of rocket engine data which we're trying to unravel.

LAPP - The complications become more severe if you have non-equilibrium effects in addition.

MCGREGOR - I might comment on the non-equilibrium on OH. The ground state appears to stay pretty much in the equilibrium. The excited electronic states, of course, are far out of equilibrium because they got there by some chemilumi-nescence reaction. Thus, one generally can't use emission for diagnostics very well, but absorption seems to work well.

PENNER - The question is really one of the uniqueness of the inversion. It is a tricky matter.

GOULARD - Uniqueness would exist if one could dispose of an infinite number of error-free measurements in space - or frequency-scanning, depending on the technique chosen. This is hard to approximate experimentally.

MCGREGOR - Right. I think it's a hard problem, no question about that. But the fact that you can get data when no other method gives it to you is an incentive to try to resolve it.

BILGER, University of Sydney - In a turbulent situation, you've really got to be very skeptical. Every small volume above 100 micron in size is non-uniform in temperature T and concentration C, and therefore the average of the absorption function $\overline{\phi(C,T)}$ (over time or space) cannot be expected to be equal to the function of the average temperature and concentration $\phi(\overline{C},\overline{T})$ as inversion theories usually assume. This error is all the more serious because the properties ϕ are often strongly nonlinear in C and T.

---

[2]B. Krakow "Spectroscopic Temperature Profile Measurements in Homogeneous Hot Gases" Applied Optics, 5, No. 2, p. 201, 1966.

*Since the Workshop, and stimulated in part by these discussions, a multiangular scanning approach has been proposed by Chen and Goulard (see AIAA Paper 76-108). It yields concentration and temperature profiles in non-symmetrical flames. (Ed.)

WRAY, Physical Sciences, Inc. - Is your work on resonance line absorption technique done with filters or with a dispersing instrument?

MCGREGOR - That work so far has been done with the low resolution ultra-violet spectrometer. We've done work in the lab with high resolution where you could isolate the lines but field work has all been done with low resolution spectrometers. There is a very good reason for that: one gets the 0,0 band, the 1,1 band and the 2,2 band. So that you can correct your data for any kind of extraneous absorption by looking at the 1,1 and 2,2 band.

Another interesting question has to do with the effect of turbulence on the radiation emitted in various parts of the flow and on its propagation across the flow. Rhodes has made an evaluation of this effect, using essentially the same kind of band models that we used for diagnostics. In those calculations the effect of turbulence had a fairly drastic effect in some cases on the radiation fluxes. So one would expect it to do the same kind of thing to our diagnostics. Would you agree, Bob?

RHODES, ARO - Turbulence-induced variations of the temperature-dependent index of refraction tend to scatter a beam passing through the flow. Still, this scatter could be accommodated by larger aperture optics for instance. However, the absorption coefficient and - in emission - the Planck function are more nonlinear in temperature than the index of refraction; I would expect some serious problems in the already unstable inversion procedure.

SESSION 6
Scattering Measurements
Chairman: MARSHALL LAPP

RAMAN SCATTERING
STUDIES OF COMBUSTION

MARSHALL LAPP
General Electric Corporate Research and Development
DANNY L. HARTLEY
Sandia Laboratories

ABSTRACT

The need for spatially and temporally well-resolved non-perturbing measurement techniques for modern combustion systems has stimulated development of various new methods for sensing the physical and chemical conditions. Strong requirements are imposed upon new potential techniques by the specific

---

This talk corresponds largely to the following manuscript by M. Lapp and D. L. Hartley, which has been prepared for a special issue of Combustion Science and Technology devoted to turbulent reactive flows. This article is reprinted here with the permission of the Editor of this journal, for which the authors are grateful. Additional material relevant to the talk is contained in the following references, which are currently in press:

Hartley, D. L., Hardesty, D. R., Lapp, M., Dooher, J., and Dryer, F. (Eds.) (1975), "The Role of Physics in Combustion." In Efficient Use of Energy, American Institute of Physics Conference Proceedings No. 25, American Institute of Physics, New York. An abbreviated version of some of this material is also contained in Hartley, D. L., Lapp, M. and Hardesty, D. R. (1975), "The Role of Physics in Combustion Modeling and Diagnostics," to be published in Physics Today.

Lapp, M. (1975). Optical Diagnostics of Combustion Processes. In Proceedings of the Society of Photo-Optical Instrumentation Engineers, Vol. 61. Optical Methods in Energy Conversion.

needs of combustion science and technology, such as the necessity to probe
rapidly fluctuating, luminous hot flows which can contain more than one
phase. Furthermore, the proven capabilities for some of these applications
of present-day probes, such as thermocouples, hot wires, gas sampling apparatus,
etc, lead to the conclusion that new combustion sensors should be developed
for experimental conditions for which present probes are clearly not adequate
or for which they cannot survive. Raman scattering is described here as a
candidate measurement technique which can satisfy many of the needs of combustion experiments, but which must be applied with care and discretion in
order to avoid disadvantages arising from its low intensity. Basic measurement principles and applications are described, and alternate stronger
scattering processes are mentioned.

I. INTRODUCTION

The impact of increased knowledge of the combustion and fluid mechanic
properties of modern technological systems for power generation and propulsion
(such as internal combustion engines, jet engine combustors, gas turbines,
etc.) can be strong from many points of view - most prominently today from
the competing demands of increased engine (i.e., fuel) efficiency and increasingly stringent emission controls.

During a recent study sponsored by the American Physical Society (Princeton,
July 1974) on the application of physics to problems in combustion (Hartley,
et al., 1974), the importance of enlarging our basic knowledge of combustion
processes to meet increasingly difficult technological needs was stressed,
as opposed to patchwork solutions to individual problems on an ad hoc basis.
From an overall viewpoint, we need to be able to design more effectively our
combustion systems to accomplish specific goals.

A major thrust to emerge from this APS Study was the extreme importance
of coupling effectively combustion analytical treatments (termed "modeling"
in the sense of predicting overall combustion system properties from a knowledge of the detailed component processes, via an approximate description
or "model" of the combustion system) with available experimental data. This
seemingly simple approach is of great importance and high potential utility
(Osgerby, 1974), but is also, in fact, rather difficult to accomplish. One
of the largest obstacles to this coupling of efforts is the often inherent
separation of research workers into those developing new physical measurement
techniques and those involved in advancing the understanding of a particular
field to which the new techniques can be conceivably applied. This article
was prepared for a special issue of Combustion Science and Technology devoted
to a discussion concerning advances in the theoretical and experimental aspects
of turbulent reactive flows. This issue will hopefully help to bridge the
gap between the workers developing some of the current new measurement schemes
and the combustion scientists and engineers who can make strong potential use
of the new methods.

The selection of the coupling between light-scattering diagnostic methods and turbulent reacting flows as a priority area in combustion research is suggested by a variety of recent publications of a generalist nature. For example, Glassman and Sirignano (1974) point out that new advances in understanding of the methodology required to describe the basic chemical and transport properties of turbulent flows have now created the need for new experiments to provide data for evaluating the theoretical models. Swithenbank (1974) emphasizes the need for more data on the turbulent properties of combustion, briefly mentioning some of the new light-scattering methods, and stating the opinion: "It is easy to justify the cost of work in combustion fluid mechanics, since astronomical sums of money may be saved by very modest improvements in the efficiency of combustion systems." Oppenheim and Weinberg (1974) have discussed wide-ranging research programs in combustion science related to our current technological needs, and have pointed out that laser techniques--with their concomitant good spatial and temporal resolution--should be of high benefit to this work.

We believe that these and many other current references and sources lead to the conclusion that well-conceived applications of light-scattering diagnostics to important combustion problems are critical to the advancement of combustion technology. Foremost on this list of new optical diagnostic techniques is Raman scattering (RS), which can determine gas density, composition, and temperature with excellent spatial and temporal resolution. Specifically, we feel that the application of RS to problems in turbulent combustion, even using equipment available today, can provide much needed information for the theorists. We note also that the combination of Raman scattering with laser velocimetry introduces the capability for providing a comprehensive set of data for characterizing a combustion system in an experimentally compatible configuration. Both experimental schemes utilize laser sources and the optics and signal processing apparatus for light scattering data. Furthermore, experiments can be designed so that both schemes correspond to essentially the same temporal and spatial resolution.

Not only do we feel that it is worthwhile to attempt the use of RS in a variety of combustion experiments, but that now is the proper time to implement this work in a large-scale, comprehensive fashion. Raman scattering applications to laboratory scale experiments in turbulent flames as well as full-scale system experimentation in complex combustion devices should be conducted for the dual purpose of providing useful experimental data now, and for identifying and solving potential limitations for future applications (such as background fluorescence interference, weak overall signals, electronic detection and data handling complications, etc.). From the point of view of Raman sensor development, momentum exists in the sense that progress in RS diagnostics for combustion and fluid mechanics has been rapid during the past few years, and has involved work

from a substantial number of laboratories.* Thus, it is our belief that this
current strong activity in Raman scattering research directed toward sensors
useful for these fields of interest required dedicated application by workers
in combustion to demonstrate the capabilities and limitations of Raman
scattering and to determine what course the future diagnostic research and
development programs should take.

II.   RAMAN SCATTERING FUNDAMENTAL CONCEPTS

Raman scattering is an inelastic process in which the scattered light
undergoes a change in frequency characteristic of the internal energy modes of
the irradiated molecule. (For our purposes here, these internal modes are
the vibrational and rotational energy level manifold, although RS also occurs
for suitably-spaced electronic levels of molecules as well as atoms.) The
resultant RS signature (i.e., spectral line intensities and distribution)
from a multicomponent hot gaseous combustion system thus depends on the
population distribution in various internal energy modes for all polyatomic
species present and is thus indicative of the system composition and
temperature. However, this relative spectral structure also depends on the
associated Raman line strengths and statistical weights of the internal
energy modes. Thus, molecular information of a fundamental nature - scattering cross sections, internal mode energy level structure, etc. - is required
for proper analysis of RS signatures. For many molecules of major importance
for combustion research, such information is available. For others, further
experimental work is required.

Since the RS signature for any given species can, in principle, be
related to the populations of the various vibrational and rotational energy

---

*Among these laboratories are: Aerospace Research Laboratories (USAF);
Air Force Aero Propulsion Laboratory; ARO, Inc.; AVCO Everett Research
Laboratory; Block Engineering, Inc.; Calspan Corp.; General Electric
Corporate Research and Development; General Motors Research Laboratories;
NASA Langley Research Center; Naval Ordnance Laboratory; New York Polytechnic
Institute; Office National d'Etudes et de Recherches Aerospatiales,
Chatillon, France; Sandia Laboratories; University of Southampton, England;
and United Aircraft Research Laboratories. If we add to this list those
laboratories supporting Raman work in other fields of fluid physics or
chemistry which have strong application to the type of work we are dis-
cussing here, we then increase our list of active workers significantly.
For example, Raman studies directed toward atmospheric science applications have been carried out at a number of locations. Some examples not
already included in the above listing are Drexel University;
Hydrometeorological Service of the USSR, Moscow; Impulsphysik GmbH,
Hamburg, W. Germany; NASA Lewis Research Center; National Oceanic and
Atmospheric Administration, Boulder; and Tohoku University, Japan.

levels, these scattering profiles are related to the temperature in the most
fundamental fashion, i.e., by means of ratios of energy level populations.
Thus, for a system in thermal equilibrium, a temperature corresponding to
the usual Maxwell-Boltzmann definition is obtained. For a system not in
thermal equilibrium, a set of internal mode excitation temperatures is
obtained from the ratios of the appropriate adjacent energy level populations
derived from the experimental Raman intensities.

The basic theory of RS has appeared in a number of volumes, too numerous
to list in detail. The recent two-volume work edited by Anderson (1971, 1973)
is representative of the high quality books available. From the point of
view of combustion measurements, an overview of Raman scattering gas-phase
diagnostics was obtained at the Project SQUID Laser Raman Workshop on the
Measurement of Gas Properties, held in Schenectady in May 1973. (Lapp and
Penney, 1974a). The proceedings (Lapp and Penney, 1974b) contains a broad
cross section of the type of work relevant to our discussion, including both
introductory comments and descriptions of combustion applications.

A description of Raman diagnostics applied specifically to combustion
problems was given in the APS Study on the Role of Physics in Combustion
(Hartley, et al., 1975a, 1975b). In the section of this study which concerned diagnostics, attention was focused upon optical probes and, in
particular, upon light scattering techniques. Thus, the description of
RS given there is accompanied by descriptions of laser velocimetry, Rayleigh
scattering, fluorescence, and particle scattering--giving an overall perspec-
tive of this general class of non-perturbing measurements as they might
be applied to combustion experimentation.

A brief survey of light scattering measurements for combustion has also
been given by Penner and Jersky (1973), and Goulard (1974) has surveyed the
Raman literature in order to determine the practicality of RS measurements
for several types of applications in combustion, fluid mechanics, and
atmospheric science. We will refer to the "feasibility index" defined by
Goulard for the application of RS for these purposes in Section III.

In the light of the references just described, we wish to stress here
mainly the high usefulness of many of the properties of RS for combustion
studies--especially those associated with its non-perturbing character and
its high degree of temporal and spatial resolution.

A. MAGNITUDES OF SCATTERING PROCESSES

In Fig. 1 we show a general schematic for light scattering in order to
depict the relevant geometrical factors and to aid in visualizing the
wide geometrical variations that are possible (i.e., right-angle scattering,
requiring different windows or apertures for the incident laser beam and
the scattered Raman radiation; backscattering, for which one window or

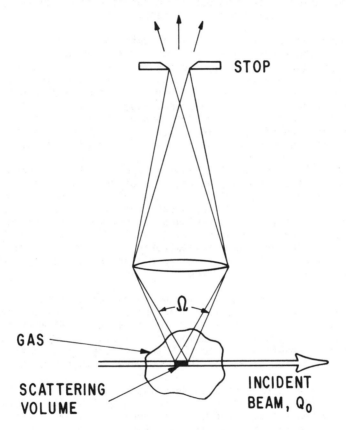

OBSERVED
SCATTERED LIGHT, Q_S

STOP

Ω

GAS

SCATTERING
VOLUME

INCIDENT
BEAM, Q_0

Fig. 1 - A general experimental configuration for light scattering.
The lens collects the scattered light, Q_S, from the solid
angle, Ω, and focuses it through a stop that defines the
length, L, of the observed segment of the incident beam, Q_0.

aperture can serve for both the laser and scattered light, but for which the depth resolution for the probed volume is somewhat diminished; etc.).

The relative magnitudes of various scattering processes are indicated by the differential scattering cross section, denoted as σ in units of cm^2/sterad, and defined by the relation

$$Q_S = Q_0 \ \sigma \ L \ \Omega \ N \ (v,J) \tag{1}$$

Here, $N(v,J)$ is the number density of the observed molecules in the initial vibrational and rotational energy levels v and J; $Q_S \equiv Q_S[v(\Delta v, \Delta J)]$ is the observed energy of the scattered light at a frequency shift v from the incident laser frequency corresponding to the changes Δv and ΔJ in the vibrational and rotational energy levels, and over the time duration corresponding to the incident laser beam energy Q_0; L is the length of the observed segment of the incident beam measured along its propagation direction; and Ω is the solid angle over which scattered light is collected. An implicit dependence upon temperature exists for Eq. (1) through the dependence of the population factor $N(v,J)$ upon temperature. The magnitudes of some typical cross sections of interest here are given in Table I. [Cross sections quoted for vibrational RS usually refer to the value integrated over the rotational structure of the band, and usually refer to the Q-branch intensity ($\Delta J = 0$) for cross sections used in combustion studies. For rotational RS, individual line cross sections are often quoted. These comments apply to the cross section given in Table I.]

The challenge in successfully employing Raman scattering is boldly illustrated in the tabulated scattering cross sections quoted here; i.e., Raman scattering is several orders of magnitude weaker than Rayleigh scattering, and ten to twenty orders of magnitude weaker than other scattering processes. The unique information available from Raman signals, however, makes this challenge irresistible. Quantitative values of RS cross section for many species of combustion interest are available in the literature. Some examples of useful vibrational RS cross sections are given by Fouche and Chang (1972, 1971), Leonard (1970), Murphy, et al. (1969), Penney, et al. (1972), and Stephenson (1974).

The relative spectral positions of Rayleigh, rotational Raman, and vibrational Raman scattering for N_2 and O_2 are shown in Fig. 2. We see that Rayleigh scattering (which is essentially elastic) and pure rotational scattering (which corresponds to only small energy exchanges) occur very near the incident laser wavelength. Thus, for a hot, multi-component gas (whose composition can be a strong function of the temperature), the signatures of Rayleigh and rotational Raman scattering processes overlap to a serious extent for many of the flame species.

In particular, Rayleigh scattering is several orders of magnitude stronger than RS, but species cannot in general be differentiated from Rayleigh observations, and the total gas density is also not easily available from the integrated Rayleigh line intensity. This fact arises because the Rayleigh

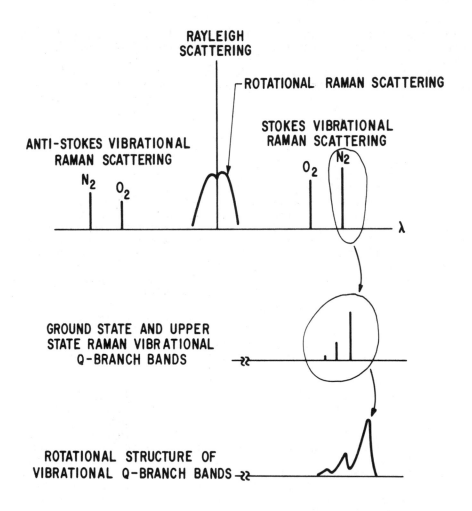

Fig. 2 - Schematic of Rayleigh, rotational Raman, and vibrational Raman scattering. Shown in detail is the splitting of the vibrational Stokes Raman signal for nitrogen into a series of fundamental vibrational bands (ground state band; first upper state, or "hot" band; second upper state band; ....) and the spreading out of this fundamental vibrational band series due to rotational structure, as would be viewed by a typical laboratory double monochromator.

cross sections for the various flame gas components differ, and the composition itself changes with temperature. Therefore, without knowing relative component concentrations a priori it is not possible to determine total density from the Rayleigh signal (unless, of course, all species present have nearly equal cross sections). Additionally, particulate scattering as well as stray laser light can obscure the results. Temperature measurements are possible with high resolution apparatus, but these also do not differentiate between species (and so cannot respond to non-thermal equilibrium conditions in an easily quantifiable way), and are also highly subject to interferences from particle scattering. Ways to remove the effect of the particle scattering (whose signature is spectrally coincident with but narrower than the Rayleigh scattering) have been devised, but this problem remains a major limitation to the utility of the method.

Rotational Raman scattering is stronger than vibrational RS by over an order of magnitude for the integrated spectra, but again the rotational RS spectral signatures of many of the flame species overlap to a serious extent. Thus, the net signature in a given wavelength passband is a sensitive function of the temperature via changes in the gas composition with temperature. Furthermore, spectral proximity to the central laser wavelength means that some interference from particle scattering is possible, although recent experiments (Hilliard and Hunter, 1974) indicate that this may not be a serious limitation. A way of observing the integrated rotational RS signature while simultaneously rejecting the Rayleigh and particle scattering signature at the laser wavelength has been described by Barratt (1974). This interferometric "comb" method has shown promise in pure gases, but has not yet been applied to combustion gases, i.e., to variable-composition gas mixtures containing some species (such as N_2, O_2, CO, etc.) with rotational constants which are not very dissimilar. Thus, it is not yet clear that sensitive measurements of combustion product gas species will be obtained. In a recent computer simulation of interferometric detection of rotational lines from the reaction zone of a chemical laser, Hill et al. (1975) demonstrated the ability to distinguish HF, DF, and F_2 in non-reacting mixtures, but were unable to distinguish HF in the reacting mixture. They have not yet attempted to reproduce these results experimentally. The greatest promise appears to exist for monitoring rotational RS lines which are relatively free from spectral interferences. This mitigates in favor of H_2 and other species containing H atoms, because of their relatively large rotational constants.

Vibrational Raman scattering, though weaker than rotational RS, does not in general suffer from overlapping spectral signals. As schematically indicated in Fig. 2, the vibrational line for one gaseous constituent, in this case N_2, is well displaced from that of the other gaseous species shown, in this case O_2. In the expanded view of the N_2 vibrational Raman signal, we observe closely-spaced, but easily-resolvable lines corresponding to "hot" bands, or upper state bands. Calculation of these lines based on simple harmonic oscillator theory would predict their coincidence, but molecular vibrational anharmonicities cause them to be slightly shifted (about 0.8 nm for N_2). This displacement allows individual "hot" band detection, the relative intensities of which can be used to predict the vibrational temperature of that gaseous species. When we also consider the vibration-rotation interaction possessed by the molecule, we find the net spectral signature gives us a "sawtooth" structure shown in the lower part of Fig. 2. Thus, the rotational temperature of the molecule can also be calculated by a curve fit to the sawtooth signature. These "hot" bands are sparsely populated at room temperature for many diatomic molecules, such as N_2, and therefore provide a poor temperature measurement there for such species. At characteristic flame temperatures (1000°K - 4000°K), however, they are sufficiently populated to provide very accurate measurement of temperature--often, to within several percent. (See, for example, Lapp 1974a). For common polyatomic molecules, such as CO_2 and H_2O, the vibrational and rotational energy level structure is such that temperatures can be determined from band peak ratios or contour fits down to room temperature. (Lapp, et al., 1973; Leonard, 1972).

Clearly, the information which can be deconvoluted from the rotation-vibration Raman spectra of hot gases is of great importance. One must be reminded, however, of the weakness of these scattering processes and the resultant limitations in its application to trace species detection with good temporal resolution. An alternative approach to enhance Raman capabilities is provided by the recent exploitation of coherent anti-Stokes Raman spectroscopy and resonant Raman spectroscopy.

Coherent anti-Stokes Raman spectroscopy (CARS) is a nonlinear optical process offering enormously greater intensities than the ordinary RS or Rayleigh scattering discussed so far. This process has been applied successfully to flame studies, (Regnier and Taran, 1974; Moya, et al., 1975) and has lately been receiving wide general attention. (Begley, et al., 1974). It has certain limitations along with its advantages--principally, that the process is an "in-line" one for gases (i.e., observation is made along the propagation direction of the incident laser beam), that a high peak-power laser is required for its implementation, and that refractive effects in turbulent flows make its application for our purposes somewhat complicated because of nonlinear effects arising from focusing and defocusing of the laser beams. It possesses, at the same time, a number of great advantages as a diagnostic probe, and should be strongly monitored for consideration as the techniques associated with the method are developed.

Finally, we mention resonant RS as another candidate measurement technique. (See, for example, Penney, 1974). We refer here to scattering processes in which the incident (tunable) laser beam wavelength is adjusted to be near but not coincident with a strong allowed dipole transition of the gas being probed. Far from this spectral region, we have ordinary RS, while coincidence between the wavelength of the incident laser beam and the absorption line in question leads to fluorescence. Under the conditions of near resonance, strong enhancement of the RS signal is expected, with the characteristics of the scattered light exhibiting many of the desirable qualities or ordinary RS, and with only small influence of the undesirable qualities of fluorescence. Of particular value in the use of resonant RS is substantial freedom from the quenching effects that usually accompany fluorescence. This freedom is critical in applications for combustion under conditions for which the relative concentrations of quenching molecules vary spatially and with temperature, and for which the various quenching cross sections (and their variations with temperature) are often poorly known or unknown.

The strength of resonant RS strongly suggests that serious consideration be given to its potential use as a diagnostic probe, especially in
situations in which minor species cannot otherwise be detected. Thus, we
consider of particular importance situations in which we enhance by resonant
RS the signal from a minority species, as compared to the signal from a
majority species with a similar Raman shift but without the presence of an
allowed dipole absorption nearby the incident laser wavelength.

Problems exist, however, in implementing resonant RS for combustion systems.
In some situations, it has been difficult to clearly determine the distinction
between resonant RS and fluorescence (St. Peters, et al., 1973). It is
expected that these difficulties would be reasonably strong for complicated
combustion systems, i.e., those containing a number of gas-phase visible or
uv absorbing species as well as particles. Well-controlled tunable lasers of
narrow bandwidth are required for proper implementation of this technique,
as are good estimates of the absorbing properties of the probed gas. Much
effort has been put into studies of resonant RS for gases of general interest
to combustion, but further development work is still required for this technique to be fully utilized in this way. A result which is encouraging for
the future utility of resonant RS for combustion work was obtained by Penney,
et al. (1973), who found that the cross sections for N_2 and O_2 at 300 nm are
several times larger than would be predicted by $(1/\lambda_{scatter})^4$ extrapolations
from the visible.

Based upon the preceding discussion of the various possible types of
Raman processes which might be suitable for combustion diagnostics, we conclude that ordinary vibrational RS is the prime immediate candidate for
exploitation in the study of turbulent reactive flows. Vibrational RS, in
spite of its weakness, has the virtue of relative simplicity and has been
used extensively in the laboratory. CARS, followed by rotational RS techniques,
are however, potential candidates currently undergoing active development.
They have already-demonstrated capabilities which can be of great value if
they can be implemented on the systems of interest here. The utilization of
resonant RS or Rayleigh scattering for combustion measurements presently
appears to be somewhat further removed than the other techniques mentioned,
but may well prove to be of substantial use in specialized instances after
further development.

B. EXPERIMENTAL APPARATUS

In principle, because of the effectively instantaneous nature of the Raman
scattering process, the temporal resolution is as short as could be desired--
limited only by the duration of a laser pulse or of a signal detector gating
pulse. In practice, the weakness of Raman scattering necessitates experimental
time durations long enough to accumulate sufficient photons to give a useful
signal-to-noise ratio. Thus, from the _scattering_ point of view (and neglecting considerations about background luminosity, fluorescence, time resolution,
etc. for the present), it is equivalent to gather a signal from a 1-watt cw

laser for one second or from a 1-joule pulsed laser at one-pulse-per-sec
repetition rate, whatever the time duration of the pulse. However, from
the experimental combustion system point of view, temporal combustion
fluctuations, luminosity, stray laser light, fluorescence, etc. are often
present to a degree sufficient to cause significant difficulties. These
problems can often be ameliorated by use of pulsed laser sources, for which
the pulse duration is sufficiently short (often, of the order of magnitude
of 10^{-8} sec) to sharply enhance the strength of the Raman signature over
other undesired signals.

Thus, for these purposes, the relatively greater expense and complexity
of a high-energy-per-pulse, visible or uv, pulsed laser source is warranted.
A tradeoff exists, naturally, between the energy per pulse (assuming a narrow
pulse width) and the pulse repetition rate for operation of the laser, and
this tradeoff also exists in desirable attributes for our scattering purposes.
We therefore compromise between the ability to obtain a high signal-to-noise
ratio for each data point (i.e., high energy per pulse) and the ability to
rapidly follow transient phenomena (high repetition rate). In the case of
turbulent flows, it is expected that the combustion system luminosity, stray
light, etc. will dictate that the most reasonable first choice will be to
favor a relatively high energy per pulse laser, and after the results of
experiments based upon this source are properly digested, progress can be
expected to be made toward optimization in the direction of following the
time history more closely--both from the points of view of upgrading the
laser capabilities and of being better able to handle and control the actual
experiments.

The spatial resolution obtainable for RS configurations can be varied
over wide limits depending upon the optical collection system used and whether
the spectral discrimination arises from a monochromator, filter, or interferometer. In general, scattering volumes < 1 mm^3 should be attainable without undue difficulty. Volumes as small as an approximate cylinder roughly
30μm diam by 1 mm long are not unreasonable, although most laboratory experiments have utilized larger volumes in work carried out so far. Substantial
enhancement of the scattering signal can be obtained by multiple reflections
of the incident laser beam through the test zone. An optical reflecting configuration for which about two order of magnitude increase in scattered signal
intensity was obtained has been described by Hill and Hartley (1974).

The general methods used for data acquisition for the RS experiments
involve spectrometers (often, scanning double monochromators) or interference filters for spectral discrimination in vibrational, rotational, or
resonant RS. (See, for example, Lapp, 1974). Interferometers are also of
use for integrated rotational RS data (Barratt, 1974). Photomultipliers
selected for low noise and high quantum efficiency are usually selected as
detectors, and varieties of electronic signal processing systems are of potential interest (Schildkraut, 1974).

The recent explosion in the development of electronic and electro-optic
components offers newly-expanding opportunities to workers in Raman gas
diagnostics. Optical multichannel analyzers (OMA's), and even two-dimensional
OMA's which integrate and store entire spectral signatures, perform background subtraction, and provide direct digital output can now be coupled to
mini-computer data acquisition systems to provide the capability to obtain,

deconvolute, and process the kind of data necessary to successfully diagnose a complex combustion system in a quantitative fashion. Further details of specific apparatus configurations and signal processing systems which have been used in RS studies of combustion may be found in the references quoted in the next section.

III. APPLICATIONS TO COMBUSTION

The attractiveness of RS for combustion diagnostics lies in its capabilities for determining the densities of individual gas components and the gas temperature, under non-perturbing conditions and with good spatial and temporal resolution. Furthermore, as has been stated at the beginning of Section II, thermal equilibrium does not have to exist in order to interpret data taken for the purpose of defining system temperatures. Suitably defined non-thermal equilibrium excitation temperatures for a given species can be obtained from ratios of its internal mode excited and ground state populations determined from various parts of the Raman signature. In principle, a complete deconvolution of a vibration-rotation Raman band contour gives a complete relative measure of the populations of all the vibrational and rotational levels contributing to that profile. Although a complete deconvolution would be difficult to accomplish without OMA's and mini-computers, approximate treatments would still yield very useful results. Vibrational excitation temperatures (i.e., neglecting information about the rotational degrees of freedom) can be obtained relatively easily through use of appropriate ratios of the integrated intensities of the fundamental Stokes or anti-Stokes vibrational band series. (See Fig. 2). (Lapp, et al., 1973; Leonard, 1972; Nelson, et al., 1971).

To illustrate the type of Raman signatures which correspond to combustion produce gases, we consider CO_2 (Lapp, 1974c) and H_2O vapor (Lapp, 1974b), shown in Figs. 3 and 4, respectively. Both spectra were obtained from flames at comparable temperatures, and indicate overall variations in the spectral contour that are attributable to the internal molecular degrees of freedom. For CO_2, the small values of the rotational constant make each of the indicated vibrational bands of this Fermi resonance doublet series quite narrow. Thus, while rotational information is lost, greater ease results for obtaining vibrational information from these band intensities. For H_2O, the observed structure is influenced strongly by both the rotational and vibrational degrees of freedom in a considerably more complicated fashion. The large values of rotational constant here make the H_2O vapor profile very extensive--roughly, twice as broad under these flame conditions as would be the case for N_2. [Since reaction rates in H_2 flames are usually rapid, the progress of turbulent mixing in such flames should be monitored well from the produce species H_2O. Thus, H_2 flames might well be a good choice for early experiments in turbulent mixing.]

Additional combustion gas data have been obtained by Setchell (1974), who measured radial and axial profiles of temperature and recorded spectra for N_2, O_2, CO_2, CO, and H_2 in laminar methane/air flames. Comparison of

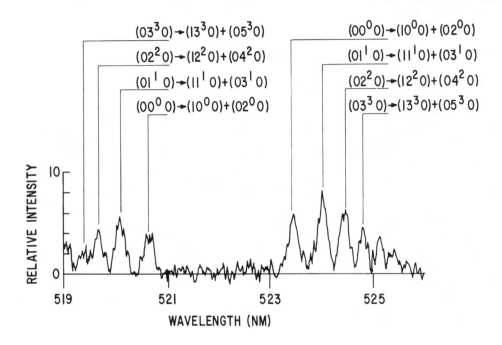

$(03^30) \rightarrow (13^30) + (05^30)$

$(02^20) \rightarrow (12^20) + (04^20)$

$(01^10) \rightarrow (11^10) + (03^10)$

$(00^00) \rightarrow (10^00) + (02^00)$

$(00^00) \rightarrow (10^00) + (02^00)$

$(01^10) \rightarrow (11^10) + (03^10)$

$(02^20) \rightarrow (12^20) + (04^20)$

$(03^30) \rightarrow (13^30) + (05^30)$

RELATIVE INTENSITY

10

0

519   521   523   525

WAVELENGTH (NM)

Fig. 3 - Experimental CO_2 Stokes vibrational Raman spectrum for CO_2 seeded into a stoichiometric H -air flame at about 1565°K, and at a partial pressure of about 1/3 atm. The data were obtained through use of a 1.1w, 488 nm argon ion laser source and a 3/4-meter double monochromator with a triangular spectral slit function full width-half maximum value of 0.163 nm. The notation for the indicated transitions, $(v_1 v_2^\ell v_3)$, corresponds to the three fundamental vibrational quantum numbers and to the vibrational angular momentum ℓ of the v_2-bending mode.

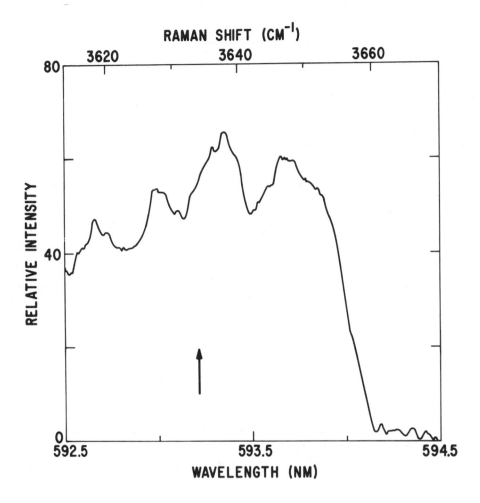

Fig. 4 - Part of the H_2O vapor Stokes vibrational Raman scattering profile for an approximately stoichiometric hydrogen/ oxygen flame at roughly 1500°K. A 488 nm argon ion laser source of 1.05 watt intensity was used in conjunction with a 3/4-meter double monochromator with a triangular spectral slit function full width-half maximum value of 0.162 nm. The data shown here are the average of two data-logged runs which have been smoothed by computer processing. The vertical arrow in the figure corresponds to the spectral vicinity in which the first upper state vibrational band is expected.

149

temperatures calculated from the N_2 and CO profiles verified thermal
equilibrium.

We next evaluate briefly the overall applicability of current Raman
scattering capabilities to measurement situations of general interest here,
and follow this with some examples of current relevant work.

Perhaps the most often asked question in these applications of RS is:
Can an RS signal be seen at all above the strong luminosity expected in many
combustion situations? Use of Eq. (1) for a simple example can help to
resolve this issue. Suppose we consider a frequency-doubled ruby laser
source at 347 nm, which produces a Stokes RS band for nitrogen at 378 nm,
with a scattering cross section value of about 2.5 x 10^{-30} cm^2/sterad. For
1-joule pulses of 30-nsec duration, a number density N for nitrogen of
10^{19} cm^{-3} (corresponding to 300°K and 0.4 atm.), a scattering length L of
0.1 cm, and a solid angle Ω of 0.05 sterad, we find a scattered power of
roughly 4 μwatts. This is approximately the same power as that radiated by
a 2270°K blackbody of 1 cm^2 area into the same solid angle about a normal to
the surface, and into a 1-nm rectangular spectral passband. Thus, the RS
signature in this example can be seen above fairly bright thermal backgrounds,
and especially so when we consider that combustion system gases often do not
possess emissivities anywhere near unity over wide ranges of the spectrum.
The example is, to be sure, for scattering in the blue, which produces a
more favorable situation than for scattering in the red (as would be the
case with a ruby source at 694 nm, for instance). However, for many combustion systems of interest (i.e., for their corresponding gas temperatures and
emissivities), the same general conclusion is obtained, viz., that RS
signatures can usually be seen above the thermal background when high power
Q-switched pulsed laser sources are used.

One method for viewing the overall performance characteristics of RS
for our purposes is through the "feasibility index" X defined by Goulard
(1974). Here, X is defined from a relation similar to that given as Eq. (1),
and includes only factors which relate to the geometry and specific nature
of the experiment. Thus,

$$X = N \sigma L \Omega e \qquad (2)$$

where e is the optical efficiency of the system, σ is calculated at some
reference wavelength (chosen here to be 488 nm), and the other factors have
been defined previously. In Fig. 5, we see a plot of the feasibility index
vs. the characteristic experimental measurement time, with zones delineated
for several measurement applications, and with representative, currently
available laser-detector system capabilities also shown. For example,
successful measurements of small scale turbulence in laboratory situations
(i.e., not remote lidar-type measurements) appear from this analysis to be
of reasonable probability. Further details of the application conditions
and laser performance characteristics concerning this figure are given by

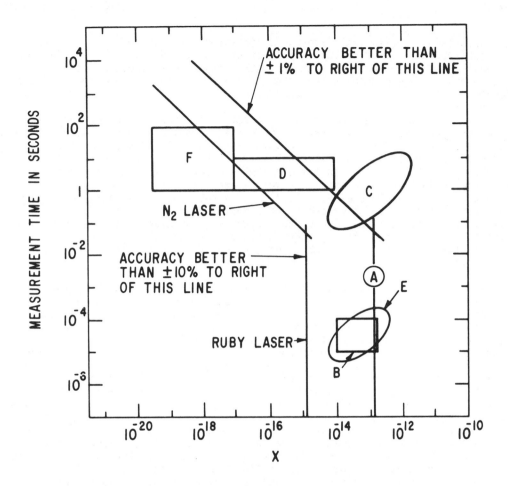

Fig. 5 - Feasibility index X [see Goulard (1974)] for
several measurement applications. Key: A - Compressor inlet.
B - Compressor blade passage. C - Laboratory temperature.
D - Remote atmospheric constituents and temperature. E - Small
scale turbulence in laboratory. F - Remote pollution sources.
The diagonal appearance of cases C and E reflect the observation
that larger turbulence scales correspond usually to longer
characteristic times. For comparison purposes, the measurement
potential of current systems is illustrated for two accuracy
figures. The N laser was chosen as typical of continuous (repeated
pulse) measurement, and the ruby laser as typical of one-pulse
measurement.

Goulard (1974). Of salient importance for interpretation here is the use of
a 1-joule/pulse, o.1-pulse/sec ruby laser at 694 nm and a 1-millijoule/pulse,
500-pulse/sec nitrogen laser, combined with measurement conditions corresponding to 10^{-5} to 10^2 sec characteristic times and 10^{-1} to 10^4 cm characteristic
lengths.

The use of RS techniques in experiments of the general type of interest
here have been reported by a variety of workers, including Arden, et al. (1974),
Hartley (1974), Hendra, et al. (1974), Hillard, et al. (1974), Lapp (1974a),
Lederman and Bornstein (1974), Leonard (1974), Regnier and Taran (1974), and
Setchell (1974). Included in these examples are results of experiments
involving simultaneous RS and laser velocimetry efforts (Lederman and
Bornstein, 1974; Hillard et al., 1974) and measurements related to turbulence
phenomena (Hartley, 1974; Lederman and Bornstein, 1974; and Hillard et al.,
1974). From the point of view of hot, reactive, turbulent flows, we note
particularly that the work of Lederman and Bornstein (1974) represents a
direct attack on laser measurements of the important perturbation and mean
system variables.

From the experimental evidence that is now available, it appears that
worthwhile contributions to combustion and fluid mechanic studies have
already been made by RS experiments. Furthermore, the potential for future
fruitful implementations is enormous. Along with our optimism, however,
we must keep in mind some of the particular difficulties introduced into the
diagnostic techniques by these proposed applications. Perhaps foremost among
these is the problem of refractive index effects upon the laser and scattered
radiation beams caused by the temperature and density gradients along the
optical paths. This is particularly true for the laser velocimeter measurements, where data acquisition rates can be reduced by orders of magnitude by
the non-intersection of the laser beams (for the dual beam method) in the
observed scattering volume. The effect on the RS data could be to poorly
define the observed scattering volume in terms of position, volume, and shape,
and to make the volume not coincident with that under simultaneous observation
by other techniques (such as laser velocimetry, fluorescence, etc.). Careful
design of the optical paths with respect to the overall experimental arrangement will be required to overcome this problem. For Raman studies reported
to date, index of refraction variations have not posed a serious problem.
This is most notably illustrated in the application of a multi-pass light-
trap by Setchell (1975) for measuring properties in an atmospheric pressure
turbulent flame. However, for large scale experiments, such as furnace
diagnostics, refractive effects could be severe.

The limitations imposed by background fluorescence, particularly from
hydrocarbon aerosols, have been investigated by Leonard (1974) and by
Aeschliman and Setchell (1975). The origin of laser-induced fluorescence
in flame gases and (when required) the experimental suppression of its
effects deserve further attention in order to apply RS techniques to the
study of combustion phenomena with confidence. This is particularly true
for the study of heterogeneous systems.

One approach to ameliorate the influence of fluorescence on vibrational
RS when it becomes a substantial experimental problem is to confine all

measurements to anti-Stokes scattering (i.e., measure density from the anti-Stokes integrated intensity and temperature from the anti-Stokes profile contour or band peak intensity ratios), since practically all of the laser-induced fluorescence occurs on the Stokes side (i.e., at wavelengths longer than the incident laser wavelength).

IV. CONCLUSION

Raman scattering diagnostics for potential application to hot, turbulent, reactive flows have been described, and brief comparisons given for several types of scattering processes. An overview of RS capabilities and limitations lead us to the conclusion that reasonable prospects exist for fruitful application to problems of interest. We therefore suggest that increased attempts at implementation of these techniques be undertaken.

It is emphasized that Raman diagnostics should be applied first by combustion workers to areas where successful implementation can be anticipated, and where other probes are not presently of substantial use. In this fashion, we can evaluate the inherent difficulties in these light-scattering techniques, such as weakness of signal or (compared with some of the simpler solid probes) the moderate complexity of the apparatus. Following this framework, it is recommended to implement initially temperature diagnostics for hot, multi-component flows, which can then evolve to include (in successive experiments) turbulence and chemical reactions. Density measurements of the major or intermediate-concentration species would then follow. A growing effort to reduce background interference, to enhance the laser intensity impinging on the sample volume, to improve data handling, and to develop innovative methods of utilizing Raman scattering measurement techniques should be encouraged.

One of the authors (ML) gratefully acknowledges partial support for this work by Project SQUID, Office of Naval Research.

REFERENCES

Aeschliman, D. P. and Setchell, R. E. (1975). Fluorescence Limitations to Combustion Studies Using Raman Spectroscopy. Sandia Laboratories Report SAND74-8688.

Anderson, A. (Ed.) (1971, 1973), The Raman Effect, Marcel Dekker, Inc., New York. Volumes 1 and 2.

Arden, W. M., Hirschfeld, T. B., Klainer, S. M., and Mueller, W. A. (1974). Studies of Gaseous Flame Combustion Products by Raman Spectroscopy. Appl. Spectroscopy 28, 554.

Barratt, J. J. (1974). The Use of a Fabry-Perot Interferometer for Study-
    ing Rotational Raman Spectra of Gases. In Lapp, M. and Penney, C. M.
    (Eds.) Laser Raman Gas Diagnostics, Plenum Press, New York, pp. 63-85.

Begley, R. F., Harvey, A. B. and Byer, R. L. (1974). Coherent Anti-Stokes
    Raman Spectroscopy, Appl. Phys. Lett. $\underline{25}$, 387; Begley, R. F., Harvey,
    A. B., Byer, R. L. and Hudson, B. S. (Nov. 1974). A New Spectroscopic
    Tool: Coherent Anti-Stokes Raman Spectroscopy, American Laboratory.

Eckbreth, A. C. (1974). Laser Raman Gas Thermometry. AIAA Paper No. 74-1144.

Fouche, D. G. and Chang, R. K. (1971). Relative Raman Cross Section for
    N_2, O_2, CO, CO_2, SO_2 and H_2S. Appl. Phys. Lett. $\underline{18}$, 579.

Fouche, D. G. and Chang, R. K. (1972). Relative Raman Cross Section for
    O_3, CH_4, C_3H_8, NO, N_2O, and H_2. Appl. Phys. Lett. $\underline{20}$, 256.

Glassman, I. and Sirignano, W. A. (Aug. 1974). Summary Report of the Workshop
    on Energy-Related Basic Combustion Research, Sponsored by NSF,
    Princeton Univ., Dept. of Aerospace and Mechanical Sciences Report No.
    1177.

Goulard, R. (1974). Laser Raman Scattering Applications. In Lapp, M. and
    Penney, C. M. (Eds.). Laser Raman Gas Diagnostics, Plenum Press,
    New York, pp. 3-14. Also, (1974). J. Quant, Spectrosc. Radiat.
    Transfer $\underline{14}$, 969.

Hartley, D. L. (1974). Application of Laser Raman Scattering to the Study
    of Turbulence. AIAA J. $\underline{12}$, 816.

Hartley, D. L., Hardesty, D. R., Lapp, M., Dooher, J. and Dryer, F. (Eds.)
    (1975a). The Role of Physics in Combustion. In Efficient Use of
    Energy, American Institute of Physics Conference Proceedings No. 25,
    American Institute of Physics, New York.

Hartley, D. L., Lapp, M., and Hardesty, D. R. (1975b). The Role of Physics
    in Combustion: Modeling and Diagnostics. To be published in Physics
    Today.

Hendra, P. J., Vear, C. J., Moss, R. and Macfarlane, J. J. (1974). Raman
    Scattering and Fluoescence Studies of Flames. In Lapp, M. and Penney,
    C. M. (Eds.), Laser Raman Gas Diagnostics, Plenum Press, New York,
    pp. 153-160.

Hill, R. A., Aeschliman, D. P., and Hackett, C. E. (1975). Optical Flow
    Diagnostics for Plasmas. AIAA Paper 75-178.

Hill, R. A. and Hartley, D. L. (1974). Focused Multiple-Pass Cell for Raman
    Scattering. Applied Optics, $\underline{13}$, 186.

Hillard, Jr., M. E., Hunter, Jr., W. W., Meyers, J. F., and Feller, W. V.
    (1974). Simultaneous Raman and Laser Velocimeter Measurements.
    AIAA J. $\underline{12}$, 1445.

Lapp, M. (1974a). Flame Temperatures from Vibrational Raman Scattering.
    In Lapp, M. and Penney, C. M. (Eds.), Laser Raman Gas Diagnostics,
    Plenum Press, New York, pp. 107-145.

Lapp, M. (1974b). Raman Scattering Probe for Water Vapor in Flames. AIAA
    Paper No. 74-1143.

Lapp, M. (1974c). Unpublished data.

Lapp, M., Penney, C. M., and Goldman, L. M. (1973). Vibrational Raman
    Scattering Temperature Measurements. Optics Comm. 9, 195.

Lapp, M. and Penney, C. M. (1974a). Measurement of Gas Properties: Comments
    on the Laser Raman Workshop. Appl. Opt. 13, A14.

Lapp, M. and Penney, C. M. (Eds.) (1974b). Laser Raman Gas Diagnostics,
    Plenum Press, New York.

Lederman, S. and Bornstein, J. (1973). Temperature and Concentration Measure-
    ments on an Axisymmetric Jet and Flame. Project SQUID (Office of
    Naval Research) Technical Report PIB-32-PU.

Lederman, S. and Bornstein, J. (1974). Application of Raman Effect to
    Flowfield Diagnostics. In Fuhs, A. E. and Kingery, M. (Eds.),
    Instrumentation for Airbreathing Propulsion, Progress in Astronautics
    and Aeronautics Series, MIT Press, Cambridge, Mass., Vol. 34, pp.
    283-296.

Leonard, D. A. (1970). Measurement of NO and SO_2 Raman-Scattering Cross
    Sections. J. Appl. Phys. 41, 4238.

Leonard, D. A. (1972). Development of a Laser Raman Aircraft Turbine
    Engine Exhaust Emissions Measurement System. AVCO Everett Res. Lab.
    Research Note 914.

Leonard, D. A. (1974). Measurement of Aircraft Turbine Engine Exhaust
    Emissions. In Lapp, M. and Penney, C. M. (Eds.), Laser Raman Gas
    Diagnostics, Plenum Press, New York, pp. 45-61.

Moya, F., Caumartin, S. A. J., and Taran, J. -P. E. (1975). Gas Spectroscopy
    and Temperature Measurement by Coherent Raman Anti-Stokes Scattering.
    (Submitted to Optics Communications).

Murphy, W. F., Holzer, W. and Bernstein, H. J. (1969). Gas Phase Raman
    Intensities: A Review of "Pre-Laser" Data. Appl. Spectroscopy 23, 211.

Nelson, L. Y., Saunders, Jr., A. W., Harvey, A. B. and Neely, G. O. (1971).
    Detection of Vibrationally Excited Homonuclear Diatomic Molecules by
    Raman Spectroscopy. J. Chem. Phys. 55, 5127.

St. Peters, R. L., Silverstein, S. D., Lapp, M., and Penney, C. M. (1973).
    Resonant Raman Scattering of Fluorescence in I_2 Vapor? Phys. Rev.
    Lett. 30, 191.

Schildkraut, E. R. (1974). Electronic Signal Processing for Raman Scattering
    Measurements. In Lapp, M. and Penney, C.M. (Eds.), <u>Laser Raman Gas
    Diagnostics</u>, Plenum Press, New York, pp. 259-277

Setchell, R. E. (1974). Analysis of Flame Emissions by Laser Raman Spectros-
    copy. Sandia Laboratories Energy Report No. SLL74-5244; also Paper
    No. WSS/CI 74-6, Western States Section, The Combustion Institute.

Setchell, R. E. (1975). Initial Raman Spectroscopy Measurements in a Turbu-
    lent Hydrogen Diffusion Flame. Submitted as Brief Communication to
    Combustion and Flame.

Stephenson, D. A. (1974). Raman Cross Sections of Selected Hydrocarbons and
    Freons. J. Quant. Spectrosc. Radiat. Transfer, <u>14,</u> 1291.

Swithenbank, J. (May 1974). The Unknown Fluid Mechanics of Combustion.
    Keynote Address, ASME Fluid Mechanics of Combustion Conference,
    Montreal. Also, University of Sheffield (UK) Report HIC 220.

DISCUSSION

<u>NASH - WEBBER</u>, MIT - Which combustion species can be effectively mea-
sured by this technique?

<u>LAPP</u> - The easiest experiments are for the highest concentrations -
usually, nitrogen. But many species of intermediate concentration in combustion systems also can be observed. Distinguishing among similar hydrocarbons appears to be rather difficult, but an effective measurement of hydrocarbon concentration should be possible by observing the Raman signature
of a common bond, such as that for CH. With sufficient sensitivity, Raman
signals from free radicals should be observed.

APPLICATION OF
LASER DIAGNOSTICS TO
COMBUSTION

SAMUEL LEDERMAN

Polytechnic Institute of New York

ABSTRACT

The applicability of laser measurement techniques to combustion
diagnostics is explored. Specie concentration and temperature measurements were obtained in a methane-air and methane-CO_2-air open flame using
laser Raman scattering techniques. Velocity profiles of the flame and
turbulence intensity profiles were obtained using a laser Doppler velocimeter.
Finally, an attempt was made to obtain concentration and temperature, using
the Raman scattering diagnostic technique in a simulated, limited accessibility
internal combustion chamber, with very encouraging results.

I.   INTRODUCTION

The development of new diagnostic techniques applicable to fluid dynamic
research has been an ongoing task in our laboratory for many years. In the
last several years, our efforts were directed to the development of flow
field diagnostics using laser Raman scattering and laser Doppler anemometry.
Specifically our attention was focussed on the applicability of Raman
scattering towards the determination of specie concentration and temperature,
among others, in a mixing axisymmetric jet and air-methane flame. The
applicability of a single pulse technique towards the acquisition of the
pertinent data was of special interest in this context. It was shown that
this particular technique can provide not only the concentration and temperature of a number of species in a flow field, instantaneously, simultaneously
and remotely, but it is also capable of providing information on the type of
flow at hand, i.e., turbulent, or laminar. The incorporation into the
diagnostic system of a laser Doppler velocimeter resulted in a very useful
and self-contained apparatus which could be of major importance in the diagnostics of a number of flow fields, including internal combustion.

II.   EXPERIMENTAL APPARATUS

As indicated in Refs. 1 and 2, the basic groundwork for obtaining
concentration and temperature data using laser Raman diagnostics by
means of a single laser pulse has been essentially established. The basic
data acquisition and processing system has been designed and constructed.
Preliminary data obtained on an axisymmetric mixing jet and a methane-
air and CO_2 flame have indicated that the basic system is capable of providing the concentration and temperature of a flow field with a resolution
exceeding the resolution possible with mechanical probes, without the
disturbance introduced into the measurement by mechanical probes. An additional advantage, as pointed out previously (Refs. 3 and 4), is the ability
to determine the pointwise values of a number of variables of interest
instantaneously and simultaneously. This kind of measurement permits one
also to draw conclusions as to the character of the flow field in question.
In order to obtain a more complete picture of a flow field, a velocity
measuring device has been incorporated using a laser Doppler velocimeter.
The complete experimental apparatus is described in detail in Ref. 5.

To enlarge the scope and applicability of these techniques, an attempt
was made to obtain measurements of temperature and specie concentration in an
internal combustion chamber. For that purpose a metal cylinder was equipped
with quartz windows and installed on the methane-air-CO_2 flame holder. The
cylinder could be moved along its axis and thus temperature and concentration
measurements could be made on the axis of the flame or jet, respectively.
(Ref. 5). The diagnostic system is based on backscattering through a small
diameter window which could generally be accommodated on most practical combustion engines, and temperature and specie concentration could be measured.
Some preliminary data on this system have been taken and they will be discussed
later.

III.   EXPERIMENTAL DATA AND DISCUSSION

As mentioned in the Introduction, one of the primary aims of this work
was to demonstrate the use of the single pulse Raman scattering technique
to measure the temperature and/or specie concentration in a gas dynamic system.

Here, a 150 MW Q-switched Ruby laser, capable of providing 6 pulses per
minute, with a pulse half-width of less than 20 nsec, was utilized as the
illuminating source. Measurement of the Raman scattered signal was
accomplished, in this case, through the use of narrow bandpass filters and
photomultiplier tubes. A detailed description of the measurement procedure
is given in Refs. 5 and 6.

Due to the "instantaneous" nature of the Raman signal, it is obvious
that by carrying out a large number of tests at each point in the flow field,
it is possible for one to obtain both the average value of the temperature
and/or specie concentration and a measure of the turbulent intensity of
these quantities.

Concurrent with the Raman scattering experiments, a series of velocity measurements were made using three separate techniques: (1) Pitot pressure probe; (2) Hot wire anemometer; and (3) Laser Doppler velocimeter. The emphasis of this work was on the temperature and specie concentration measurements in a flow field. The flow field in this case was a free circular mixing jet, a methane-air-CO_2 flame, and a simulated combustion chamber. The temperature of the flame was measured using the ratio of the Stokes to Anti-Stokes intensity of a given specie in the flame. Since in the methane-air-CO_2 flame the temperature could be obtained from the CO_2 Stokes to Anti-Stokes ratio and from the N_2 Stokes to Anti-Stokes ratio, both measurements were conducted. The spectroscopically measured temperature at a given point was always compared to the temperature as indicated by a thermocouple. These two indicated temperatures were plotted in the next several figures. Thus, Fig. 1 presents the temperature at a position of X/D=2, as obtained with a thermocouple and from the Stokes and Anti-Stokes Raman intensity ratios of N_2. Fig. 2 presents the N_2 concentration in the flame. Fig. 3 presents the temperature profile at X/D=7. Here the scattered Raman intensities were monitored from the CO_2 lines.

It is evident from these examples that the data points exhibit excessive scattering. This problem can be viewed from the general point of data acquisition accuracy and measurement precision, or one can look at the properties of the flow field and interpret the data in the light of these properties. If one examines the obtained data, it is seen that in the center of the flame the temperatures as obtained by thermocouple and Raman scattering agree very well indeed. As one moves out towards the edges of the flame the scattering of the data becomes relatively large.

At this point it must be stated that the agreement between a static calibration of this spectroscopic method of temperature measurements and thermocouple measurements was within 5%, in a range of temperatures where the Stokes to Anti-Stokes intensity ratio is very high, which in itself may contribute to some errors in measurements. In view of the above, the large scattering of the data points at the outside edges of the flame cannot be attributed to the lack of accuracy and precision of the Raman technique. As was pointed out in Ref. 3, this scatter of the data points indicates the presence of turbulent fluctuations in the flow field. Those turbulent fluctuations picked up by the Raman technique seem to decrease substantially with the thermocouple as the sensor. This apparent decrease of turbulence when the thermocouple is used can be explained by the integrating effect of the thermocouple and the time constant of the thermocouple E.M.F. recording instrument. In order to visualize the flow field, some Schlieren photographs were taken of the flame. These indicated that the center of the flame at the exit of the jet at X/D=2 appeared laminar; thus the temperature T_{Raman} and $T_{thermocouple}$ agree very well. As X/D was increased, the laminar portion of the flame disappeared and so did the agreement between T_R and T_{th}. The same is true as the measurements are made closer to the edges of the flame. Better agreement could be obtained by averaging a large number of data points as shown in Ref. 3.

As indicated above, in order to completely describe the flow field, a laser Doppler velocimeter was introduced into the experimental apparatus.

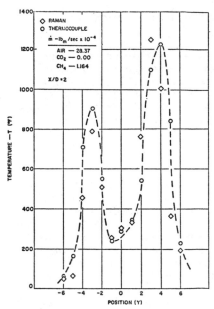

Fig. 1 - Temperature Distribution of Flame, Monitoring N_2.

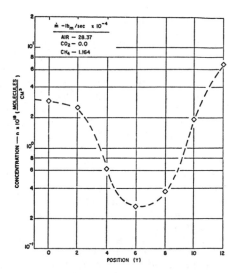

Fig. 2 - Variation of N_2 Concentration with Position at X/D = 2.

160

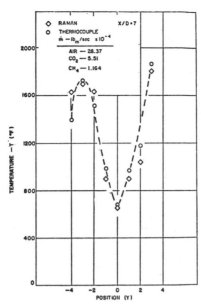

Fig. 3 – Temperature Distribution of Flame, Monitoring CO_2.

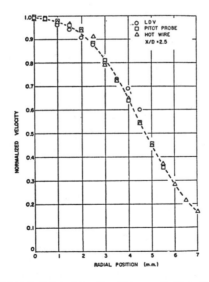

Fig. 4 – Normalized Velocity Profiles of the Jet.

Figs. 4 and 5 show two velocity profiles on the cold jet. The profiles as
obtained using the L.D.V. were compared to velocity profiles as obtained
using pitot pressure probes and a constant temperature hot wire anemometer.
As can be seen, the agreement between the three measurement techniques is
quite reasonable. With this in mind, velocity profiles and turbulent
intensity measurements were taken in a flame. These are shown in Figs.
6 and 7. It is thus evident that the apparatus described is capable of
providing most of the experimental information necessary to describe a flow
field, including combustion.

As part of this effort an attempt was made at applying the Raman
scattering technique to the measurement of concentration and temperature
in a confined internal combustion chamber. The apparatus was described
previously. Since in some cases of combustion, and in particular rocket
propulsion, aluminum oxides appear which could interfere with the Raman
measurements, an experiment was conducted in the combustion chamber by
introducing into the chamber an unknown quantity of powdered Al_2O_3. As can
be seen, the Al_2O_3 appears to increase the CO_2 concentration by about 5%.
At this point it is difficult to make definite conclusions about the effect
of Al_2O_3. A more systematic experiment is required and in particular the
amount of Al_2O_3 introduced into the chamber must be monitored and the degree
of attenuation of the bandpass filter in the stop region must be measured
more precisely.

Fig. 9 presents some preliminary temperature measurement in this
simulated combustion chamber using thermocouple measurements and CO_2 Stokes
and Anti-Stokes lines. The decrease in temperature of the flame with the
introduction of CO_2 should be noted. The agreement between the Raman
temperature and thermocouple temperature is in most cases very good. Based
on those results it is believed that a more comprehensive experimental program
is warranted to determine unambiguously the scope and limitations of the Raman
diagnostic technique in the field of internal combustion and rocket and jet
exhausts. At this point it must be remarked that the data as presented here,
while processed to a certain extent, have not been subjected to a rigorous
data processing routine as generally applied to experimental data. In that
sense the data as shown may still be considered raw data. As such they do
exhibit a reasonable agreement with data obtained by conventional means.

IV. CONCLUSIONS

The experimental apparatus thus far constructed, and the techniques
developed and described, are capable of providing most of the information
necessary for the diagnosis of a flow field. The problem of accuracy and
precision of the measurements is still to be investigated. While the data
as obtained, using the single pulse Raman scattering technique, appear to
exhibit some scatter, it is believed that the Raman technique is capable of
providing data of higher accuracy. The contributing factors to the inaccuracies
of the measurements may be traced to a number of sources. Most of the sources

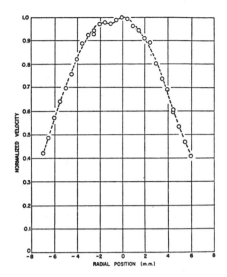

Fig. 5 - LDV Velocity Profile at X/D = 2.5.

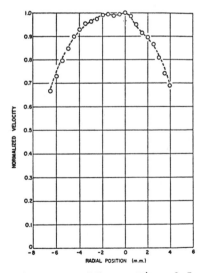

Fig. 6 - LDV Velocity Profile at X/D = 2.5

163

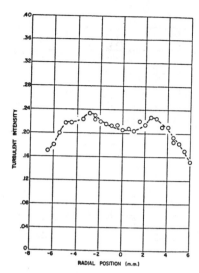

Fig. 7 - Turbulent Intensity at X/D = 2.5

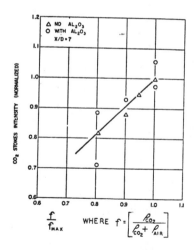

Fig. 8 - Combustion Chamber Normalized Centerline CO_2 Specie Concentration Measurement.

164

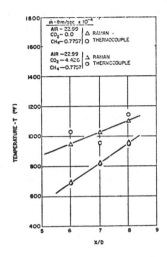

<u>Fig. 9</u> - Combustion Chamber Centerline Temperature Determination at Three Weights (CO_2).

of error must be identified and properly dealt with. Provisions must be made in the instrumentation to make the identification easy and part of a routine periodic check.

The Raman scattering technique provides a means of identifying regions of strong turbulent fluctuations in a flow field. This technique may also provide a means of measuring the frequency spectrum of the turbulent field.

ACKNOWLEDGMENT

This research was sponsored by the Office of Naval Research and is administered by Purdue University through Contract No. N00014-67-A-0226-0005, NR 098-038, Subcontract No. 4965-36.

REFERENCES

1. G. F. Widhopf and S. Lederman, "Specie Concentration Measurements Utilizing Raman Scattering of a Laser Beam," Polytechnic Institute of Brooklyn, PIBAL Report No. 69-46, November 1969.

2. S. Lederman, "Raman Scattering Measurements of Mean Values and
    Fluctuations in Fluid Mechanics," Laser Raman Gas Diagnostics, ed by
    M. Lapp and C. M. Penney, Plenum Press, pp. 303-310, 1974.

3. S. Lederman and J. Bornstein, "Specie Concentration and Temperature
    Measurements in Flow Fields," Project SQUID Technical Report No. PIB-
    31-PU, March 1973.

4. S. Lederman and J. Bornstein, "Temperature and Concentration Measure-
    ments on an Axisymmetric Jet and Flame," Project SQUID Technical Report
    No. PIB-32-PU, December 1973.

5. S. Lederman, et al, "Temperature, Concentration, and Velocity Measure-
    ments in a Jet and Flame," Project SQUID Technical Report No. PIB-33-PU,
    November 1974.

6. J. Bornstein and S. Lederman, "The Application of L.D.V. to Combustion
    Diagnostics," Polytechnic Institute of New York (in preparation).

DISCUSSION

BILGER, Univ. of Sidney - On your Fig. 4, the LVD data stops half way
across the jet. Why?

LEDERMAN - Since it was a small coaxial jet with the seeded particles
being fed through the center core, the particle population density was not
as great at the edge as it was at the center of the jet, due to their finite
rate of turbulent diffusion with the ambient air. Therefore, we could get
the data in a very short time at the center, whereas at the edge it would
have taken much more time to obtain a Raman signal of acceptable accuracy.
This is why we did not take those data points.

GOULDIN, Cornell University - What kind of uncertainty did you obtain
on temperature and concentration?

LEDERMAN - I would say that if you have a laminar flow in a closed
chamber, you could get around 5% accuracy for concentration and temperature.
With special care, we should be able to do better than that.

GOULDIN - What was your turbulence level?

LEDERMAN - We did measure it. It was on the order of 10% in the core
to 15% at the edges.

GOULDIN - Spatial resolution?

LEDERMAN - In our experiments, the spatial resolution was of the order of 1 mm^3. Since the Raman signal is directly proportional to the number of scatterers, and to the incident laser light intensity, and to the f number of the receptor lenses, it follows that with better optics, more power, higher concentration, etc., one could have better resolution. Our power source gives 6 pulses per minute each pulse lasts about 20 nanoseconds and delivers between 2-3 joules. Thus the power is of the order of 100 to 150 megawatts per pulse.

LAPP, GE - I can't resist two comments, one of which is that pretty soon you will be bringing your thermocouples to us and we will calibrate them for you! (laughter). Also, I am sure that most of us would agree that the accuracy Sam gave is a good estimate of what we can produce easily right now. But don't forget that when we collect N photons, the error is $N^{-1/2}$. As you get more photons by multiple passing of your laser beam, by larger pulses of energy, by multiple pulsing and storing, etc..., $N^{-1/2}$ will go quickly down and you can improve the accuracy considerably.

Right now some of us are working on understanding the fundamental limits and tradeoffs for the Raman experiments - for example, the balance between gas breakdown by the strong fields of the light pulses and the spatial and temporal resolution of the resultant data. These studies should give us a better understanding of attainable accuracies for diagnostic data.

GAS CONCENTRATION AND TEMPERATURE MEASUREMENTS BY COHERENT ANTI-STOKES RAMAN SCATTERING

F. MOYA
S. A. J. DRUET
J-P E. TARAN

*Office National d'Etudes
et de Recherches Aerospatiales (ONERA)*

Coherent anti-Stokes Raman Scattering is a parametric-like process at optical frequencies which shows promise for the measurement of gas concentrations and temperatures in flames (Eqs. 1-4). It can be observed in a gas with two intense, collinear optical beams of frequencies ω_1 and ω_2 such that $\omega_1 - \omega_2 \approx \omega_v$, where ω_v is the frequency of a Raman active vibrational transition. An anti-Stokes sideband at $\omega_a = 2\omega_1 - \omega_2$ is then generated in the same direction as the incoming beams.

If focused beams are used, the anti-Stokes power at ω_a is approximately independent of f-number; it is generated from a narrow region about the focus and in the same cone angle as the pump beams. It is given by:

$$P_a \simeq (\frac{4\pi\omega_1^2}{c^3})^2 \ |\chi|^2 P_1^2 P_2 \qquad (1)$$

where we assume $\omega_1 \simeq \omega_2 \simeq \omega_a$, P_1 and P_2 are the powers at ω_1 and ω_2 respectively, and χ is the susceptibility of the gas. One has: $\chi = \chi^{res} + \chi^{nr}$, where χ^{res} is a resonant contribution from the nearby Raman active vibration-rotation resonances and χ^{nr} a nonresonant term independent of $\omega_1 - \omega_2$, contributed by the electrons and the remote resonances.

In a pure gas and on resonance, χ^{res} is 3 to 5 orders of magnitude larger than χ^{nr}. A specific, homogeneously broadened Raman transition j gives a contribution:

$$\chi_j^{res} \simeq \frac{2c^4}{h\omega_1^4} N\Delta_j g_j \ (\frac{d\sigma}{d\Omega})_j \ \frac{\omega_j}{\omega_j^2 - (\omega_1 - \omega_2)^2 + i\,\gamma_j\,(\omega_1 - \omega_2)} \qquad (2)$$

Here, Δ_j is the average population difference per molecule between the lower vibration rotation level (v_j, J) and the upper one, $\hbar\omega j$ is the energy jump between these levels, y_j the transition linewidth, g_j the weighting factor (e.g., $g_j = v_j + 1$ for a Q-line in a non degenerate mode), and $(\frac{d\sigma}{d\Omega})_j$ the spontaneous Raman scattering cross section of the mode; N is the molecular number density. The actual resonant susceptibility of the gas is the algebraic sum $\Sigma_j \chi_j^{res}$ of all the terms such that $\omega j \simeq \omega_1 - \omega_2$. The other resonances in the gas, which are too far to produce an appreciable variation of χ over the spectral domain of interest, are small and real; therefore they can be lumped together in the constant χ^{nr}.

With modest pump pulses in the (1 kW to 1MW range), the anti-Stokes signal P_a is typically 5 to 10 orders of magnitude larger than the spontaneous Raman signal would be in a well designed experiment and using the same gas as a scatterer. In addition, the scattered light is well collimated (10^{-3} sr cone angles are typical), which reduces the interference from stray light and fluorescence.

The experiments are done with a single mode ruby laser pumping a tunable dye laser in a near longitudinal configuration (Fig. 1). Its linewidth is 0.01 cm^{-1}, its stability better than 0.1 cm^{-1}. The beam is amplified to the 5-10 MW level in an amplifier. The dye laser has a spectral width of about 0.2 cm^{-1}.

The advantages of the laser assembly are peak power and frequency stability, the drawback is low repetition rate. The lenses used are achromatic lenses. The reference cell contains 10 atm of N_2 (non resonant at the H_2 vibration frequency). Color and interference filters transmit the anti-Stokes pulses to the photomultipliers while blocking the laser and room light. The signal read in the sample leg P_a is divided by that in the reference leg (P_a^{ref}). The ratio P_a/P_a^{ref} is proportional to N^2 for a given gas under the same temperature and pressure conditions; therefore, if the system is calibrated on a sample of known composition, N can be obtained directly. If the temperature varies, a correction must be made to account for the Boltzmann population changes.

The experiments were done on a Bunsen flame. The next Figs. present typical data on the H_2 formed by the pyrolysis of hydrocarbon molecules. Fig. 2 shows the distribution in the flame (which then was mounted horizontal, the R axis pointing downward). These results were obtained with the first set-up built, and were not compensated for temperature; however, a large maximum in the H_2 concentration is observed in the blue reaction zone; H_2 vanishes in the diffusion zone.

Fig. 3 is the spectral profile of pure H_2 at room temperature, showing a characteristic interference between the real parts of the susceptibilities of the weaker lines and that of the stronger Q(1) line. The solid curve is theoretical; it was calculated assuming a dye laser linewidth of 0.3 cm^{-1} (the results were obtained at an early stage of the set-up, with mediocre resolution). Fig. 4 presents the line maxima obtained at one location in the flame, near the tip of the cone. The lines Q(1) through Q(5) are distributed according

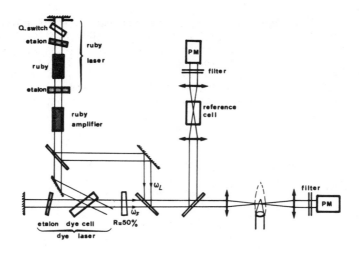

Fig. 1 - Experimental set-up.

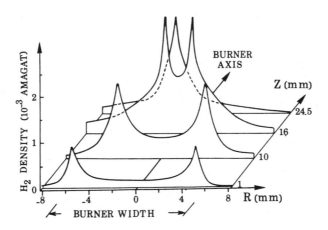

Fig. 2 - H_2 distribution in a horizontal gas flame; R is the distance from the burner axis, Z the distance along the axis; the results are deduced from the Q(1) line intensity.

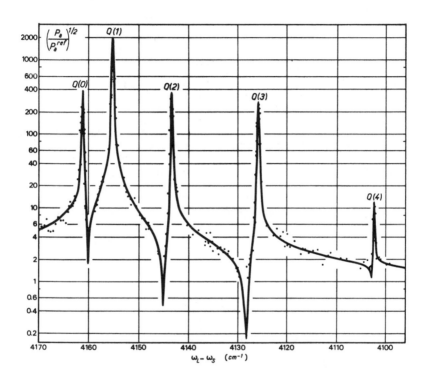

Fig. 3 - Spectrum of the susceptibility of pure H_2 at STP; the quantity $(P_a/P_a^{ref})^{1/2}$ is plotted in arbitrary units.

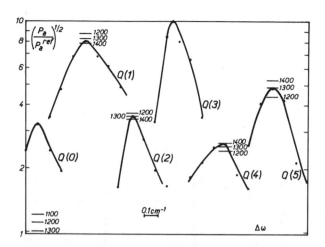

Fig. 4 - Lines Q(0) through Q(5) near their maxima, in the flame. The points at line centers represent averages over 10 laser shots; fewer shots were fired for the other points; this and the flame fluctuations explain the irregular shapes and widths. The lines are normalized with respect to Q(3).

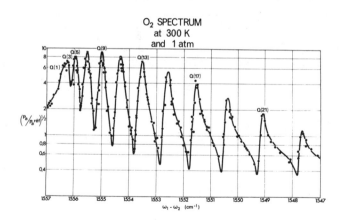

O₂ SPECTRUM
at 300 K
and 1 atm

Fig. 5 - CARS spectrum of O_2 at room temperature

to the Boltzmann factors for a temperature of 1350 ± 30 K, in good agreement with a crude thermocouple measurement (1340 ± 20 K). The Q(0) line is stronger than expected; this result may be an artefact, but it can also be explained by the endothermal nature of the production of H_2 from CH_4, continuously supplying colder molecules into the hot bath. From the absolute Q(1) line intensity, we can also deduce a number density of (5.4 ± 0.5) 10^{16} cm^{-3} for H_2 at that point.

Other gases have also been studied. Fig. 5 is the spectrum of the Q branch of 0 from Q(1) to Q(23), taken with a slightly different dye laser (shortened cavity, with narrowband interference filter plus 75 μm and 500 μm etalons for fine tuning). Simple monochromators made of three highly disper- sive Brewster angle prisms were also added in front of the photomultipliers; this was done in order to prevent the ruby light from entering the colored filters and causing fluorescence at the anti-Stokes frequency. A good fit is seen with the theoretical curve, calculated for a Boltzmann distribution among rotational levels, with a monochromatic ruby laser and a Gaussian 0.14 cm^{-1} dye laser line.

In conclusion, this method seems capable of handling difficult problems such as concentration and temperature measurements in flames, plasmas, etc., because of its considerable brightness. One can also envision turbulence measurements with high repetition rate lasers of moderate peak power; temperature measurements could be made in a single shot by using a broadband dye laser on the Stokes side instead of a narrow line tunable one and record- ing the anti-Stokes spectrum with a spectrograph and a photographic plate. One should not overlook the fact, however, that all these applications require a delicate apparatus and that some development will be needed before they can be set-up in the vicinity of large engines.

REFERENCES

1. P. R. Régnier and J-P E. Taran, Appl Phys. Letters, 23, 240 (1973).

2. P. R. Régnier and J-P E. Taran, in Laser Raman Gas Diagnostics, edited by M. Lapp and C. M. Penney, Plenum Publishing Corporation, New York, London (1974), p. 87.

3. P. R. Régnier, F. Moya and J-P E. Taran, AIAA J. 12, 826 (1974).

4. F. Moya, S. A. J. Druet and J-P E. Taran, Optics Com. 13, 169 (1975).

COHERENT ANTI-STOKES RAMAN SPECTROSCOPY (CARS) IN LIQUIDS AND NRL'S PROGRAM ON GASES

A. B. HARVEY

Naval Research Laboratory

What I am about to discuss does not directly involve combustion, but rather concerns the CARS technique in liquids. Nonetheless, the work underscores the very fine results of Dr. Taran in the previous talk. This research was conducted at Stanford University under the NRL sabbatical program in Professor Byer's laboratory. Also involved in the research were Dr. R. Begley now at Allied Chemical and Professor B. Hudson of the Chemistry Department at Stanford.

Fig. 1 illustrates our first apparatus utilizing a Chromatix Nd:YAG laser (532 nm) which in turn pumps a tunable dye laser. The two beams (532 nm + dye laser) are then focused and crossed in a cell of benzene. They are crossed in condensed media for reasons of phase matching, which I do not have enough time to discuss in this short talk. The Raman emission is evolved in the form of a very intense, coherent, laser-like beam at the corresponding anti-Stokes frequency when the dye laser is tuned across the 992 cm^{-1} Raman resonance in benzene (dye laser tuned to 562 nm). The next slide* is a color photograph of the apparatus schematically illustrated in the previous slide. Note the extremely intense laser-like Raman emission appearing in the lower right hand corner. With 1 kw input beams about 0.1% of the dye laser is converted into Raman emission or about 5 orders of magnitude greater than conventional Raman efficiencies. Note that no monochromator is needed in these experiments.

---

*Most of the slides used in this presentation were color slides and could not be reproduced here. We apologize for not making available these vivid illustrations of the CARS method potential (Ed.).

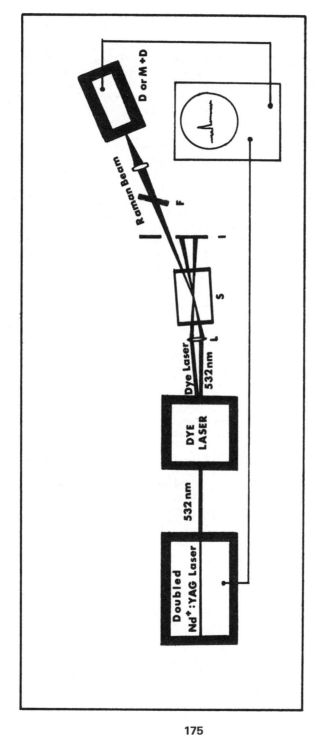

Fig. 1 - CARS Apparatus in Benzene Concentration Measurement

175

Fig. 2 schematically shows an experiment utilizing a nitrogen laser pumped dye laser system. Here a Molectron laser pumps two dye lasers each of which can be independently tuned. The following slide is a color photograph of this system. Again, notice the intense blue, laser-like Raman emission from the benzene sample in the anti-Stokes region. With 15-50 kw beams, Raman emission as high as 135 watts had been observed.

The next two slides are close-up color photos of the Raman beam and cell with and without Rhodamine-6G added to the benzene sample. Since the Raman signal by CARS is in the anti-Stokes region, the blue beam is clearly observed even though the fluorescence emanating from the cell, doped with Rhodamine-6G, is quite enormous. This strong fluorescence would swamp all conventional Raman measurements. Fig. 3 shows the intensity of the fluorescence emission particularly at the 992 cm^{-1} Stokes region and the absence of fluorescence at the anti-Stokes region. This reduction in laser induced fluorescence has obvious advantages for combustion analysis especially for the new fuels derived from shale oil and coal, since these are known to contain high concentrations of aromatics which will probably fluoresce very strongly.

Fig. 4 schematically shows our CARS apparatus for analyzing gases, which is currently under construction. It consists of a 10 MW Nd:YAG laser (10 pps, at 532 nm) which in part pumps a dye laser. The tunable dye laser and YAG pulse are focused and mixed in the sample chamber. The laser beams plus the anti-Stokes signal are split into two fractions. The smaller is filtered to remove the anti-Stokes signal and the resulting laser beams are then mixed in a reference cell of high pressure argon. The larger fraction is spectrally filtered to detect the anti-Stokes signal in the sample. The sample and reference signals are then ratioed to compensate for amplitude fluctuations in the laser beams, displacement of the beam, perturbations during dye laser tuning, dye laser efficiency, optics transmission, etc.

The last slide shows the application of the CARS technique to a problem of population distribution and temperature measurements in D_2 for an ARPA supported D_2/HCl transfer laser. After these experiments have been made, the apparatus will be used for making combustion studies in simple laboratory flame systems prior to a more ambitious program in combustion diagnostics.

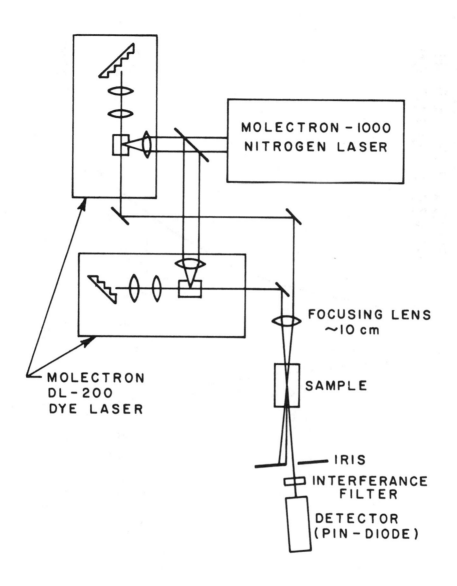

MOLECTRON – 1000
NITROGEN LASER

MOLECTRON
DL – 200
DYE LASER

FOCUSING LENS
~10 cm

SAMPLE

IRIS

INTERFERANCE
FILTER

DETECTOR
(PIN – DIODE)

Fig. 2 – CARS Experiment (N$_2$ + Dye Laser)

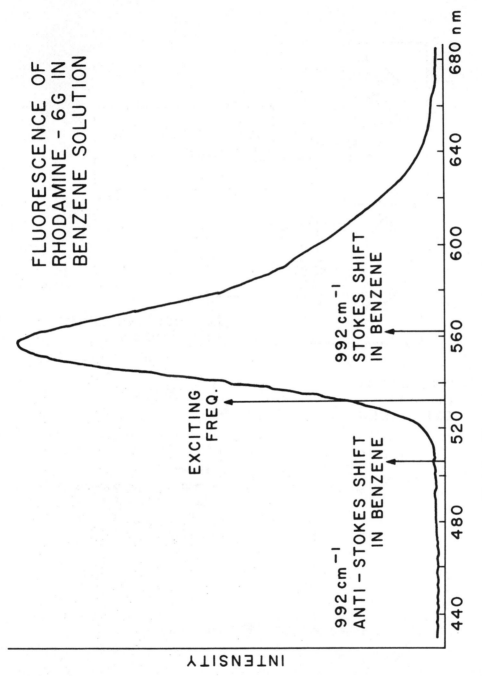

Fig. 3 - Fluorescence vs Stokes and Anti-Stokes Signals

178

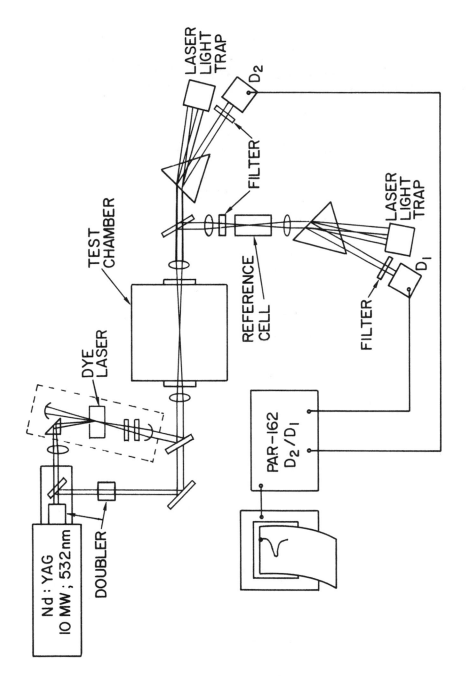

Fig. 4 – The New NRL CARS Apparatus

179

COMPARISON OF DENSITY AND TEMPERATURE MEASUREMENT USING RAMAN SCATTERING AND RAYLEIGH SCATTERING

FRANK ROBBEN

Lawrence Berkeley Laboratory

ABSTRACT

Calculations of the performance, in terms of the signal-to-noise ratio, for vibrational Raman and Rayleigh scattering measurements under combustion conditions are made. The assumed conditions are 1) a hydrogen-air flame at a temperature of 2000°K, 2) a 1 watt argon-ion laser, and 3) f/5 scattered radiation collection optics. Based on photo-electron counting statistics, the Rayleigh scattering technique has much larger signal-to-noise ratio than Raman scattering, but either technique will give good results for mean temperature and density. However, for the measurement of a nominal 10% rms turbulent intensity, only Rayleigh scattering is capable of a satisfactory measurement. It is noted that by use of higher powered lasers and lower f/number light collection optics, Raman scattering should give some information on turbulent fluctuations, especially in the case of turbulent diffusion flames where the fluctuations are of a very large amplitude.

INTRODUCTION

The use of Raman scattering for species concentration and temperature determination in gases, in particular during the combustion process, has been developed and tested by several investigators (Refs. 1-3). The principal difficulty of the technique is the low scattered signal intensity,

Supported by the Air Force Office of Scientific Research.
Work done in the Mechanical Engineering Department- University of California- Berkeley.

which requires (1) relatively long integration times for an adequate signal-to-noise ratio, (2) very low background flame radiation, and (3) low fluorescence from either gases or particulates.

Rayleigh scattering offers a competing technique for measurement of the temperature and the density of gases (Ref. 4). The Rayleigh scattering cross section is about 1000 times larger than the vibrational Raman cross section, resulting in a much larger intensity. This considerably alleviates the first 2 problems mentioned above. However, scattering from particulates, if present in the gas, is a very serious problem with Rayleigh scattering measurements.

In this discussion, a comparison of the expected signal-to-noise ratios for vibrational Raman and Rayleigh scattering measurements of density and temperature in a premixed combustion zone, without either background radiation or particulates, is made. As has already been shown, satisfactory time averaged measurements under these conditions can be obtained with either technique.

Our conclusions are that (1) satisfactory turbulent density measurements, and marginally satisfactory turbulent temperature measurements, can be obtained with Rayleigh scattering, and (2) satisfactory turbulent density and temperature measurements are not possible with Raman scattering unless much larger laser powers are used, or the turbulence intensity is much larger than assumed. It is possible that either of these latter requirements may be satisfied, so we do not by any means rule out the possibility of obtaining significant information on the turbulent properties of combustion by use of Raman scattering. However, Rayleigh scattering may offer significant advantages for combustion measurements in certain cases.

RAMAN DENSITY AND TEMPERATURE MEASUREMENTS

Fig. 1 is intended as a reminder of the principles of Rayleigh and Raman scattering. Incident photons (from a laser in this case) excite the molecule to a virtual state which essentially instantaneously decays back either to the original state or to a state with different vibrational and rotational quantum numbers. Decay to the original state results in Rayleigh scattering, where the only change in wavelength is due to the Doppler shift associated with scattering from a moving target (molecule). Decay to a state of higher vibrational energy results in a longer wavelength, called Stokes Raman scattering, while decay to a lower vibrational energy results in a shorter wavelength, called anti-Stokes Raman scattering. The cross section for Rayleigh scattering is typically 10^3 times larger than that for vibrational Raman scattering.

Fig. 2 shows schematically the measurement of scattering at right angles from a focussed laser beam. Our evaluation of scattering takes place in the post-combustion region of a hydrogen-air mixture at atmospheric pressure

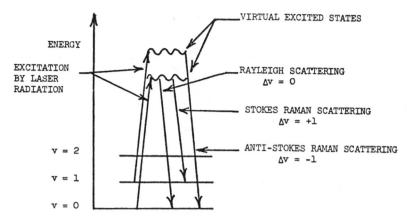

<u>Fig. 1</u> - Schematic energy level diagram of a molecule indicating the low vibrational levels and Raman and Rayleigh scattering by excitation to a virtual state.

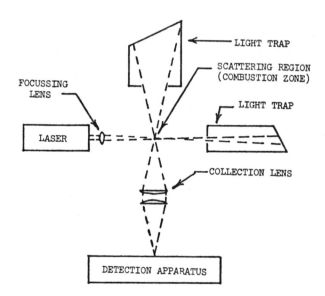

<u>Fig. 2</u> - Schematic diagram for the measurement of Raman and Rayleigh scattering from the combustion zone of a hydrogen-air flame. The parameters common for both Raman and Rayleigh scattering are:

<u>Laser</u>: Argon-ion, 1 watt continuous power at 4880 Å.
<u>Combustion</u>: Premixed hydrogen-air, near stoichiometric with I = 2000°K, p = 1 atm, no background radiation, fluorescence, or particles.
<u>Scattering volume and detector optics</u>: Scattering at 90° of a 1 mm length of laser beam is collected by f/5 optics (0.03 steradian).

and 2000°K temperature. The parameters which are common to both Raman and Rayleigh scattering are listed below the figure. The 1 mm length scattering volume is fixed by the requirement of turbulent measurements, as larger lengths would certainly average the turbulent fluctuations spatially. (Due to the optical properties of grating and Fabry-Perot spectrometers, longer lengths can be used to increase the Raman signal, but not the Rayleigh signal for temperature measurements.)

All signal strengths will be calculated in photoelectron counts per second, as this represents the optimum detection system and also leads to simple estimation of the noise (Ref. 5). The Raman counting rate for the nitrogen Stokes vibrational band is given by

$$C_N = P \; N_S \; f_N \; \sigma_R \; \Omega \; \ell \; \varepsilon \; \eta \qquad (1)$$

where:

C_N = photoelectron counts for nitrogen, integrated for 1 second

P = 2.5 x 10^{18} photons/sec   (1 watt)
   laser power

N_S = 3.7 x 10^{18} molecules/cm^3
   density of molecules in the flame

f_N = 0.67
   fraction of nitrogen molecules

σ_R = 5.4 x 10^{-31} cm^2/sr
   vibrational Raman cross section for nitrogen

Ω = 0.03 sr
   solid angle of collection for scattered light

ℓ = 0.1 cm
   length of laser beam observed

ε = 0.15
   photomultiplier quantum efficiency

η = 0.35
   optics and filter (or spectrometer) transmission

From Eq. (1) we find that the average number of photoelectron counts in 1 sec will be

$$< C_N > \; = \; 520 \text{ counts} \qquad (2)$$

These counts will have a Poisson error distribution, with the variance equal to the average,

$$Var(C_N) = < C_N >$$ (3)

and thus the realtive standard deviation of the measurement, for 1 second integration time, will be

$$\frac{\sigma(C_N)}{< C_N >} = 1/< C_N >^{1/2}$$

$$= 0.044$$ (4)

Severa methods of determining the temperature from Raman scattering have been discussed and evaluated by Lapp et. al. (Ref. 4). In this temperature range the measurement of the ratio of the Stokes to the anti-Stokes vibrational intensities is within a factor of 2 of being optimum, and is a good, practical technique. As seen from Fig. 1, it is simply a measurement of the ground and first vibrational densities, and thus gives the vibrational temperature. If we let $R = C_S/C_A$, where C_S is the Stokes intensity and C_A is the anti-Stokes intensity, then we find that the temperature is a function of R which is easily calculated. At T = 2000°K we find that an error in the ratio R results in a temperature error given by

$$\frac{\Delta T}{T} = 0.6 \frac{\Delta R}{R}$$ (5)

The standard deviations of C_S and C_A are found as dicussed previously, and for a 1 second integration time, the relative standard deviation of a temperature measurement is given by

$$\frac{\sigma(T)}{< T >} = \frac{1.7}{< C_N >^{1/2}}$$

$$= 0.072$$ (6)

RAYLEIGH DENSITY AND TEMPERATURE MEASUREMENTS

Unlike Raman scattering, where the density of a single constituent such as nitrogen could be measured, Rayleigh scattering sums the scattered radiation from all the constituents. For isotropic molecules, the Rayleigh scattering cross-section σ_{Ri} for species i can be obtained from the index of refraction N_i for species i by (Ref. 6)

$$\sigma_{Ri} = \frac{4\pi^2}{\lambda^2} \frac{N_i-1}{n} \sin^2\theta$$ (7)

where n_i is the corresponding number density and θ is the direction of scattering as measured from the E vector of the incident radiation. Here we assume that the incident radiation is polarized and that $\theta = 90°$.

The total scattered signal is then given by

$$C_{Ray} = K \, n_s \, \Sigma \, \mu_i \, \sigma_{Ri} \quad , \tag{8}$$

where μ_i is the mole fraction of species i, n_s is the total number density, and K is a constant. Thus it appears that one needs to know the actual composition of the gas in order to interpret the results. However, it turns out that the contribution of each atom to the Rayleigh scattering is roughly independent of the molecular bonds (Ref. 7), which means that the scattered intensity will be approximately independent of the degree of reaction and, in premixed combustion, approximately proportional to the mass density. Further, since nitrogen accounts for some 60% of the mixture, this approximation is quite good.

The scattered Rayleigh signal C_{Ray} can be calculated from Eq. (1), using an average Rayleigh cross section

$$\sigma_{Ray} = 8 \times 10^{-28} \, cm^2/sr \quad . \tag{9}$$

We find that

$$< C_{Ray} > \; = 1.2 \times 10^6 \, counts/sec \tag{10}$$

and for a 1 sec integration time,

$$\frac{\sigma(\rho)}{< \rho >} = 1/< C_{Ray} >^{1/2}$$

$$= 0.001 \tag{11}$$

The measurement of temperature from Rayleigh scattering is more complex than from Raman scattering, and will be described briefly. The technique consists of measuring the Doppler broadening of the scattered radiation with a Fabry-Perot interferometer. The spectrum of the scattered radiation has a Gaussian distribution as shown in Fig. 3a. The spectral intensity is given by

$$I(\nu) = K \, \Sigma_i \, \mu_i \, \sigma_{Ri} \, f_i \, (\nu) \tag{12}$$

where $f_i(v)$ is the molecular velocity distribution for species i,

$$f_i(v) = \left(\frac{m_i}{2\pi kT}\right)^{1/2} e^{-\frac{m_i v^2}{2kT}} \quad . \tag{13}$$

The light frequency υ is obtained from the molecular velocity v by the Doppler shift relation

$$\frac{\upsilon - \upsilon_o}{\upsilon_o} = \left(\frac{v}{c}\right) 2 \sin(\phi/2) \quad , \tag{14}$$

where υ_o is the laser frequency, c is the velocity of light, and ϕ is the angle of scattering, measured from the incident radiation direction. The composition of the gas must be known in order to interpret the Doppler broadening and obtain the temperature; however, for most combustion species the line width does not depend too strongly on the degree of reaction, further, nitrogen is the principle constituent.

The line width at 1/2 the peak density, for a hydrogen-air flame at 2000°K, is approximately 0.13 cm^{-1}, or 3.9 GHz. We have shown that this line width can be measured to a high degree of accuracy with a Fabry-Perot interferometer and that temperatures in a hydrogen-air flame can be measured satisfactorily (Refs. 4 & 8). The measured line profile $S(\upsilon)$ is a convolution of the instrument function $g(\upsilon - \upsilon_o)$, shown in Fig. 3b, with the scattered radiation $I(\upsilon)$ from Eq. (12),

$$S(\upsilon) = K' \int I(\upsilon')g(\upsilon - \upsilon')d\upsilon' \tag{15}$$

The width at 1/2 maximum of the instrument function, $\Delta\upsilon_a$, is determined, among other things, by the solid angle of light accepted by the Fabry-Perot interferometer (Ref. 9). The étendue E, the product of the area of the interferometer mirrors A_M and the solid angle of acceptance at the mirrors, is related to $\Delta\upsilon_a$ by the equation

$$E = 2\pi A_M \Delta\upsilon_a/\upsilon_o \quad . \tag{16}$$

This approximation to the instrument function width based on the aperture (or étendue) is valid when it is larger than the width determined by the reflectivity and roughness of the mirrors. Since the etendue is a constant of the interferometer optical system, the area of the entrance aperture of the interferometer as imaged at the laser beam is given by

$$A_b = E/\Omega_b \tag{17}$$

where Ω_b = 0.03 sr as determined by the f/5 light collection optics. Thus
we want $\Delta\upsilon_a$ to be large to give good intensity, but at the same time small
so that the de-convolution of Eq. (15) will not introduce large errors into
the Doppler line width. A somewhat conservative compromise is to let

$$\Delta\upsilon_a = \Delta\upsilon_D/3 \quad ,$$

and using this value with 5 cm diameter interferometer mirrors in Eqs. (16)
and (17) gives a laser beam length of 0.1 cm from which scattering will be
measured.

The fraction of light transmitted at the peak of the line is estimated
as follows. Filter transmission, 0.4; optics transmission, 0.6; Fabry-Perot
transmission as determined by the bandpass, $\Delta\upsilon_a/\Delta\upsilon_D$ = 0.3; and Fabry-Perot
transmission as determined by the losses in the mirror coatings, 0.6. This
gives an overall light transmission of 0.05. The peak photoelectron counting
rate given by Eq. (1) is

$$C_{RT} = 8 \times 10^4 \text{ counts/sec} \tag{18}$$

with

$\quad$ P $\quad$ = 1.25×10^{18} photons/sec
$\quad\quad\quad$ (1/2 watt power for single mode of laser)

$\quad$ f_N $\quad$ = 1.0

$\quad$ σ_{Ray} = 8×10^{-28} cm^2/sr

$\quad$ η $\quad$ = 0.05 .

The other parameters in Eq. (1) are unchanged from the values used for Raman
scattering.

To measure width, and hence the temperature, of the scattered radiation
in an optimum manner a line tracking feedback technique has been used. The
scattered radiation is split into 3 parts of equal intensity (by time division
in our previous experiments) and these parts are used to tract the half
intensity points on the left and right sides of the line and the peak intensity.
Under these conditions we have shown that the standard deviation in the line
width measurements is given by Ref. 8.

$$\frac{\sigma(W)}{<W>} = \frac{2k_c}{<C_p>^{1/2}} \tag{19}$$

where W is the measured line width, k_c is a constant of the line shape and equal to 0.36 for a Gaussian line, and C_p is the total number of photoelectron counts measured at the line peak. Since the temperature is proportional to the line width squared, we have

$$\frac{\sigma(T)}{< T >} \cong \frac{4 \times 0.36}{< 0.3 \times C_{RT} >^{1/2}}$$

$$\cong \frac{2.6}{< C_{RT} >^{1/2}} \qquad (20)$$

Using the previous value for C_{RT}, we find for a 1 sec integration time,

$$\frac{\sigma(T)}{< T >} = 0.01$$

The relative standard deviations for density and temperature measurements with 1 second integration times are summarized in Table I.

TURBULENCE MEASUREMENTS

In order to evaluate the feasability of turbulence measurements, estimations are based on Gaussian turbulence with parameters typical for free shear and boundary layers. The turbulence will be assumed to have simple statistical properties.

Let $\xi(t)$ represent a parameter of the fluid. The fractional rms turbulence f_t of ξ is defined by

$$f_t^2 = < \xi^2 > / < \xi >^2 \qquad (21)$$

where the brackets indicate time averaged values. The frequency spectrum of ξ^2, the turbulent power spectrum $G(\omega)$, is shown approximately in Fig. 4 and is normalized so that

$$\int_0^\infty G_t(\omega) d\omega = f_t^2 \qquad (22)$$

As shown in Fig. 3, $G_t(\omega)$ has a maximum value at low frequencies equal to G_{to}, and a width at 1/2 maximum of $\Delta\omega_t$. Reasonable values for the parameters are as follows:

	Density	Temperature
Raman Scattering	0.044	0.072
Rayleigh Scattering	0.001	0.01

Table 1. Comparison of the relative standard deviations for temperature and density using Raman and Rayleigh scattering. Assumed conditions: 1 sec integration time; 1 watt argon-ion laser; atmospheric pressure hydrogen-air flame at 2000°K.

	Density	Temperature
Raman Scattering	0.6	1.7
Rayleigh Scattering	3×10^{-4}	0.03

Table 2. Comparison of the expected standard deviations of the turbulent power spectrum to the maximum of the turbulent power spectrum. Assumed conditions: atmospheric pressure hydrogen-air flame at 2000°K; 1 watt argon-ion laser; 10% rms turbulent intensity; Gaussian turbulent spectrum with width (at 1/2 intensity) of 160 Hz; spectrum analyzer resolution of 48 Hz and 200 sec integration time.

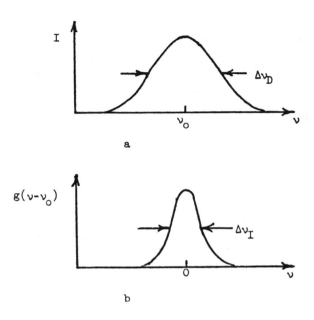

Fig. 3 - a) Intensity of Rayleigh scattered light plotted versus frequency ν, showing Doppler broadening by random molecular velocities with half-intensity width $\Delta\nu_D$.

b) Instrument function $g(\nu-\nu_0)$ with half-height width $\Delta\nu_I$.

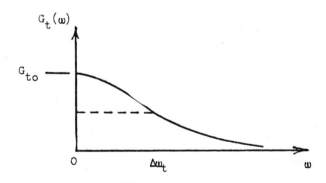

Fig. 4 - Assumed turbulent power spectrum $G_t(\omega)$ with half-intensity width $\Delta\omega_t$.

$$f_t = 0.1$$

$$\Delta\omega_t = 1000 \text{ rad/sec}$$

$$G_{to} \cong f_t^2/\Delta\omega_t$$

$$\cong 10^{-5}\text{sec} \tag{23}$$

In a measurement of the turbulent power, or spectrum, we will assume that the only important additional fluctuations will come from the Poisson counting statistics of the photoelectrons, which may be called "shot noise". The photoelectron counting power spectrum G_p is a constant, independent of frequency (Ref. 5) (white noise), and is given by

$$G_p = \frac{1}{\pi r} \tag{24}$$

where r is the average counting rate.

The measured power spectrum $G(\omega)$ is just the sum of these two sources,

$$G(\omega) = G_t(\omega) + G_p \tag{25}$$

In order to determine if we can make a satisfactory measurement of $G_t(\omega)$, we will compare the expected standard deviation in $G(\omega)$ with the maximum of the turbulence, G_{to}.

The standard deviation in $G(\omega)$, $\sigma(G)$, when measured with an ideal spectrum analyzer, (Ref. 10) is independent of ω and is given by

$$\frac{\sigma(G)}{<G>} = \frac{2\pi}{\Omega T}^{1/2} \tag{26}$$

where Ω is the bandpass of the analyzer and T is the measurement time. If we let

$$\Omega = 300 \text{ rad/sec}$$

$$T = 200 \text{ sec}$$

then we find that

$$\frac{\sigma(G)}{<G>} = 0.01 \tag{27}$$

This value, based on rather coarse resolution and long measurement time, represents about the minimum practical standard deviation of G.

Let us now consider the measurement of the turbulence power $G_t(\omega)$, using Eq. (25). Since G_p is independent of frequency, it can be determined from the asymptotic value of $G(\omega)$ at high frequencies, where $G_t(\omega)$ approaches zero. This value must then be subtracted from $G(\omega)$. If for simplicity we let the error in G_p be negligible, then $\sigma(G_t)$ will be equal to $\sigma(G)$. Thus $\sigma(G)/G_{to}$, the standard deviation divided by the maximum value, will give a measure of the signal-to-noise ratio for the detection of 10% rms turbulence in the combustion zone.

TURBULENCE IN DENSITY BY RAYLEIGH SCATTERING

Using Eq. (24) and the counting rate from (10), we find that

$$G_p = 3 \times 10^{-7} \text{ sec} \qquad (28)$$

and, from (23) that

$$G_p/G_{to} = 0.03 \quad , \qquad (29)$$

indicating that the photomultiplier shot noise is negligible for low frequency turbulence measurement. Thus $\sigma(G_t)/< G_t >$ will be limited to 1% by (27) at low frequencies. At high frequencies where $G_t(\omega)$ becomes small compared to G_p, we find that

$$\sigma(G) = 10^{-2} G_p \quad ,$$

and thus

$$\sigma(G)/G_{to} = 3 \times 10^{-4} \qquad (30)$$

for high frequencies. We conclude that very satisfactory measurement of the turbulence in density can be made by Rayleigh scattering.

TURBULENCE IN DENSITY BY RAMAN SCATTERING

Using Eq. (24) and the counting rate from (2), we find that

$$G_p = 6 \times 10^{-4} \text{ sec} \qquad (31)$$

and, from (23), that

$$G_p/G_{to} = 60 \quad . \tag{32}$$

Thus the photomultiplier shot noise is dominant even at low frequencies. In this case $\sigma(G)$ is obtained from Eq. (27) using the average value of G_p, and we find that

$$\sigma(G)/G_{to} = 0.6 \tag{33}$$

which is too large for a satisfactory measurement. We conclude that the detection of 10% rms turbulence level by Raman scattering is not possible unless a more powerful laser is used. Since we have neglected all background radiation, a pulsed laser is limited by the same general considerations.

TURBULENCE IN TEMPERATURE BY RAYLEIGH SCATTERING

For the temperature measurement a more complex analysis of the photon counting signal is required, as described previously, and Eq. (24) cannot be used directly to find the photon noise contribution to the noise power spectrum of the temperature. However, the photon noise contribution will still be independent of frequency, and from (20) results in a relative standard deviation of 1% for 1 sec measurement time. This corresponds to the standard deviation of a counting rate of 10^4 sec^{-1}, which can be used in Eq. (24) to find

$$G_p = 3 \times 10^{-5} \text{ sec} \tag{34}$$

for the photon noise power contribution to the temperature power spectrum. This gives us

$$G_p/G_{to} = 3 \quad , \tag{35}$$

and the photon noise is dominant even at low frequencies. As in the case of Raman scattering just considered,

$$\sigma(G)/G_{to} = 0.03 \quad , \tag{36}$$

from which we conclude that 10% rms turbulence in temperature is marginally measurable by the method of Rayleigh scattering.

TURBULENCE IN TEMPERATURE BY RAMAN SCATTERING

The photon noise results in relative standard deviation of 7.2% for 1 second measurement time, as given by (6). This corresponds to the standard deviation of a counting rate of 190 sec^{-1}, giving

$$G_p = 1.7 \times 10^{-3} \text{ sec} \qquad (37)$$

and

$$G_p/G_{to} = 170 \qquad . \qquad (38)$$

Thus the photon noise power is much larger than the turbulent power. The spectrum analyzer will have a relative standard deviation given by

$$\sigma(G)/G_{to} = 1.7 \qquad , \qquad (39)$$

much too large for the measurement of turbulence.

The relative standard deviations for turbulence measurements are summarized in Table II.

CONCLUSIONS AND DISCUSSION

The conclusions of this analysis, which are apparent from Tables I and II, are that both Raman and Rayleigh scattering can be used for satisfactory time averaged measurements of density and temperature, but that only Rayleigh scattering can be used for measurement of the turbulence intensity. These conclusions apply to the rather restricted case of a premixed hydrogen-air flame with 10% rms turbulence intensity, a 1 watt argon-ion laser, and f/5 light collection optics. The following speculations on other conditions are offered.

Flame: The conclusions are probably applicable to most premixed combustion systems, as there will be essentially no particulates, and proper choice of the laser frequency will, at least for Rayleigh scattering, keep the background radiation from the combustion region small. We have found that it is not difficult to eliminate particulates from the air.

For diffusion flames the situation is quite different. The interpretation of Rayleigh scattering is complicated because of the mixing of the fuel and oxidizer, which in general will have different mean scattering cross sections and molecular weights. It may be possible to prepare special fuel and oxidizer mixtures with the same mean cross sections by addition of various inert gases. Any formation of free carbon in the flame zone would greatly increase the Rayleigh scattering, which actually could be the basis of interesting measurements.

Diffusion flames are normally non-steady and turbulent in nature. Raman scattering, while unaffected by the mixing phenomena and relatively insensitive to particulates, is generally not capable of following the real time fluctuations. However, since the turbulence levels are very large (100%), powerful pulsed lasers should be capable of providing meaningful instantaneous Raman measurements. However, it will be quite difficult to obtain accurate temperature measurements.

For particulate-free diffusion flames Rayleigh scattering may give more accurate instantaneous temperature measurements than Raman scattering. The uncertainty in molecular weight may not be a serious source of error, depending on the fuel composition, as one will also have the total scattering intensity at the same time. Further, Raman measurements could be combined with the Rayleigh measurement. However, the Rayleigh scattering would be complex to apply, both in experimental instrumentation and in analysis.

Laser: More powerful continous laser, approaching 10 watts, are available and can be used. Pulsed lasers with average powers of 1 watt or more are also available, and are particularly important for Raman measurements where background radiation is a problem.

The turbulent frequency spectrum can in principle be measured with low pulse rate lasers by using them in a double-pulse mode, so that the auto-correlation function can be measured.

Light Collection Optics: We have assumed f/5 optics in order to minimize the access problem to an experimental combustor. Lower f/number optics require quite large windows and other special considerations. By going to an f/1.2 system, the solid angle of light collection is increased about 16 times to 0.5 sr. This results in 16 times improvement in the Raman intensity. However, the signal for temperature measurement by Rayleigh scattering would be increased only a factor 4, because of the limitation on the aperture size of the Fabry-Perot interferometer. Thus, larger light collection optics is particularly advantageous for Raman scattering measurements.

At this time both Raman and Rayleigh scattering should be useful instrumentation techniques for basic research in combustion phenomena. Their application to some forms of model combustors should also be possible, but will require careful consideration of the operating conditions.

ACKNOWLEDGEMENT

I would like to thank R. Pitz, and in particular Dr. R. Cattolica, for their collaboration in the experiments which form the background of this paper.

REFERENCES

1. M. Lapp, C. M. Penney, and J. A. Asher, "Applications of Light Scatter-
    ing Techniques for Measurements of Density, Temperature, and Velocity
    in Gas Dynamics," ARL 73-0045 (1973) Aerospace Research Laboratories,
    Wright-Patterson Air Force Base.

2. M. Lapp, C. M. Penney, R. L. St. Peters, "Laser Raman Probe for Flame
    Temperature," Project SQUID Report GE-1-PU (1973).

3. M. Lapp and C. M. Penney, editors, Laser Raman Gas Diagnostics, Plenum,
    New York, 1974.

4. F. Robben, R. Cattolica, and R. Pitz, "Premixed Hydrogen-Air Flame
    Temperature from Rayleigh Scattering," Technical Digest Applications
    of Laser Spectroscopy, Optical Society of America Topical Meeting,
    Anaheim, CA. (March 1975).

5. F. Robben, "Noise in the Measurement of Light with Photomultipliers,"
    Applied Optics, 10, 776 (1971).

6. C. M. Penney, "Light Scattering in Terms of Oscillator Strengths and
    Retractive Indices," J.O.S.A. 59, 34 (1969).

7. M. Born and E. Wolf, Principals of Optics, Pergamon Press (New York, 1965)
    p. 88.

8. R. Cattolica and F. Robben, "Computer Controlled Fabry-Perot Spectrometer,"
    unpublished manuscript.

9. P. Jacquinot, "New Developments in Interference Spectroscopy," Report.
    Prog. Phys. 23, 267 (1960).

10. A. A. Kharkevich, Spectra and Analysis, Consultants Bureau, New York,
    1960.

LASER MODULATED PARTICULATE INCANDESCENCE NOISE EFFECTS IN LASER RAMAN SCATTERING DIAGNOSTICS

ALAN C. ECKBRETH

United Technologies Research Center

The following problem has been encountered in the laser Raman thermometry experiments we are conducting on model combustors at the United Technologies Research Center (formerly United Aircraft Research Laboratories). Particulates in the flow, whether naturally occurring, purposively introduced (LDV), or, what I believe to be true in our situation, produced by combustion (soot), absorb laser radiation and heat to temperatures far in excess of the local stream value. Due to the strong sensitivity of incandescence with temperature, the particulate radiation increases substantially, and, unfortunately, nearly in phase with the laser pulses (2 μsec full width at half height) precluding the use of phase sensitive discrimination techniques. In our work with pulsed, focal flux levels on the order of 10^6 watts/cm^2, the particulate noise was typically an order of magnitude more intense than the anticipated Raman signal levels.

* * * * * * * * * *

Before describing the problem in greater detail and indicating possible solutions to it, I would briefly like to describe our program and approach in laser Raman thermometry. Our program is aimed at demonstrating the feasibility of laser Raman scattering diagnostics for use in practical combustion situations, and, accordingly, we are attempting Raman measurements in actual model combustors. As an adjunct to the program, we are attempting to simplify and to reduce the expense as much as possible of laser Raman techniques. This is reflected in our approach, schematically illustrated in Fig. 1, which utilizes a high energy dye laser, narrowband interference filters, and a coaxial collection geometry. Pulsed dye lasers are technically attractive for

a number of reasons and, in addition, are relatively inexpensive. High
energy laser pulses are attainable so that hundreds to thousands of Raman
photons are collected in a single pulse minimizing photocathode statistical
errors and obviating the need for photon-counting electronics. Attendant
with the high pulse energy are short pulse durations leading to the high peak
powers necessary to discriminate against luminous combustor backgrounds.
High average powers, i. e., watts to tens of watts, are or soon will be available to minimize averaging times or expedite measurements of probability distribution functions. By monitoring vibrational scattering from N_2, the Raman
signatures are spectrally well displaced from the exciting laser line permitting the use of narrowband interference filters for signal detection. In
addition by viewing coaxially only one optical access port is required on the
combustor.

Temperature information is obtained by the well-known technique of
ratioing the vibrational anti-Stokes to Stokes Raman intensities. In Fig. 2,
a ten trace overlay of Stokes and anti-Stokes scattering from a laboratory
air-methane flame is shown. The shot-to-shot variation seen is completely
explainable from photocathode statistics. By averaging these signals and
ratioing, temperatures are obtained which are in reasonable agreement with
radiation-corrected thermocouple readings.

However, in experiments on model combustors, several problems were
uncovered which are not generally encountered in laboratory flames, which
are relatively benign vis-a-vis combustor flames. Background luminosity
levels are found to be comparable to or greatly exceed Raman peak signal
powers. In the experiments conducted, burner fluctuations caused the luminosity to vary by over an order of magnitude but at a relatively slow rate
(~ 150 hz) as seen in Fig. 3. Since the Raman signatures have a frequency
spectrum characteristic of the fast dye laser pulses electronic filtering
techniques have been successfully employed to discriminate against the luminous background excursions as shown in Fig. 3.

More serious, and more difficult to circumvent, is the laser modulated
particulate incandescence problem alluded to earlier and illustrated in
Fig. 4. In the top oscillogram, Stokes scattering from ambient air in the
combustor is shown for reference; in the temperature range about $1500^{\circ}K$, the
Stokes and anti-Stokes intensities are approximately 14% and 21% respectively
of this signal level. With the combustor operating (the lower traces),
"signals" are recorded which are actually much larger than the reference
signal level and obviously not Raman scattering. Based upon the temporal
behavior of these signals (note that they are not identical with the top
Raman trace), their variation with interference filter bandwidth and the
spectral rejection capabilities of the spectrometer, we are fairly well convinced that we are seeing laser modulated particulate incandescence. The
severity of the problem can be appreciated by noting that only ten, one-
micron diameter particles at the vaporization temperature of carbon in the
Raman sample volume emit enough radiation (into the appropriate spectral
regions, solid angle, etc.) to equal the anticipated Raman signal levels.

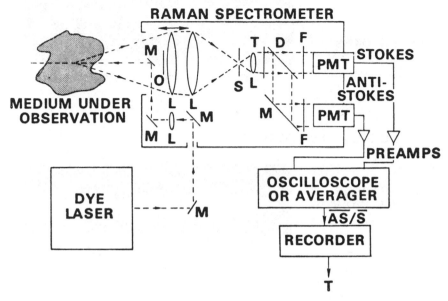

Fig. 1 - Schematic of Two Channel Raman Spectrometer.

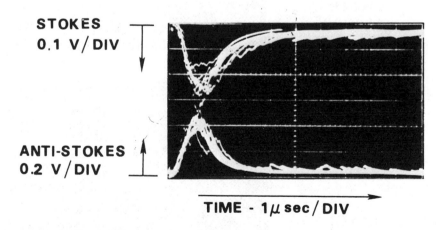

STOKES
0.1 V/DIV

ANTI-STOKES
0.2 V/DIV

TIME - 1μ sec/DIV

Fig. 2 - Raman Signatures in CH_4 - Air Flame (Ten Trace Overlay).

NO FILTERING

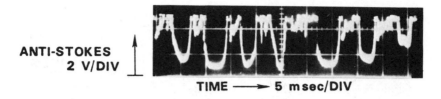

ANTI-STOKES
2 V/DIV

TIME ——→ 5 msec/DIV

20 KC HIGH PASS FILTERING

ANTI-STOKES
0.5 V/DIV

TIME——→ 5 msec/DIV

Fig. 3 - Luminosity Suppression with Electronic Filtering.

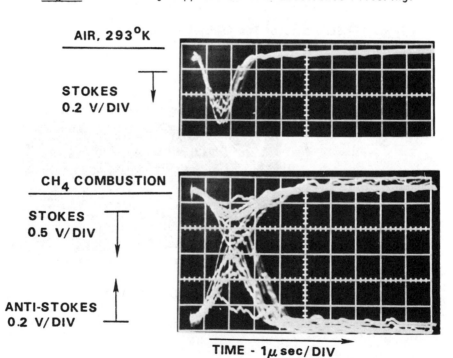

AIR, 293°K

STOKES
0.2 V/DIV

CH_4 COMBUSTION

STOKES
0.5 V/DIV

ANTI-STOKES
0.2 V/DIV

TIME - 1μ sec/DIV

Fig. 4 - Particulate "Noise" in JBTS Combustor (Multiple Trace Overlays).

Analyses of the laser irradiated particulate heating problem predict the behavior exhibited in Fig. 5, where particle temperature rise is shown as a function of laser flux, q, and particle radius, a. The factor $\alpha\beta$ depends on the optical properties of the soot and is approximately equal to 0.8 for plane wave illumination of a hydrocarbon soot. Fig. 5 is not only helpful in predicting where potential problems may occur but also in suggesting a possible solution to the problem. Namely, due to the vaporization kinetics, the noise can be anticipated to exhibit saturable behavior with increasing laser flux levels. For example, S/N improvements from 50 to 60 are predicted for a two order of magnitude flux increase from 10^6 to 10^8 Watts/cm^2. Other improvements in S/N can be made by decreasing the interference filter spectral bandwidth and by using polarized laser beams. Ten Angstrom wide filters are currently employed; the use of narrower filters would require environmental controls to ensure spectral stability and are not being considered at this time. A doubling of S/N can be effected through use of polarized laser fluxes. This arises since the vibrational scattering from N_2 is only slightly depolarized ($\sim 2.2\%$) whereas the particle incandescence is unpolarized. It is hoped that the use of high flux, polarized laser beams will permit sufficiently high S/N levels to be attained to permit accurate single pulse temperature measurements. Single pulse measurements with good accuracy are required to assemble temperature probability distribution functions and for turbulence studies.

For average temperature measurements, the possibility exists of sampling the noise in adjacent spectral regions to the Raman bands as shown in Fig. 6, averaging, and then subtracting from the Raman channel "signals." However, such an approach probably cannot be implemented for single pulse measurements where subtraction in real time at high frequencies (10^5 hz and above) would be required.

We are currently pursuing experimental programs to examine the feasibility of both the increased focal flux and noise sampling approaches.

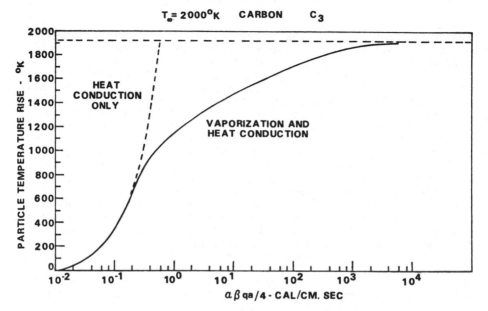

Fig. 5 - Laser Irradiated, Volatile Particle Temperature Rise.

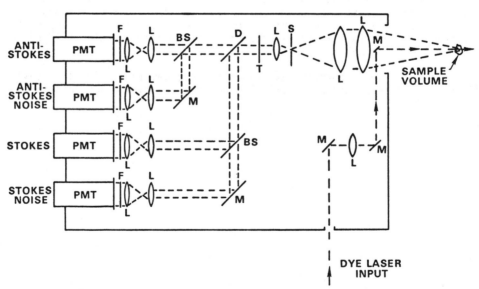

Fig. 6 - Four Channel Laser Raman Spectrometer.

RAMAN MEASUREMENTS OF SPECIE CONCENTRATION AND TEMPERATURE IN AN AIRCRAFT TURBINE EXHAUST

DONALD A. LEONARD

Computer Genetics Corporation

I would like to describe some laser Raman field experiments in which measurements of specie concentration and temperature were made on the exhaust gases of a gas turbine aircraft engine. Fig. 1 and Fig. 2 are schematics which show the optical system that was used.

The Raman scattering collected is essentially backscatter and the measurement volume was approximately 0.1 cm x 1.0 cm x 1.0 cm with the short dimension of the volume parallel to the direction of flow. The laser transmitter was a pulsed nitrogen laser operating at 337 nm at a one-half watt average power level with 100 kw pulses at a pulse repetition rate of 500 pulses per second. The receiver consisted of a fully computer controlled double 1 meter scanning spectrometer with combined photon counting and synchronous detection.

The aircraft combustor utilized in this work was an Avco Lycoming T-53 Gas Turbine Combustor, the emissions from which had previously been thoroughly investigated by conventional methods. In addition, the emissions from a small piston engine and a simulated exhaust gas generator were also utilized for purposes of instrument development and calibration.

Raman data were obtained which could be used to accurately measure the mole fractions of the major species in the flow, i. e., N_2, O_2, CO_2 and H_2O over the entire range of engine operating conditions from idle to full power. These Raman measurements were compared with the expected values of the specie concentrations as calculated from the measured fuel/air ratio at the various operating conditions.

Initial tests were conducted to prove and demonstrate the basic performance of the laser Raman field unit using normal atmospheric air as a test gas.

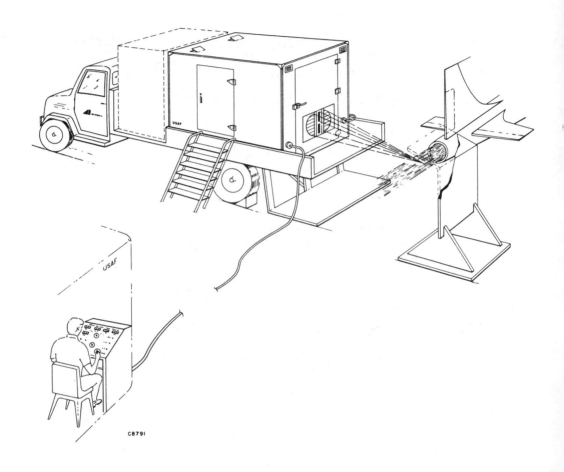

Fig. 1 - Schematic of Laser Raman Aircraft Engine Exhaust Emissions
Measurement System

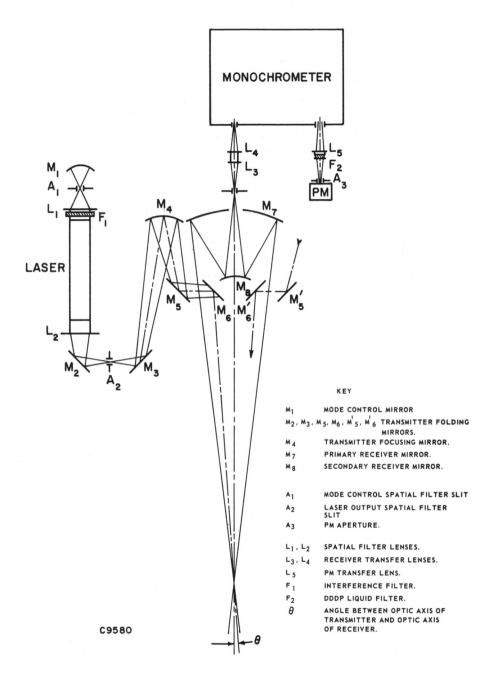

MONOCHROMETER

LASER

KEY

M_1 MODE CONTROL MIRROR

M_2, M_3, M_5, M_6, M'_5, M'_6 TRANSMITTER FOLDING
                                              MIRRORS.

M_4 TRANSMITTER FOCUSING MIRROR.

M_7 PRIMARY RECEIVER MIRROR.

M_8 SECONDARY RECEIVER MIRROR.

A_1 MODE CONTROL SPATIAL FILTER SLIT

A_2 LASER OUTPUT SPATIAL FILTER
               SLIT

A_3 PM APERTURE.

L_1, L_2 SPATIAL FILTER LENSES.

L_3, L_4 RECEIVER TRANSFER LENSES.

L_5 PM TRANSFER LENS.

F_1 INTERFERENCE FILTER.

F_2 DDDP LIQUID FILTER.

θ ANGLE BETWEEN OPTIC AXIS OF
               TRANSMITTER AND OPTIC AXIS
               OF RECEIVER.

C9580

Fig. 2 – Schematic of Transmitter and Receiver System Details

205

With the laser Raman van in place at the Lycoming test site, the Raman spectrum in Fig. 3 was obtained before the exhaust stream generator was operated. This spectrum shows, on a semi-logarithmic scale, the usual atmospheric nitrogen vibrational line with its rotational side-bands at about 1% of the peak value and an ambient noise level at about the 0.01% level compared to the nitrogen peak. The spectrum in Fig. 3 thus showed that the laser Raman gas analyzer system was working in accordance with its design.

When the exhaust stream generator was operated under simulated 'idle' conditions and produced several hundred ppm of hydrocarbon emissions, spectra such as shown in Fig. 4 were obtained. This figure shows a scan of the same spectral region as previously shown in Fig. 3 which includes the nitrogen Raman line. What is most apparent in Fig. 4 is that with the high concentration of hydrocarbon emissions present, a laser induced continuum level is produced which is the same order of magnitude as the nitrogen Raman peak.

Very accurate Raman measurements of the O_2 and CO_2 concentrations in the T-53 engine exhaust were however possible. These measurements showed excellent agreement when compared with the expected values of the concentration of these species on the basis of the measured fuel/air ratio of the actual operating engine.

Typical data of the O_2 vibrational Raman line in the hot exhaust gases is shown in Fig. 5. This data corresponds to an engine power setting of 30%. The dip in the data trace at the right side shows the results of temporarily turning off the laser during the scan, indicating that the background level is entirely a laser induced signal. The value of the O_2 signal is obtained by subtracting the average fluorescence level from the observed signal at the O_2 Raman line position. In a similar manner, the N_2 Raman signal for the same engine condition is obtained after suitable corrections for the fluorescence level in that spectral region. The ratio of the corrected O_2 to N_2 Raman signals can then be used to obtain the O_2/N_2 mole ratio. The system was calibrated in the field for effective cross-section and system transfer function by using the O_2/N_2 ratio obtained from ambient air and assuming that air is 21% O_2 and 79% N_2 on a mole basis.

The results of the oxygen measurements are shown in Fig. 6 where the O_2/N_2 mole ratio as obtained from the Raman measurements is plotted as a function of the O_2/N_2 mole ratio as calculated from the fuel/air (F/A) ratio of the actual operating Engine. Perfect agreement would cause the data to fall on the line connecting zero with the point labelled 'air.' It can be seen that the data fall generally along that line to within the accuracy of the respective measurements. The error bars are mainly due to the shot noise in the Raman signal at the high power conditions and are dominated by the fluctuations in the higher fluorescence levels at the lower power conditions. A minor temperature correction due to the difference in the vibrational Boltzmann factor between O_2 and N_2 should be applied to such data if higher precision is required. Similar Raman measurements were made of the CO_2 concentration in the T-53 turbine engine exhaust.

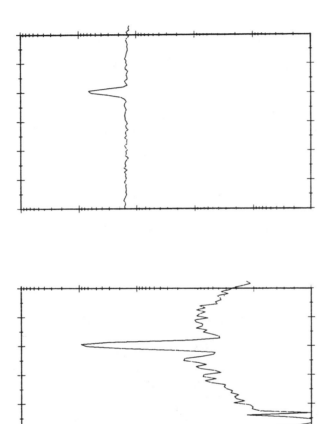

Fig. 4 - Nitrogen Vibrational
Line in Combustor Exhaust.

Fig. 3 - Ambient Air Nitrogen
Vibrational Raman Line.

207

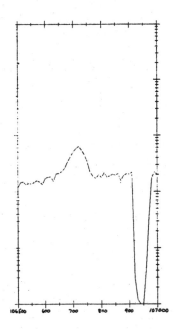

Fig. 5 - Oxygen Vibrational Raman Line with Fluorescence Background.

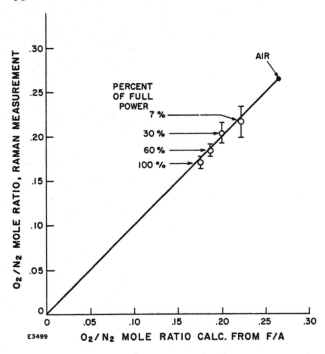

Fig. 6 - O_2/N_2 Mole Ratio for T-53 Engine Raman Measurement Vs. Calculated Value.

The temperature of the T-53 exhaust was measured by means of the N Raman density method whereby the temperature is assumed to be inversely proportional to the density of nitrogen, with a constant static pressure. If the dynamic pressure is a significant portion of the total pressure, i. e., in a high velocity flow, then a correction for this must be made to the data.

The density of nitrogen was measured at six points in the exhaust stream of the T-53 for each of the four operating conditions. The temperature and density of the ambient air was used as a scale factor to obtain the temperature profiles shown in Fig. 7. Thermocouple measurements in the exhaust stream gave 725°K for the 7% power point and 890°K for the 100% power point. The lower power points are in reasonable agreement. The higher power Raman data is somewhat higher than the thermocouple, but a thermocouple radiation correction is probably necessary.

This work was sponsored by the Aero Propulsion Laboratory of the United States Air Force, Wright-Patterson Air Force Base. A more complete description of the work may be found in the Technical Report AFAPL-TR-74-100, "Field Tests of a Laser Raman Measurement System for Aircraft Engine Exhaust Emissions," October 1974.

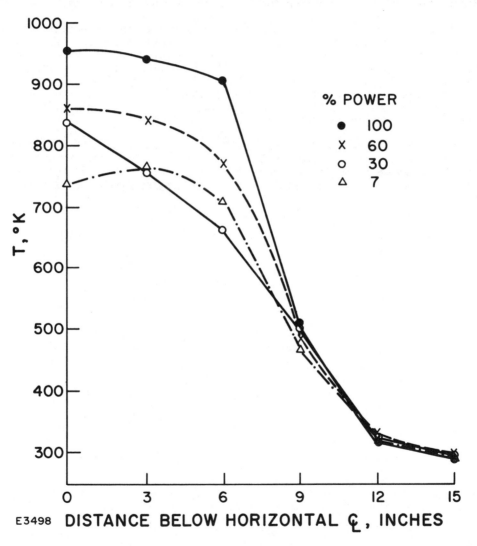

TEMPERATURE PROFILES T-53 ENGINE
RAMAN DENSITY METHOD

% POWER
● 100
✕ 60
○ 30
△ 7

T, °K

E3498 DISTANCE BELOW HORIZONTAL C_L, INCHES

Fig.7

210

RAMAN SCATTERING FROM LAMINAR AND TURBULENT FLAME GASES

ROBERT E. SETCHELL

Sandia Laboratories

I. INTRODUCTION

I would like to briefly summarize some results that have been obtained
in laboratory flames using a sensitive Raman spectroscopy system. The
experimental configuration is shown schematically in Fig. 1. An important
feature of this system is an ellipsoidal mirror cavity used to repeatedly
reflect the beam from a CW argon-ion laser through a small scattering
volume (Ref. 1). For a scattering volume having dimensions no larger than
1 mm, the ellipsoidal cavity can provide a gain of more than 20 over the laser
power available from a single laser pass. A second feature of the system is
a commercial two-channel photon counter whose channels can be gated syn-
chronously with the chopping of the laser beam. This feature permits an
emission background to be continuously subtracted from the total signal,
allowing weak Raman signals to be recorded from luminous flames.

II. LAMINAR FLAMES

Figs. 2-5 show Raman vibrational spectra from the post-flame gases of a
premixed, laminar, methane/air flat flame. (Ref. 2). The Stokes Q-branch
bands of nitrogen are shown in Fig. 2. The temperature of the flame gases is
easily determined from the ratio of the peak of the first upper-state band
($v = 1$) to the peak of the ground-state band ($v = 0$). The predicted nitrogen
spectrum shown in Fig. 2 is based on the experimentally determined temperature.
Fig. 3 shows Stokes vibrational bands of CO_2 and O_2 in the post-flame gases
of a lean flame. Fig. 4 shows Stokes Q-branch bands of carbon monoxide in
the post-flame gases of a rich flame. The temperature can be determined from

This work was supported by the U.S. Energy Research and Development
Administration, Contract Number AT(29-1)789.

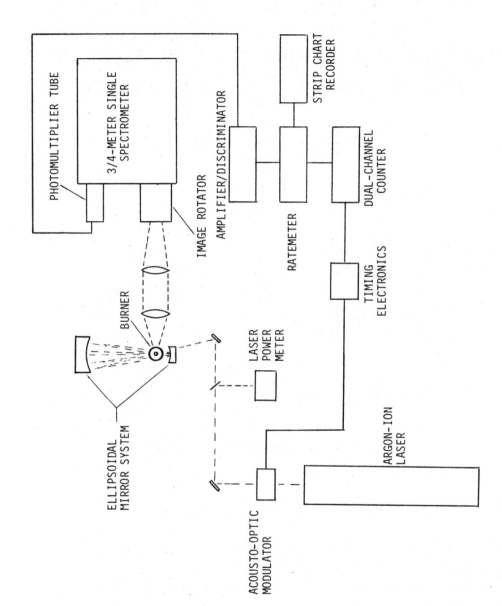

Fig. 1 – Raman Spectroscopy System for Laboratory Flame Studies.

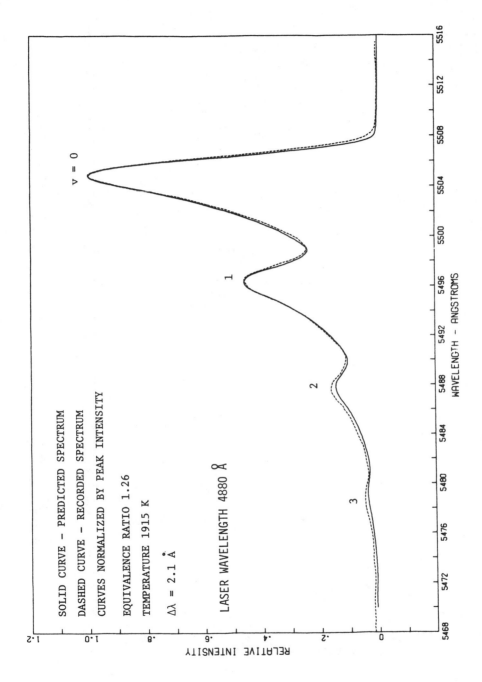

SOLID CURVE - PREDICTED SPECTRUM

DASHED CURVE - RECORDED SPECTRUM

CURVES NORMALIZED BY PEAK INTENSITY

EQUIVALENCE RATIO 1.26

TEMPERATURE 1915 K

$\Delta\lambda = 2.1$ Å

LASER WAVELENGTH 4880 Å

v = 0

RELATIVE INTENSITY

WAVELENGTH - ANGSTROMS

Fig. 2 - Comparison between a Predicted Nitrogen Stokes Q-branch Spectrum and a Spectrum Recorded from a Laminar Methane/Air Flat Flame.

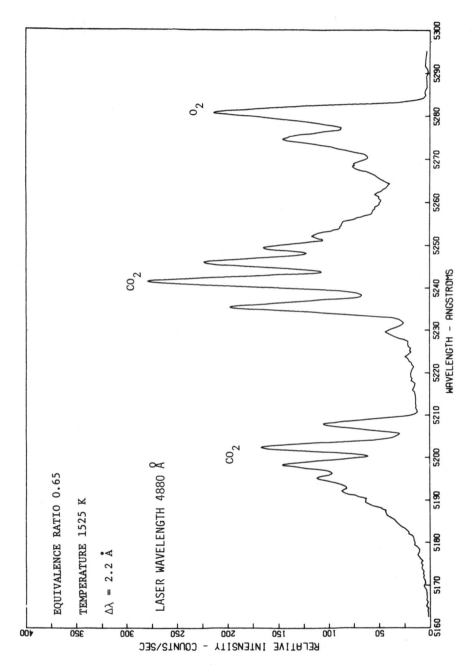

Fig. 3 - Stokes Vibrational Spectra of Carbon Dioxide and Oxygen in the Post-Flame Region of Lean Methane/Air Flame.

214

the recorded spectrum in the same manner as with nitrogen, and the predicted
spectrum shown in Fig. 4 is based on this temperature. Temperature values
obtained from both N_2 and CO spectra recorded at the same position within the
flame gases have been found to agree within 5%. Nitrogen provides the most
sensitive measurement, with errors resulting from statistical variations in
the Raman signals typically less than 1% when 10-second signal accumulation
times are used. Fig. 5 shows the complex Raman spectra of water vapor in
the post-flame gases of a lean flame.

The initial goal for this experimental system was to demonstrate the
capabilities of Raman spectroscopy in laminar, non-luminous, particulate-
free flame gases. As a demonstration of these capabilities, Fig. 6 shows
axial distributions of temperature and absolute concentrations of carbon
monoxide and hydrogen in the post-flame region above a rich methane/air flat
flame. The dotted lines on this figure show concentrations predicted by
equilibrium calculations using the initial fuel/air mixture ratio and locally
measured temperatures.

III. TURBULENT FLAMES

In order to evaluate the capabilities of the Raman system in turbulent
flame gases, the burner for producing laminar flat flames was replaced by
a simple circular nozzle for generating turbulent hydrogen diffusion flames.
Since the system incorporates a CW laser, the goal was to demonstrate that
time-averaged temperatures and species concentrations could be obtained.
However, a time-averaged value obtained by accumulating a Raman signal
generated by a CW laser introduces undesirable correlations. (Ref. 3). The
expression relating the Raman signal I_i observed at an appropriate wave
number ω_i and the number density n_i of the scattering species in the ith
internal energy state can be written:

$$I_i(\omega_i) = \text{constant} \cdot I_o \cdot \chi(\Delta\omega; T) \cdot n_i \qquad (1)$$

where I_o is the laser power within the scattering volume and $\chi(\Delta\omega; T)$ is
a function which depends on both the temperature T and the spectral resolu-
tion $\Delta\omega$ at the save number ω_i. The factor $\chi(\Delta\omega; T)$ is a consequence of the
near coincidence of successive Q-branch lines corresponding to particular
vibration-rotation energy states, and of the overlapping of successive
A-branch bands corresponding to complete vibrational levels, in the spectra
of most molecules. If Equation (1) is written in terms of the total
number density n_t of the scattering species, we have:

$$I_t(\omega_t) = \text{constant} \cdot I_o \cdot F(\Delta\omega; T) \cdot n_t \qquad (2)$$

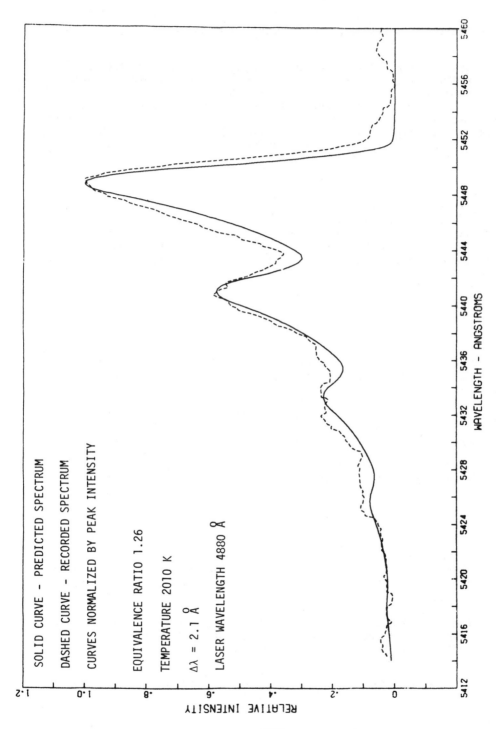

SOLID CURVE - PREDICTED SPECTRUM

DASHED CURVE - RECORDED SPECTRUM

CURVES NORMALIZED BY PEAK INTENSITY

EQUIVALENCE RATIO 1.26

TEMPERATURE 2010 K

$\Delta\lambda$ = 2.1 Å

LASER WAVELENGTH 4880 Å

RELATIVE INTENSITY

WAVELENGTH - ANGSTROMS

Fig. 4 - Comparison between a Predicted Carbon Monoxide Stokes Q-Branch
Spectrum and a Spectrum Recorded from a Rich Methane/Air Flame.

216

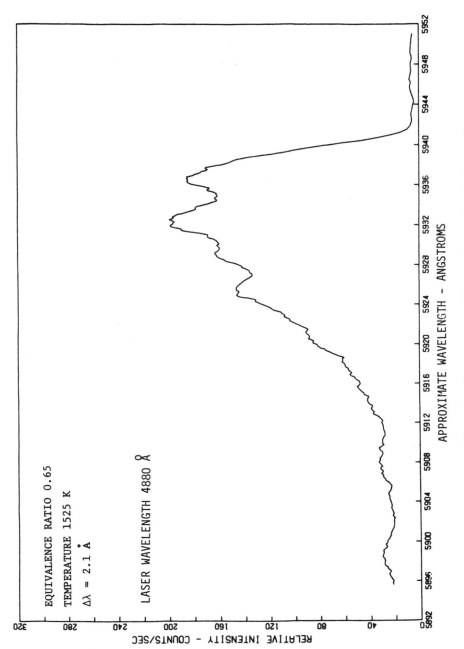

Fig. 5 – Stokes Vibrational Spectrum of Water Vapor in the Post-Flame Region of a Lean Methane/Air Flame.

217

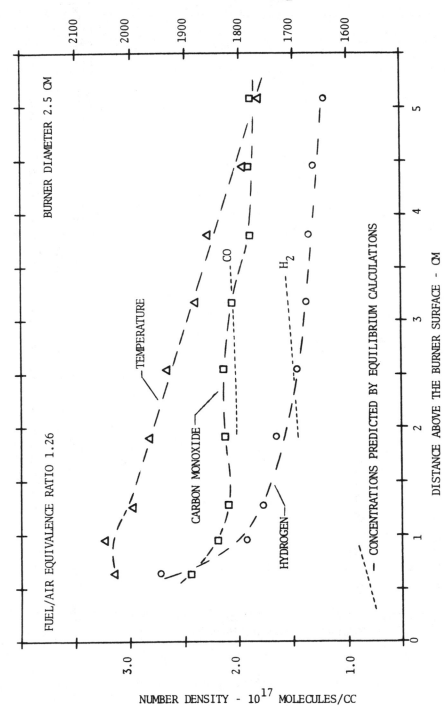

Fig. 6 – Profiles of Temperature, Hydrogen Concentration and Carbon
Monoxide Concentration Along the Flow Axis in the Post-Flame
Region of a Rich Methane/Air Flat Flame.

where $F(\Delta\omega; T) = \chi(\Delta\omega; T) \quad n_i/n_t$. If the laser power within the scattering volume can be assumed constant, then time-averaging Equations (1) and (2) gives:

$$\overline{I_i} \propto \overline{\chi(\Delta\omega; T) \cdot n_i} \qquad (3)$$

and

$$\overline{I_t} \propto \overline{F(\Delta\omega; T) \cdot n_t} \qquad (4)$$

The correlations shown between number densities and functions of temperature represent limitations to the accuracy that can be achieved in temperature and concentration measurements. However, for a spectral resolution ≤ 5 cm^{-1}, the individual Q-branch lines of hydrogen can be completely resolved, and approximate rotational temperatures can be obtained from ratios of time-averaged intensities of these lines. Fig. 7 shows the Stokes Q-branch lines of hydrogen recorded in a turbulent hydrogen diffusion flame at a position along the flame axis and 12 nozzle diameters downstream from the nozzle exit. The predicted spectrum shown in this figure was obtained using the temperature inferred from the ratio of the $J = 3$ and $J = 1$ Q-branch lines.

Time-averaged concentrations can be found in two ways. One approach is to obtain approximate values by measuring:

$$\overline{F(\Delta\omega; T) \cdot n_t} \Big/ [F(\Delta\omega; T^*)]$$

where T^* is an experimentally-determined mean temperature. An analysis of the errors introduced by this approach has not yet been done. A second approach is to make the factor $F(\Delta\omega; T)$ insensitive to temperature fluctuations by an appropriate choice of $\Delta\omega$. Calculations for N_2, CO and O_2 have shown that $F \doteq 1$ for temperatures from 300°K to over 2000°K for values of $\Delta\omega$ equal to 90 cm^{-1}, 85 cm^{-1} and 70 cm^{-1}, respectively.* These are relatively large values for the spectral resolution and their use could introduce serious spectral interferences. No practical choice of $\Delta\omega$ is possible to restrict $F(\Delta\omega; T)$ to values near unity for H_2.

Figs. 8 and 9 show radial distributions of time-averaged temperature, relative concentrations and normalized Raman signal levels measured at different axial positions downstream from the nozzle exit. The concentration measurements are ratios of the quantities $F n/F(T^*)$ for the species H_2, O_2 and N_2, while the normalized Raman signal levels are values of $F n/(F n)_{max}$ for H_2O and (in Fig. 9) the hydroxyl radical OH. (Ref. 4).

---

*In these calculations, $\Delta\omega$ represents the full width at half maximum of a triangular spectrometer slit function.

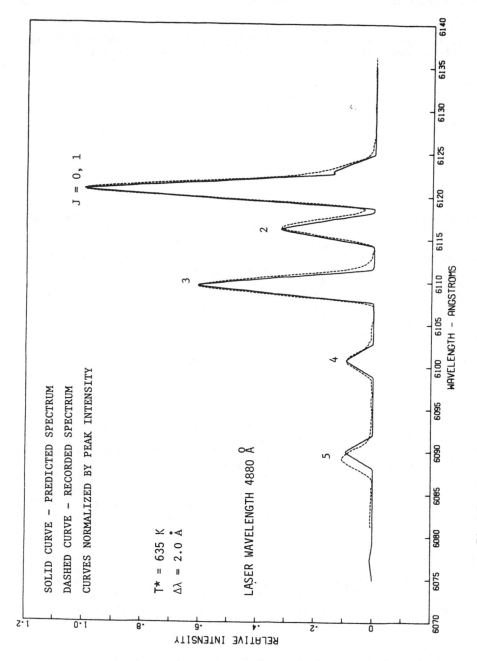

SOLID CURVE - PREDICTED SPECTRUM

DASHED CURVE - RECORDED SPECTRUM

CURVES NORMALIZED BY PEAK INTENSITY

$T^* = 635$ K

$\Delta\lambda = 2.0$ Å

LASER WAVELENGTH 4880 Å

Fig. 7 - Comparison between a Predicted Hydrogen Q-branch Spectrum and a
Spectrum Recorded in a Turbulent Hydrogen Diffusion Flame.

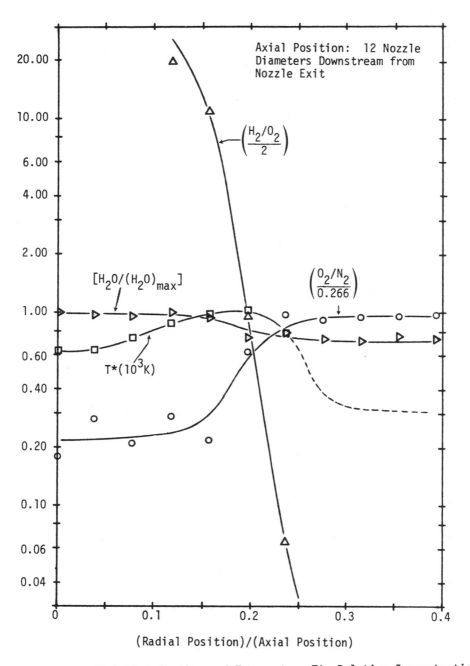

Fig. 8 - Radial Distributions of Temperature T*, Relative Concentrations (in parentheses) and Normalized Raman Signal Levels (in brackets) in a Turbulent Hydrogen Diffusion Flame. Nozzle Exit Velocity 370 m/sec, Nozzle Diameter 2.1 mm.

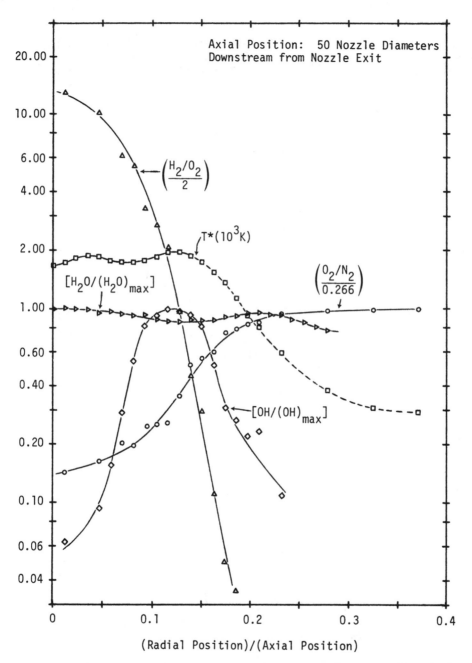

Fig. 9 – Radial Distributions of Temperature T*, Relative Concentrations (in parentheses) and Normalized Raman Signal Levels (in brackets) in a Turbulent Hydrogen Diffusion Flame. Nozzle Exit Velocity 840 m/sec, Nozzle Diameter 0.9 mm.

222

Concentrations have not been determined for these species because the scattering cross-sections are unknown, or the factors $F(\Delta\omega; T)$ have not been calculated. The Raman signal for the hydroxyl radical is difficult to obtain accurately, with corrections necessary for contributions due to coincident Raman bands from H_2O and H_2 (O-branch). The H_2O contribution is quite large, and present undertainties in properly accounting for this signal make it necessary to classify the OH data as preliminary. Identification of the observed spectra was based on good agreement with predicted spectral details for OH at comparable temperatures, (Ref. 5) and on distinct differences with the details of water spectra at comparable temperatures recorded in the post-flame gases of a premixed methane/air flame.

REFERENCES

1. R. A. Hill and D. L. Hartley, "Focused Miltiple-Pass Cell for Raman Scattering," Appl. Opt. 13, 186, 1974.

2. The laminar flame measurements reported in this summary are presented in more detail in: R. E. Setchell, "Analysis of Flame Emissions by Laser Raman Spectroscopy," Western States Section/The Combustion Institute Paper No. 74-6, 1974. Also SLL-74-5244.

3. The analysis and experiments for turbulent flames are discussed in more detail in: R. E. Setchell, "Time-Averaged Measurements in Turbulent Flames using Raman Spectroscopy," submitted to the AIAA 14th Aerospace Sciences Meeting, Washington, D.C., Jan. 1976.

4. When Fig. 8 was presented during the Workshop, data identified as OH was included with the other radial profiles. However, the source of this data was recently determined to be the J = 3 O-branch line of hydrogen. The measurements shown in Fig. 9 were obtained since the Workshop, but are included with this summary to reaffirm that Raman spectroscopy--with some difficulty--is capable of observing the OH radical in flames.

5. M. Lapp, private communication.

INTERFEROMETRIC FLAME
TEMPERATURE MEASUREMENTS

M. M. El WAKIL
University of Wisconsin

I. INTRODUCTION

Flame temperatures may be accurately measured without physically inter-fering with the flame or altering its state by the use of interferometry. An interferometer, to be sure, has inherent problems such as end effects and refraction errors, but these are predictable and with some care, can be accounted for. It has the added advantage of obtaining almost continuous temperature profiles, nearly up to the limiting surface, thus permitting the accurate evaluation of temperature gradients and heat transfer.

An interferometer is usually used to study two-dimensional systems, though its extension to three dimensions is possible under certain conditions. Care should be taken, therefore, to build physical models that are two-dimen-sional. Even so, it presents a map of two-dimensional events that is superior to such other methods as thermocouples.

An interferometer does not measure temperature directly, but the index of refraction of the medium in question. In the ordinary heat transfer problem, where only one component exists in the medium, the index of refraction is eas-ily related to density and in turn, through the gas laws, to temperature. The same is true if there exist two species, such as the case of diffusion of one pure component into another in an isothermal field. Where both heat and mass transfer coexist, and where exist a large number of components, as in a flame, the composition of the medium must be evaluated, or measured independently. If done accurately, this not only helps evaluate the index of refraction and heat transfer, but also the mass transfer.

II. BASIC PRINCIPLES

The basic components of an ordinary M-Z interferometer, (Fig. 1) are a light source and slit, an interference filter, a collimating lens, two beam splitters, two plane mirrors, an objective lens, and a screen.

The light beam emerging from the collimating lens, nearly monochromatic and parallel, is divided into two separate beams by beam splitter BS1, the reference beam and the test beam. They traverse paths that take them to beam splitter BS2 where they recombine.

Electromagnetic light travels in the form of wave trains. Two light waves that meet at a surface interfere with each other. If they are in phase, they reinforce one another and the surface appears bright. If they are 180° out of phase, their effects cancel and the surface appears dark. Phase relationships between 0° and 180° result in shades between bright and dark.

Let us consider the case wherein the media in the reference and test paths are identical (room air, for example). When formed at beam splitter BS1, the two light beams are in phase, depending upon the difference in the path lengths they traveled. If these were equal, the recombined beams would be in phase and the screen would be uniformly bright. If they differed by $\lambda/2$ where λ is the wavelength of the monochromatic light, they would be 180° out of phase and the screen would be uniformly dark. The path length of the reference beam may be changed by translating mirror M2. If the reference beam is further made to differ from the test beam by an amount λ, the screen brightens again.

III. THE FRINGE SHIFT

The change from bright to dark to bright is said to result in a *fringe shift* of one. Two fringe shifts occur when the path lengths differ by 2λ, etc. The fringe shift is therefore defined as the change in path length L divided by the wavelength, λ.

$$\text{Fringe shift, } S = \frac{\text{Change in path length } L}{\text{wavelength, } \lambda} \qquad (1a)$$

We have considered the case wherein the media in the reference and test sections were identical. If not, the above concepts would still hold. The term "path length," however, is changed to "optical path length." The *optical path length* is defined as the path length light travels in a vacuum in the same time it took to travel a specified distance in the medium. Path length L and optical path length D are related by the *index of refraction* of the medium n, defined as

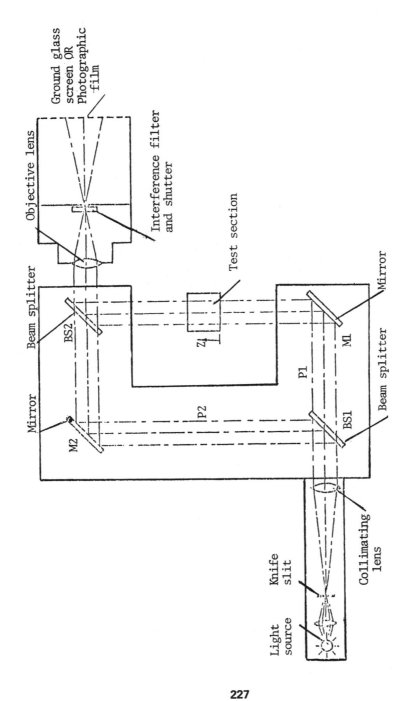

Fig. 1 - Schematic of the Mach-Zehnder Interferometer.

227

$$n = \frac{\text{velocity of light in vacuum}}{\text{velocity of light in the medium}} \qquad (2)$$

Thus $D = nL$ (if n is constant)   (3a)

or $D = \int_0^L n(z)dz$ (3b)

where z is a coordinate along the light path. Equation (1a) now modifies to

$$\text{Fringe shift, } S = \frac{\text{change in optical path length } D}{\text{wavelength, } \lambda} \qquad (1b)$$

When the index of refraction of the medium in the test section is different from that in the reference section because the temperatures are not the
same or because the fluids are different or both, the optical path lengths will
not be the same, even though the path lengths might be identical. These differences result in fringe shifts and changes in the brightness of the screen.
Changes in the index of refraction of the test medium from that of the reference
medium are commonly referred to as a *disturbance*.

If the index of refraction of the medium in the test section is not uniform
but changes in a direction perpendicular to the light path, the x-y plane
(Fig. 2), such as due to a temperature gradient, or a concentration gradient, or
both, successive light and dark fringes (of varying widths depending upon the
gradient) will appear on the screen. The result is called an *interferogram*
(Fig. 3).

The magnitude of the fringe shift at any one point in the x-y plane is
directly proportional to the change in the optical path length of the ray of
light that passes through that point in the test section. An important assumption here is that the index of refraction gradients do not *appreciably bend* the
light rays in the optical path length, i. e., they do not appreciably *refract*
them (otherwise refraction errors must be accounted for). In this case, the
change in optical path length $\Delta D(x,y)$ caused by disturbance of length L in the
test section is given by

$$\Delta D(x,y) = \int_0^L [n(x,y,z) - n_r]dz \qquad (3c)$$

where $n(x,y,z)$ = index of refraction of the test medium at a
                    point x,y,z

        n_r = index of refraction of the medium in the reference
               path.

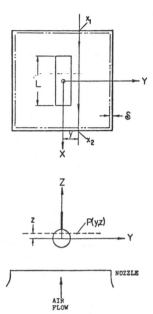

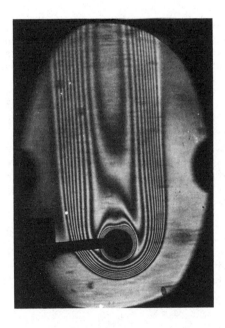

Fig. 2 - Identification of Coordinate Axes.

Fig. 3 - Infinite Fringe Interferogram.

229

The fringe shift $S(x,y)$ at the same point is the change in optical path length divided by the wavelength or

$$S(x,y) = \int_0^L \frac{[n \; x,y,z \; - \; n_r]}{\lambda} dz \qquad (1c)$$

As stated above, care is usually taken to build physical models that are two dimensional, eliminating variations in the z direction, so that

$$S(x,y) = \frac{L}{\lambda}[n(x,y) - n_r] \qquad (1d)$$

$S(x,y)$ can be quantitatively evaluated from the interferograms. For known L and n_r (usually for room air), the index of refraction $n(x,y)$ is evaluated. This is now to be related to physical properties of the test medium.

IV. INFINITE AND REFERENCE FRINGE ADJUSTMENTS

The M-Z interferometer can be operated on the so-called *infinite fringe* adjustment. This results in an interferogram as shown in Fig. 3. In this adjustment, the rays uniting at BS2 are coincident. With no disturbance the screen will appear uniformly bright or dark depending upon the difference between the two paths (0,λ,2λ, etc. for bright; $\lambda/2$, $3\lambda/2$, etc. for dark). When a disturbance is introduced, the screen will show a succession of dark and bright bands or contours where the disturbance occurs. Each of these contours is a locus of constant path difference. The path difference on an adjacent contour of like brightness is greater or less by λ.

In an infinite fringe interferogram, errors may arise due to the uncertainty of fringe position, especially when fringe spacings are wide such as at the edge of a boundary layer. While the infinite fringe interferogram gives a more realistic and pleasing visual picture of the disturbance, the second type is normally used for quantitative analysis.

In the *reference fringe* setting, one or two of the plates are rotated a very small angle from parallel. The two light paths are no longer coincident but intersect. A path difference is created, resulting in a series of parallel evenly spaced bright and dark bands, in the absence of disturbance. These are called *reference fringes*. The width of these fringes w is given by

$$w = \frac{\lambda}{2\epsilon} = \frac{\lambda}{\phi} \qquad (4)$$

where ϵ is the angle of rotation of the plate and ϕ the angle of intersection of the rays. Note that as ε approaches zero, the fringe width approaches infinity and we approach the infinite fringe setting (hence the name infinite).

The reference fringes can be adjusted to any desired spacing and location. The orientation can also be chosen since they are always parallel to the axis of rotation. A reference fringe interferogram is shown in Fig. 4.

When a disturbance is introduced, the shift of one undisturbed fringe width constitutes a path difference of λ. The values of fringe shift are obtained as shown in Fig. 5. This shows reference fringes oriented normal to a plate. The reference fringes (in the undisturbed field) are projected and fringe shifts (in terms of one undisturbed fringe width) are obtained at various positions from the plate.

V. EVALUATIONS OF DENSITY

The index of refraction n and the density ρ of a medium are functionally related by the well-known relation obtained independently by Lorenz and Lorentz (the latter based his approach on electromagnetic theory):

$$N = \frac{n^2 - 1}{n^2 + 2} \frac{M}{\rho} \tag{5a}$$

where N is a quantity known as the *molar refractivity* of the medium, and M is its molecular mass.

N is a constant for any given substance, independent of pressure or temperature (but not of wavelength). It is relatively independent of the phase of the substance differing by only a few percent between the gas and liquid phases. It can therefore be easily found, via measuring the density and index of refraction of the liquid on a refractometer.

The index of refraction $n(x,y)$ evaluated from Eq. (1d), may now be used to evaluate the densities $\rho(x,y)$ of the test medium, for known N and M.

VI. THE GENERAL CASE OF SIMULTANEOUS HEAT AND MASS TRANSFER

We shall now consider the general case of a boundary layer in which there exist a temperature gradient due to heat transfer and more than one species, each having its own concentration gradient due to mass transfer, as in the case of flames.

A further important property of the molar refractivity N, is that, for a homogeneous mixture or solution, it is expressed in terms of individual molar refractivities as

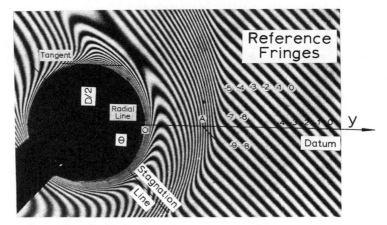

Fig. 4 - Reference Fringe Interferogram and Fringe Numbering System.

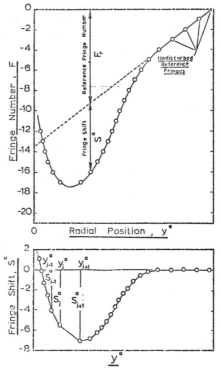

Fig. 5 - Evaluation of Apparent Fringe Shifts.

$$N = \overset{k}{\underset{}{\Sigma}} \ X_i N_i \tag{6}$$

where X is the *mole fraction*, subscript i denotes component i in the mixture and k the number of such components. Equation (5a) now becomes

$$(n-1) = \frac{(n^2+2)}{(n+1)} \frac{\rho_m}{M} \overset{k}{\underset{}{\Sigma}} \ X_i N_i \tag{5b}$$

where ρ_m and M_m are now the density and molecular mass of the mixture. The indexes of refraction of gases are very close to unity, and Eq. (5b) reduces quite accurately to

$$(n-1) = \frac{3}{2} \frac{\rho_m}{M_m} \overset{k}{\underset{}{\Sigma}} \ X_i N_i \tag{5c}$$

The gases in a boundary layer in interferometer work are usually at low pressures or partial pressures, and therefore obey the ideal gas laws. Thus

$$\frac{\rho_m}{M_m} = \frac{P}{R_0 T} \tag{7}$$

and

$$(n-1) = \frac{3}{2R_0} \frac{P}{T} \overset{k}{\underset{}{\Sigma}} \ X_i N_i \tag{5d}$$

where P is the total absolute pressure, T, the absolute temperature, and R_0 the universal gas constant. The reference medium (usually room air) has a fixed composition and temperature, so that for it

$$(n_r-1) = \frac{3}{2R_0} \frac{P}{T_r} N_r \tag{5e}$$

where the total pressure is assumed the same in the reference and test sections. The fringe shift in the general system can now be evaluated by combining Eqs. (1d), (5d) and (5e) to give

$$S(x,y) = \frac{3}{2R_0} \frac{PL}{\lambda} \left[\frac{1}{T(x,y)} \quad \overset{k}{\underset{i}{\Sigma}} \ X_i(x,y)N_i \ - \ \frac{N_r}{T_r} \right] \tag{1e}$$

VII. THE MULTIWAVELENGTH INTERFEROMETER

If the test medium were composed of only one component and variable temperature, one equation of the type (1e) is sufficient for the complete solution of the temperature gradient. If it were composed of two components in an isothermal system, again only one equation is sufficient for the complete solution of the mole fraction gradients. In the general case of heat and mass transfer, one equation is *not* sufficient.

A way to resolve the above indeterminancy is found from the implicit dependency of the index of refraction n, and hence the molar refractivity N, on wavelength. A characteristic phenomenon of transparent substances is termed *normal dispersion*. In this (a) n increases as the wavelength decreases, (b) the rate of increase is greater the shorter the wavelength, and (c) substances with larger n exhibit larger rates of change of n. Most transparent substances have normal dispersion in the visible region of light (they exhibit so-called *anomalous dispersion* near absorption bands).

If one selects more than one wavelength such that there would be an appreciable change in the value of n with wavelength, there will be sufficient change in N and S, so that independent equations may be solved. Theoretically, if one has a heat- and mass-transfer boundary layer with k components, there will be $(k-1)$ independent mole fractions. Including the temperature, there will be k unknowns for which k independent equations are needed. Thus k interferograms taken at k wavelengths may be used for the complete solution of the temperature and concentration gradients. The equations take the form

$$S_j(x,y) = \frac{3}{2R_0} \frac{PL}{\lambda_j} \left[\frac{1}{T(x,y)} \quad \overset{k}{\Sigma} \ X_i(x,y)N_{ij} \ - \ \frac{N_{rj}}{T_r} \right] \tag{1f}$$

where j denotes the wavelength employed and identifies the wavelength-dependent quantities. It is given by

$$j = 1,1,3,\ldots,k \tag{8}$$

VIII. THE TWO-WAVELENGTH INTERFEROMETER

The number of wavelengths that are practicable is limited. Besides the requirement of sufficient dispersion, the wavelength spread is limited to the transparency range of the optical components of the interferometer. For all practical purposes a maximum of two wavelengths may be used, thus allowing, in principle, the study of most heat- and mass-transfer problems which involve two components and a temperature gradient. An example is the diffusion of vapors into air.

Interferograms at two wavelengths may be taken separately with a change of filters. A difficulty immediately arises, however, caused by the fact that most flow fields are nonsteady by choice or by nature. The above is therefore not a good method, and the two interferograms are taken simultaneously by passing unfiltered, continuous (white) light from the light source through the interferometer, producing interference in all wavelengths in the transmitted spectrum, from which any number of multiple interferograms may be obtained by filtration at the exit, and by a specially designed composite prism which directs the beam of interfering rays into separate paths of equal length. This method, however, resulted in temperature profiles that were unacceptably erratic, though the concentration profiles were consistent.

It was determined that the reason was due to lack of sufficient dispersion. The errors in fringe-shift measurement, small as they are, caused the scatter because the coefficients of the independent equations were then not sufficiently different to render the equations truly independent. The use of vapors with better dispersion would improve the situation. A search was undertaken, which pointed to carbon disulphide as the best available, but even it yielded unsatisfactory results.

IX. REMEDIAL METHODS

The above deficiency of the two-wavelength technique spelled its doom. To obtain temperature and concentration gradients simultaneously, only one wavelength is used and a density gradient is obtained. To separate the density into its components, temperature and partial pressure (or concentration), independent or dependent determination of one of these is necessitated.

In the independent determination of temperature, another probe such as a fine thermocouple may be used. This obviates one of the main advantages of interferometry, that of not physically interfering with the physical model.

A dependent method relies on the theoretical premise that in heat- and mass-transfer boundary layers, the governing equations for diffusion and energy take the form (forced convection):

$$u \frac{\partial \omega}{\partial x} + v \frac{\partial \omega}{\partial y} = D \frac{\partial^2 \omega}{\partial y^2} \qquad (9)$$

and

$$u \frac{\partial T}{\partial x} + v \frac{\partial T}{\partial y} = \alpha \frac{\partial^2 T}{\partial y^2} \qquad (10)$$

where ω is the mass concentration, T the temperature, u and v the velocity components in the x and y directions, D the diffusion coefficient, and α the thermal diffusivity.

If the Prandtl (v/α) and Schmidt (v/D) numbers are equal or nearly equal (v is the kinematic viscosity), then the normalized concentration and temperature profiles would be the same. This results in a functional relation between concentration and temperature of the form

$$\frac{T - T_\infty}{T_w - T_\infty} = \frac{X - X_\infty}{X_w - X_\infty} \qquad (11)$$

where the subscripts w and ∞ denote the wall and free stream conditions respectively. Equation (11) is used, then, in conjunction with only one equation of the type of (1f) with fringe shifts taken at one wavelength only ($j=1$), to yield both concentrations and temperature profiles in a two-component system.

Finally, as in the case of flames with a large number of components, concentrations, rather than temperature may be measured in the one-wavelength method. This is done by the use of gas chromatography.

X. FLAME STUDIES

The only early work known using interferometry was by Ross and El-Wakil. In a preliminary study on droplet combustion, they used a two-wavelength Mach-Zehnder interferometer to measure temperature profiles along several radial lines of a 1 cm diameter, n-heptane wetted cylinder burning under free convection conditions. However, they had to make some assumptions regarding the composition profiles in the boundary layer, which limited the accuracy of their results. The cylinder was used to eliminate the troublesome three-dimensional effects of a sphere. More recently El-Wakil and coworkers refined this technique and extended it to obtain energy and mass balances on a diffusion-type drop flame. The drop was simulated by porous bronze cylinders through which liquid n-heptane was forced at carefully controlled temperatures and flow rates. The fuel flow rate entering the cylinder was 2 to 4 times the burning rate. The inlet and exit flows were adjusted until a thin liquid film was established on the surface of the cylinder with no excess fuel dripping from it. To prevent the fuel inside the cylinder from boiling, cooling water, at controlled temperature and flow rate, flowed through jackets built around the fuel inlet and exit tubes (Fig. 6).

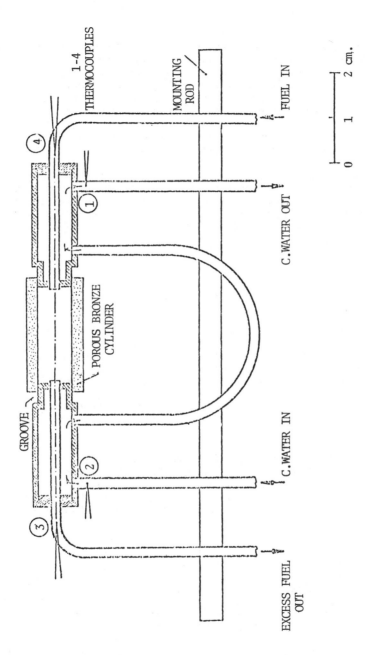

Fig. 6 - The Experimental Model.

THERMOCOUPLES 1-4

MOUNTING ROD

FUEL IN

C.WATER OUT

POROUS BRONZE CYLINDER

GROOVE

C.WATER IN

EXCESS FUEL OUT

0   1   2 cm.

237

Two different size cylinders were examined: 12.8 mm in diameter and 25.4 mm long, and 6.35 mm in diameter and 16.0 mm long. These sizes, large compared to those of drops in sprays (for which no feasible means of measuring temperature and composition profiles exist), were dictated by fabrication limitations and by disturbances of the flame by the finite-size chromatographic sampling probe. The lengths were chosen so that a compromise between light refraction errors and end effects was made. The large cylinder was equipped with 0.25 mm diameter copper-constantan thermocouples to measure the fuel and cooling water temperatures necessary for energy balance calculations.

The fuel was allowed to burn steadily at atmospheric pressure in uniform air flow fields. The latter were supplied by a convergent nozzle assembly that yielded steady, low turbulence, uniform velocity air jets over a wide range of velocities. Runs were made at 12.0, 40.0, and 64.0 cm/sec. These were well below the extinction value so that envelope flames were always established around the cyclinders.

XI. GAS CHROMATOGRAPHY

Samples were withdrawn from the flame at several points along each of the 0°, 45°, 90°, 135°, and 180° radial lines (measured from the forward stagnation line) by means of microprobes. These were made from 5 mm diameter quartz tubes tapered down to 30-50 micron diameter orifices. They were mounted on a micro-manipulator which was indexed to measure probe tip positions relative to the cylinder surface. All samples were withdrawn in a plane midway along the cylinder axis. Care was taken to prevent the condensation of the fuel vapor or other high boiling point species that might be present in the samples by heating the components of the sampling system.

The samples withdrawn from the flame were first dried, then diluted by mixing them with helium in a fixed 1:3 ratio. The mixtures were analyzed by means of a three-column gas chromatograph with a thermal conductivity detector. The sample size was 2.0 ml at 1.0 atm and 75°C. Details of the chromatographic columns and the sampling and chromatographic analysis systems may be found elsewhere (Ref. 5).

The chromatographic analysis yielded concentrations of CO_2, CO, O_2, N_2, CH_4, C_2H_2, C_2H_4, and C_7H_{16} in dried samples. Sample composition profiles at one air velocity and cylinder size are given in Figs. 7 and 8 for the 0° and 90° radial lines. Similar data are available for other angles and experimental conditions. Sample contours of constant fuel and oxygen concentrations are shown in Fig. 9. The flame on the small cylinder was studied at a single air velocity of 12.0 cm/sec. At higher velocities the flame on this cylinder was visibly disturbed by the sampling probe. The differences between the effective sampling positions and the actual positions of the probe tip depend upon the local velocity, probe diameter and probe orientation. No corrections were made to account for these differences, since they are believed to be small.

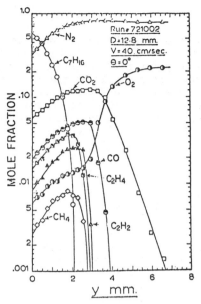

Fig. 7 - Composition Profiles of a Burning Heptane Cylinder.
D=12.8 mm, V=40 cm/s, θ=0°

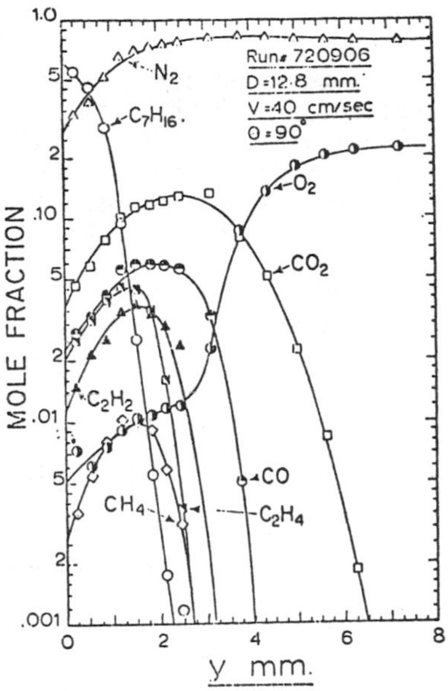

Fig. 8 - Composition Profiles of a Burning Heptane Cylinder.
D=12.8 mm, V=40 cm/s, θ=90°

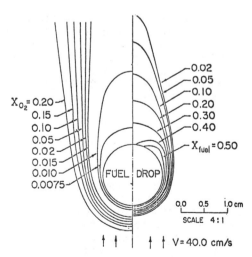

Fig. 9 - Contours of Heptane and Oxygen of a Burning Heptane Cylinder
D=12.8 mm, V=40 cm/s.

The composition profiles were used to determine the fuel vapor mass-flux distributions around the drop as well as the molar refractivities along the sampled radial lines. The latter are necessary for the evaluation of the temperature profiles from the interferograms. In calculating the molar refractivities, the gas mixture was assumed to be ideal. The water vapor concentration was taken to be a constant ratio of that of CO_2, equal to that in the equilibrium products of combustion in a stoichiometric n-heptane - air mixture. This was justified by available experimental data on diffusion flames. Small errors in water vapor concentrations have negligible effect on the molar refractivity.

XII.   INTERFEROMETRY CORRECTIONS

The burning cylinder was mounted in the test section of a 7.5 cm diameter Mach-Zehnder interferometer, with its axis parallel to the reference light path. The resulting fringe shift distributions along the sampled radial lines were used with the molar refractivity profiles to determine the temperature distributions in the flame. Severe index of refraction gradients in the radial directions necessitated refraction corrections to the apparent fringe shift data, given by

$$\Delta S = 5G^2 L^3 / 12\lambda \qquad\qquad (12)$$

$$\Delta y = (G-G_1)L^2 / 2 \qquad\qquad (13)$$

where G is the fringe shift gradient. An effective light path length L was used instead of the physical length of the cylinder because of end effects. Its value was determined from the boundary condition specifying that the temperature at a point on the liquid surface is equal to the saturation temperature corresponding to the partial pressure of the fuel vapor at that point. Sample temperature profiles along various radial lines are given in Figs. 10 and 11. Sample contours of isotherms are shown in Fig. 12.

XIII.   ENERGY BALANCE AND TOTAL RADIATION

The gradients of the composition and temperature profiles at the liquid surface were obtained by a linear least-square fit of the data near the surface for each radial line. These were then used, along with appropriate diffusivities and thermal conductivities, to determine the local and integrated convective mass and heat-flux distributions around the drop. The local values varied between 0.06×10^{-3} and 1.86×10^{-3} g_m/s cm^2 and 0.35 and 1.00 cal/s cm^2 respectively.

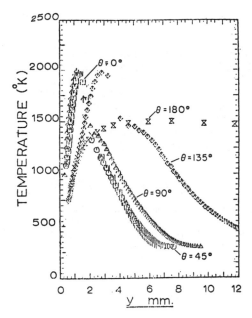

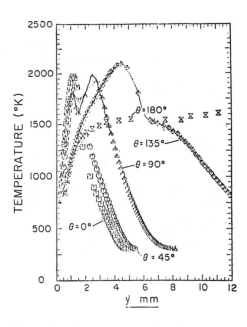

Fig. 10 - Temperature Profiles of a
Burning-Heptane Cylinder
D=12.8 mm, V=12 cm/s.

Fig. 11 - Temperature Profiles of a
Burning Heptane Cylinder
D=12.8 mm, V=40 cm/s.

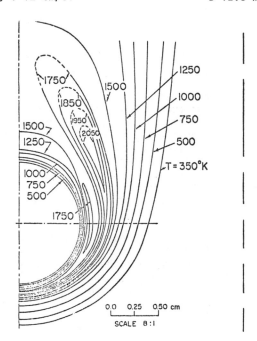

Fig. 12 - Contours of Isotherms of a Burning Heptane Cylinder
D=12.8 mm, V=40 cm/s.

241

Energy balances were made for the 12.8 mm diameter cylinder. The enthalpies of the incoming and outgoing fuel and cooling water, together with the convective heat transfer rate to the cylinder and the amount of heat necessary to evaporate the fuel burned, were used to evaluate the radiative heat transfer rate to the drop. This was found to be about 40% of the total heat transferred. The medium of the diffusion flame is composed of gaseous combustion reactants and products and solid carbon particles (soot). The total radiation was the divided up into gas and soot contributions, with the gas contribution found to account for about 10% of the total radiation.

The optical paths encountered in the drop flame geometry are generally short, and the interaction between gas and soot may be neglected. The contributions are, therefore, treated separately and are assumed to be additive. It is also assumed that the emissivity of the fuel surface is unity and that the chemical reactions in the flame have negligible effect on the absorption coefficients.

XIV. DISCUSSION AND CONCLUSIONS

1. The mole fraction of the fuel vapor at the surface varies between 40-60% depending on angular position, air velocity, and cylinder diameter. A value of 65% along the forward stagnation line in a similar diffusion flame, though under natural convection conditions, was reported. The fuel vapor concentrations are generally lower in the trailing half than those in the leading half of the cylinder. Higher fuel vapor concentrations at the surface occurred at higher air velocities for the same size cylinder, and for the smaller cylinder at the same air velocity except near the boundary layer separation point.

2. The fuel vapor concentrations drop rapidly in the radial direction becoming zero at the reaction zone. The radial position of the reaction zone, relative to the cylinder surface, is nearly constant in the leading half of the flame, but moves away from the cylinder surface in the trailing half reaching a maximum value along the $180°$ radial line. At higher air velocities, the reaction zone is closer to the cylinder surface except for the $135°$ radial line. This phenomena can be attributed to the increased recirculation down stream of the separation point. For the same air velocity, the radial positions of the reaction zone relative to the surface are nearly the same for both cylinders.

3. Considerable amounts of intermediate hydrocarbons, mainly CH_4, C_2H_2 and C_2H_4, are present in the fuel-side of the reaction zone, in agreement with other investigators. They are probably the pyrolysis products of the fuel and are believed to be the precursors of soot particles. This is supported by the higher concentrations of these species found in the more luminous trailing half of the flame. Except for the wake, they rapidly disappear near the region of maximum CO_2 concentration.

4. The oxygen concentrations do not drop to zero at the reaction zone. Small amounts are found in the fuel-side of the reaction zone up to the liquid surface.

5. The radial temperature profiles differ significantly from one another, especially in the trailing half of the flame. Therefore, theories assuming spherical symmetry should not be used to predict temperature-dependent quantities such as heat-fluxes and concentrations of nitrogen oxides.

6. The air velocity has a considerable effect on the temperature profiles in the trailing half of the flame. Double-peaked temperature profiles occurred along the 90° and 135° lines for the 12.8 mm cylinder at high air velocities. This is believed to be a result of the flow patterns behind the separation point in which free carbon and pyrolysis products are transferred from the inner flame zone by the stationary vortices in the wake to regions outside the reaction zone. There, the higher oxygen concentrations cause them to burn immediately resulting in a second temperature peak.

7. As expected, higher air velocities cause the point of maximum temperature to approach the surface and thin the boundary layer in the leading half of the drop. In general, the point of maximum temperature is closer to the cylinder surface than the point of zero fuel concentration. The extent of the difference, however, depends upon the uncertainty of the correspondence of sample and probe tip positions.

8. The maximum flame temperature in the leading half of the flame is about 1950°K. It increases slightly with the cylinder diameter, but not with air velocity within the examined range. (Values as low as 1500°K along the forward stagnation line were reported. Such low values are now believed to be due to unrealistic molar refractivity distributions.) Maximum temperatures along the 90° and 135° radial lines are slightly higher than in the leading half probably due to the rising hot gases.

9. The maximum flame temperatures along the 180° radial lines are 300-500°K lower than other radial lines, probably due to the large wake and lower energy density. This temperature, however, increases with air velocity which increases turbulence, reduces the size of the wake, and makes oxygen more accessible to the fuel vapor and other combustible species.

REFERENCES

1. P. A. Ross and M. M. El-Wakil, A Two-Wavelength Interferometric Technique for the Study of Vaporization and Combustion of Fuels. Progress of Astronautics and Rocketry, Vol. II, Liquid Propellants and Rockets, 1960.

2. C. L. Jaeck and M. M. El-Wakil, A Two-Wavelength Interferometer for the Study of Heat and Mass Transfer. Journal of Heat Transfer, August 1964.

3. M. M. El-Wakil, G. E. Myers and R. J. Schilling, An Interferometric Study
    of Mass Transfer from a Vertical Plate at Low Reynolds Numbers. <u>Journal
    of Heat Transfer</u>, Nov. 1966, pp. 399-406.

4. G. G. Hardel and M. M. El-Wakil, An Interferometric Study of Combined
    Forced and Free Convective Mass Transfer on a Vertical Flat Plate. Sympos-
    ium of the International Seminar on Heat and Mass Transfer, Herceg Novi,
    Yugoslavia, 1969.

5. S. I. Abdel-Khalik, An Investigation of the Diffusion Flame Surrounding a
    Simulated Liquid Fuel Droplet, PhD Thesis, University of Wisconsin-Madison,
    1973.

6. S. I. Abdel-Khalik, T. Tamaru and M. M. El-Wakil, Presented at the Sixth
    International Seminar-Heat Transfer from Flame, Trogir, Yugoslavia, 1973.

7. T. Tamaru, Radiative Heat Transfer in a Diffusion Flame, M.S. Thesis,
    University of Wisconsin-Madison, 1973.

PROGRESS ON IMPROVED FLOW VISUALIZATION BY RESONANCE REFRACTIVITY

DANIEL BERSHADER
S. GOPALA PRAKASH
Stanford University

This report covers an initial phase of the current study of flow visualization enhancement by use of resonant refractivity. The motivation for the
effort stems from a continuing interest in making visible a class of fluid
dynamic phenomena which require refractive sensitivities outside the range
of present methods. They include vortex phenomena in low-speed flow, aerody-
namically generated sound at or near the audible range, and turbulence in low-
speed flows, including meteorological flows.

It is the large increase of the polarizability or specific refractivity
at the half-power points of resonant lines in gases which underlies the present
study, a phenomenon which was predicted by the classical Lorentz electron theory (Ref. 1), and further verified by the quantum mechanics (Ref. 8). However,
the results of these earlier studies were obtained only for the case of natural
line broadening, whereas it is well known that pressure (collision) and particularly Doppler broadening mechanisms are much more prominent under conditions
that obtain in typical fluid-dynamic investigations. This progress report
presents some quantitative results of the effect of line broadening on the
resonant refractivity of sodium vapor, a principal additive candidate for future applications. Information in the literature on this problem is quite
scarce, the work of Marlow (Ref. 2) being an exception. In any case, adequate
quantitative data is not available.

Additionally, discussion of preliminary work on the experimental phase is
included. That program deals with a dispersion interferometric experiment on
sodium vapor.

Work supported by a Grant from the Air Force Office of Scientific Research.

245

I.   LORENTZIAN BROADENING MECHANISMS

The classical Lorentz theory gives the index of refraction, near an isolated line at frequency ω_0 with radiation damping only, as a function of
frequency ω, (Ref. 1):

$$(n-1) = (n_0-1) + \frac{\pi e^2 Nf}{m\omega_0} \frac{\omega_0 - \omega}{(\omega_0-\omega)^2 + \gamma_N^2/4} \tag{1}$$

where n_0 = refractive index in non-resonant region
           f = oscillator strength of resonance line at ω_0
           N = number density of atoms
           e = electronic charge
           M = electron mass
           γ_N = classical damping constant

The constant γ_N represents the natural line broadening which, in practice,
is equalled or overshadowed by other broadening mechanisms associated, respectively, with collisional interruptions in the electromagnetic wave trains and
with the Doppler effect. With respect to their effects on the refractive dispersion, the collision broadening between neutral particles, also called pressure broadening, and that due to Coulomb collisions of charged particles (Stark)
may be combined into a single Lorentzian formula identical to Eqn. 1, except
that γ_N is replaced by a total γ given by

$$\gamma = \gamma_N + \gamma_{Collision} + \gamma_{Stark} \tag{2}$$

Note that 2γ is the full frequency width in radians at half the resonant amplitude. In the range of conditions of interest here, Stark broadening may be
neglected.

II.   DOPPLER BROADENING AND REFRACTIVE DISPERSION

Doppler broadening is due to frequency shifts associated with the atomic
thermal velocities. Here, one replaces the frequency ω by the well-known
Doppler-shifted frequency $\omega(1 + v/c)$, where v is the line-of-sight component
(positive or negative) of the atom velocity. Now, assuming a Maxwellian velocity distribution, we have

$$\frac{dN}{N} = \left(\frac{M}{2\pi kT}\right)^{\frac{1}{2}} e^{-\frac{Mv^2}{2kT}} dv = e^{-\frac{M v^2}{2kT}} \frac{dv}{(\pi/2)\bar{v}} \tag{3}$$

where $\bar{v}$ denotes the mean thermal velocity of the particles, and where dN is the number of atoms per unit volume with velocity components between v and v + dv and where the other symbols have their usual definitions. The reduced refractivity n - 1 is obtained as

$$(n-1) = (n_0-1) + \frac{\pi e^2 Nf}{m\omega_0} \int_{-\infty}^{+\infty} \frac{\omega_0-\omega(1+v/c)}{\{\omega_0-\omega(1+v/c)\}^2+\gamma^2/4} \frac{e^{-\frac{Mv^2}{2kT}}}{(\pi/2)} \frac{dv}{\bar{v}} \qquad (4)$$

Next, it is useful to recast the above relation by introducing non-dimensional variables in terms of ω_0, c and a characteristic molecular thermal velocity $\sqrt{\frac{2kT}{M}}$. Define

$$x = \frac{(\omega_0/c)v}{(\omega_0/c)\sqrt{2kT/M}} = v\sqrt{M/2kT} \qquad (5)$$

$$a = \frac{\gamma/2}{(\omega_0/c)\sqrt{2kT/M}} \qquad * \qquad (6)$$

$$u = \frac{\omega_0-\omega}{(\omega_0/c)\sqrt{2kT/M}} \qquad (7)$$

The refractive index, Eqn. 4, now appears in a form more adaptable for evaluation:

$$(n-1) = (n_0-1) + \sqrt{\frac{M}{2\pi kT}} \frac{\pi e^2 Nfc}{m\omega_0^2} \int_{-\infty}^{+\infty} \frac{u-x}{(u-x)^2+a^2} e^{-x^2} dx \qquad (8)$$

We have evaluated Eqn. 8 as a function of the normalized frequency u. The integral was computed by interpolating the tabulated values in Ref. 4 in the range $0 < u < 3.9$ and a computer program was employed for $u > 3.9$. The profile is shown in Fig. 3. It should be noted that, apart from a constant, u represents the frequency difference normalized with respect to the Doppler shift, $\Delta\nu_D$:

$$u = \frac{\omega_0-\omega}{(\omega_0/c)\sqrt{2kT/M}} = \frac{2\sqrt{\ln 2}\,(\nu_0-\nu)}{\Delta\nu_D} \qquad (7a)$$

where $\Delta\nu_D$ is the Doppler half width given by

---

*See Appendix.

$$\Delta \nu_D = \frac{2\nu_0}{c} \sqrt{\frac{2kT}{M}} \sqrt{\ell n 2} \qquad\qquad (7b)$$

III. LINE-BROADENED REFRACTIVE INDEX

Equations 1 (as modified by Eqn. 2) and 8 are the pertinent relations to be examined for present purposes. The former is a Lorentzian (see Fig. 1), but heavily scaled with respect to the damping coefficient. Indeed, pressure broadening is approximately four times the natural broadening for the case discussed in the Appendix, and would be appreciably larger still for application in a seeded air flow.

The refractive behavior of sodium vapor depends in particular on the number density of atoms as determined by the vapor pressure-temperature relation for this gas (Ref. 5). Fig. 2 gives a plot of atomic number density vs. temperature. It is seen that at $700°K$ ($427°C$), $N = 10^{16}/cm^3$. This is approximately 10^{-3} of atmospheric density and is considered as a target value for the proposed experimental designs. The corresponding temperature is not overly high from the point of view of experimental feasibility.

Figure 3 gives $n-n_0$, modified by line broadening, as a function of non-dimensional frequency u, but also shows an Angstrom scale representing wavelength distance from line center. The curve shown corresponds to the conditions just mentioned (sodium vapor density of $10^{16}/cm^3$). A peak value of nearly .24 is reached around .08 Å from line center, falling to a value of about .02 at approximately 1 Å spectral displacement.

It is instructive to use values from Fig. 3 in the usual non-resonant Gladstone-Dale formula, in order to compare with values as used in conventional refractive studies. The GD expression relates refractive index and mass density and reads

$$n - 1 = K\rho \qquad\qquad (9)$$

In studies with air at typical wavelengths in the visible region, $K = .23cm^3/gm$. The corresponding value for sodium vapor at the peak of refractive curve in Fig. 3 is

$$K = \frac{n-1}{\rho} = 6.28 \times 10^5 \ cm^3/gm$$

This value is 2.7×10^6 larger than the value for air mentioned just above! Even at 1 Å displacement from center, K assumes a value of more than 10 times the conventional value. While it is true that the Gladstone-Dale form of the dispersion relation as given by Eqn. (9) is not appropriate for describing the detailed resonant dispersion behavior, the point of this discussion is to show the enormous increase in the sensitivity factor K.

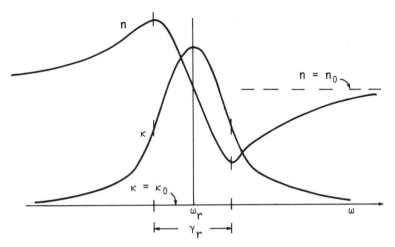

Fig. 1 - Dispersion and Absorption Characteristics Near Resonance for a
Bound Electron Oscillator in the Presence of Electromagnetic Radiation.
Note that the Extreme Values of n Occur at Values of ω Corresponding
to the Half-maximum Values of $\kappa - \kappa_o$. The Two Curves are not Scaled with
Respect to Each Other..

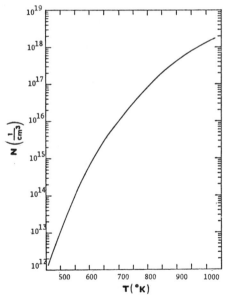

Fig. 2 - Atomic Number Density of Sodium Vapor as a Function of Temperature,
in Accordance with Vapor Pressure-Temperature Data (5).

Table I - Values and spectral locations of the maximum refractive index for the various broadening mechanisms. Symbol u is a non-dimensional frequency separation defined in Equation 7a.

Broadening Mechanisms	$(n-n_o)_{max}$	u	Displacement from Line Center, Å
Natural	27.2	0.0041	0.00037
Natural + Collision	6.47	0.0164	0.0015
Natural + Collision + Doppler	0.237	0.9	0.079

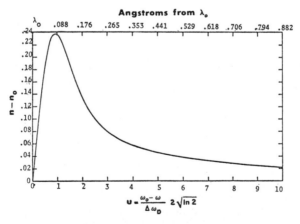

Fig. 3 - Dispersion of the D_2 Line of Sodium Vapor Including the Effects of Line Broadening: Refractive Index vs. Frequency Separation from Line Center. Note that n_o Represents the Non-resonant Refractivity Value.

Until now, the effects of line broadening on refractive sensitivity have not been ascertained numerically. Because line broadening has such an important effect on the resonant refractive dispersion (see discussion which follows) the numerics discussed in the above paragraph represent, in turn, an important and encouraging result for purposes of the present program.

Table 1 gives peak values of index and wavelength displacement for the three cases, respectively, of purely natural broadening, natural plus collision broadening, and the combined effects of all mechanisms, including Doppler broadening. It is quite clear that the line broadening produces a significant effect both on the height of the refractive peak and on its spectral location. Although the peak value itself is decreased by a large factor, we have shown that the sensitivity magnification is still enormous. What we have is a trade-off between peaking and spreading of the enhanced refractivity region. For practical design, the spreading feature will likely turn out to be an advantage which outweighs the loss in peak value.

One final comment on Fig. 3: Note that by definition of n_0, the curve asymptotically approaches zero for sufficiently large values of u or $\lambda - \lambda_0$. A plot on a compressed abscissa scale would show that the spectral region of significantly high refractivity covers several Å.

IV. PRELIMINARY EXPERIMENTAL WORK

The first phase of the experimental program consists of a set of interferometric experiments on pure sodium vapor. The wide band dispersion of sodium vapor will be measured at various wavelengths with the help of suitable light sources. That program will then be followed by similar experiments with a narrow band dye laser which is tunable over the resonant region.

A double tube heat pipe oven has been constructed to be so arranged that one tube is in each arm of a Mach-Zehnder interferometer. The tubes are of considerably different lengths so that all boundary effects cancel out, the only difference being the length of the sodium vapor column in the two tubes. The latter are made from stainless steel and 80 mesh stainless steel screen is used on the inside wall to provide the capillary action for the return flow of liquid sodium after condensation at the cold interface. A set of matched sapphire windows is available for the system, and thermocouple probes will monitor the temperature. Argon gas pressure in the cooled ends of the tubes will provide a pressure controlling mechanism for the whole system. Operating temperature will be around 700°K, corresponding to approximately 10^{16} sodium atoms per cc.

The tunable dye laser system is now being assembled. A nitrogen ultraviolet laser source will serve as the pumping laser and a dye cell within a laser cavity containing a beam expander and a rotatable grating will serve as the dye laser subsystem. Modifications of the existing interferometer configuration are now in progress to accommodate the heat pipe oven assembly and manifold.

APPENDIX

Numerical determination of parameter 'a'

The quantity 'a' represents a ratio of the combined Lorentzian-type line width to the Doppler width:

$$a = \frac{\Delta\nu_N + \Delta\nu_C}{\Delta\nu_D} \sqrt{\ell n 2}$$

where $\Delta\nu_N$, $\Delta\nu_C$, and $\Delta\nu_D$ are the halfwidths of natural, collision and Doppler profiles, respectively.

For the conditions of interest here, namely, T = 700°K, with associated Na vapor pressure p = 0.75 torr, we have (3)

$$\Delta\nu_N + \Delta\nu_C = 0.04 \times 10^9 \text{ sec}^{-1}$$

$$\text{and} \qquad \Delta\nu_D = 2.03 \times 10^9 \text{ sec}^{-1}$$

$$\therefore \quad a = 0.0164$$

REFERENCES

1. D. Bershader, "Refractive Behavior of Gases," in Modern Optical Methods in Gas Dynamic Research, D. Dosanjh (ed.), Plenum Press (1971).

2. W. C. Marlow, "Hakenmethode," Appl. Opt./Vol. 6, No. 10/1715 (1967).

3. A. C. G. Mitchell and M. W. Zemansky, Resonance Radiation and Excited Atoms, Cambridge Univ. Press, (1961).

4. M. Abramowitz and I. A. Stegun (Eds.), "Handbook of Mathematical Functions," Nat'l Bur. Stds. (1964).

5. E. von Theile, Ann. Physik 14, 937 (1932).

6. C. A. Forbrich, Jr., W. C. Marlow, and D. Bershader, "Measurement of Sodium D-Line Absolute Oscillator Strengths by the Roschedestvenskii Hook Method," Phys. Rev., Vol. 173, 150 (1968).

7. Proposal No. 4-74 to the AFOSR for a Study of Improved Flow Visualization by Use of Resonant Refractivity, for a period of one year, Jan. 1, 1974- Dec. 31, 1974.

8. A. S. Davydov, Quantum Mechanics, Chap. IX, Pergamon, 1965.

HOLOGRAPHY AND
HOLOGRAPHIC INTERFEROMETRY

G. O. REYNOLDS
Technical Operations

This talk will introduce the subjects of holography and holographic interferometry. Messrs. Vest and Sweeney will then proceed to discuss the analysis
of holographic data. The real advantage of holographic interferometry over
conventional interferometry is that all the information present in a three-
dimensional volume is recorded. However, the storage of this immense amount
of information makes the task of quantitatively analyzing the data difficult.

A hologram is a recording made in an interferometer, where one beam of the
interferometer carries the object information and the other beam is a reference
wave. The hologram, usually made on photographic film, is reconstructed by
placing the processed film into a coherent beam of light. At the appropriate
image plane a three-dimensional image of the original object is reconstructed.
If the original object is not optically visible (i. e., it consists of refractive index changes in a turbulent flow field) phase contrast techniques can be
applied to the reconstructed image to render the image visible. In such cases
holographic interferometry is used to measure the object information. I will
now indicate, by example, the types of results which can be expected from holography and holographic interferometry. Holography is only recently being considered in the combustion problem so there are very few experimental results
relating to the diagnostics of the combustion problem. However, as we have
already heard from the previous speakers, holography is potentially a very useful tool in combustion diagnostics since it is non-invasive and three-dimensional.

The first slide* (Fig. 8-11 of Ref. 1) shows an image reconstructed from a
hologram of a stained specimen of neutrons taken from the work of Van Ligten
and Osterberg.[2] The arrow in the slide indicates detail on the order of one

---

micron showing that holographic microscopy has resolution capability as well
as depth of field. Holographic microscopy has also been applied to measuring
particle size of moving aerosols in a three-dimensional volume. The hologram
is made using a pulsed laser which freezes the particles. The next slide from
the work of Thompson et al[3] (Fig. 8-4 of Ref. 1) shows that particle diameters
on the order of five to ten microns have been measured in field situations.
The problem with measuring particle size holographically is in the readout of
the data. When one scans through the reconstructed image volume the out-of-
focus particles cause a noisy background. This is shown on the next slide
(Fig. 8-6 from Ref. 1).[3] The arrows point to the in-focus particles. However
the out-of-focus particles make mensuration of large masses of data nearly impossible. The next slide (Fig. 8-9 from Ref. 1)[3] shows that some noise can be
removed by electronic clipping; however mensuration accuracy is not improved
since an observer still cannot readily distinguish between in-focus and out-of-
focus particles. The depth of focus problem can be improved by using diffusers
when making the holograms, and an example of such a result done by Leith and
Upatnieks at Michigan[4] is shown in the next slide (Fig. 3-23 from Ref. 1).[4]
The hologram (3-23a of Ref. 1) is a double exposure containing two different
scenes. When an observer looks at an image plane 14 inches from the hologram
the image of one object is observed and when observing 24 inches from the hologram the other object is observed. Since these holograms were made with diffusers the reconstructed images do not interfere with each other.

In order to demonstrate the ability of a hologram to store phase information we recall the work of Stroke[5] as seen in the next slide (Fig. 8-27 of
Ref. 1). Fig. 8-27a is a direct image of the pure phase object and as expected
no visible image is discerned. Fig. 8-27b is the hologram, and Fig. 8-27c is
the reconstructed image which again is not visible. However by applying phase
contrast techniques a visible image is obtained as shown in Fig. 8-27d. This
example illustrates that optical phase information can be retrieved from a
hologram. Another example of phase recording is shown in the next slide due
to Zinky (Fig. 8-28 of Ref. 1). The hologram of small particles was made in
the vicinity of a blast. The reconstructed particles are visible and the wavefront of the shock wave is also shown in the reconstructed image. This shock
wave physically existed in the field of view of the holographic camera at the
instant the hologram was recorded, and since it is located in a different plane
from the reconstructed particles the wavefront is visible because it is out of
focus.

In the last few years, many configurations for performing holographic
interferometry have appeared in the literature. These may be classified as
(1) double exposure technique, (2) time lapse technique, (3) real time technique, and (4) time average technique. In order to keep this discussion short
each technique will be illustrated by an example from the literature.

In the double exposure technique a hologram of the sample volume is made,
and then a second hologram is made when the volume is disturbed with a wave
front. Upon reconstruction the empty volume acts as a reference wave for the
wave front and the wave front is rendered visible. The classic example of
this technique due to Weurker et al[6] is shown in the next slide (Fig. 8-33a of
Ref. 7). The bullet in the sample volume was captured in flight with a

pulsed laser beam and upon reconstruction both the bullet and its shock wave are visible. This kind of holographic interferometry would probably be useful for the combustion process.

In time lapse holographic interferometry two holograms of the same object, but at two different times, are superimposed on the same film with a double exposure. Upon reconstruction the differences in the physical state of the object at the two times is observable as a fringe pattern. The next slide (Fig. 15-13 of Ref. 8) shows a time lapse holographic interferogram of an automobile tire where air pressure was added between exposures. The localized circular interference patterns reveal separations between the plies of the four-ply tire. This work was performed by G. C. O. Inc. of Ann Arbor, Michigan.

Real time holographic interferometry is performed by allowing the wave field of the object to illuminate a previously recorded hologram of the object such that differences in the state of the object are observed interferometrically in real time at the image plane. This is hard to demonstrate by an example since a motion picture is needed. Such examples were shown at this conference by Dr. Panknin.

Finally, long time exposure holographic interferometry has also been discussed in the literature.[1,7,8] In the next slide (Fig. 8-31 of Ref. 1) due to Powell and Stetson,[9] the nodal patterns of a vibrating object are shown for different physical driving forces. During the holographic exposure the nodes and anti-nodes are stationary whereas the rest of the vibrating object moves during the exposure thereby destroying the interference pattern in the hologram. Upon reconstruction the interference fringes appear only at the nodal positions thereby reconstructing a visual rendition of the mode pattern.

REFERENCES

1. J. B. DeVelis and G. O. Reynolds, "Theory and Applications of Holography," Addison Wesley, Reading, Mass. (1967).

2. R. F. Van Ligten and H. Osterberg, Nature 211, 282 (1966).

3. B. J. Thompson et al, J. Appl. Meterol, 5, 343 (1966).

4. E. N. Leith and J. Upatneiks, J. Opt. Soc. Am. 54, 1295 (1964).

5. D. Gabor, G. W. Stroke et al, Nature, 208, 1159 (1965).

6. R. E. Brooks, L. O. Heflinger and R. F. Weurker, Appl. Phys. Lett., 1, 248 (1965).

7. J. W. Goodman, "Introduction to Fourier Optics," McGraw-Hill, New York (1968).

8. R. J. Collier, C. B. Burckhardt and L. H. Lin, "Optical Holography," Academic Press, New York (1971).

MEASUREMENT OF ASYMMETRIC TEMPERATURE FIELDS BY HOLOGRAPHIC INTERFEROMETRY

C. M. VEST
University of Michigan

Holographic interferometry can be used to measure asymmetric temperature fields. In these remarks, the experimental and computational concepts are outlined. In the following presentation, Professor Sweeney discusses some specific experiments.

Fig. 1 is a schematic diagram of a typical holographic interferometer which could be used to measure the temperature distribution in a flame. Two sequential exposures are used to form a holographic interferogram. The film plate (hologram) is first exposed to the reference wave and to a plane wave of light which has traversed the test section while it is in a quiescent state. The flame under study is then ignited and a second exposure is made on the same film plate. As described by Dr. Reynolds in the preceding remarks, appropriate illumination of this developed hologram enables one to reconstruct the two optical waves which passed through the test section. These two waves interfere to form an interferogram, or fringe pattern, which can be related to the temperature field. This is referred to as double-exposure holographic interferometry. The interferogram is essentially identical to that which would be produced with the Mach-Zehnder interferometer which has been described above by Dr. El Wakyl. The holographic system is sometimes advantageous because it is a single-path interferometer, and therefore pathlength errors due to optically imperfect test section windows are eliminated.

The usefulness of the holographic interferometer can be extended by placing a diffuser (Ref. 1), such as a plate of ground glass or opal glass, behind the test section (see Fig. 2). Laser light will be scattered in many directions from each point of the diffuser, so no matter what direction we view the test section from, it will be back illuminated by the diffuser. Because holograms can record large amounts of optical information, it is possible to make double-exposure holographic interferograms with this diffuse illumination. When such a hologram is viewed, fringe patterns like

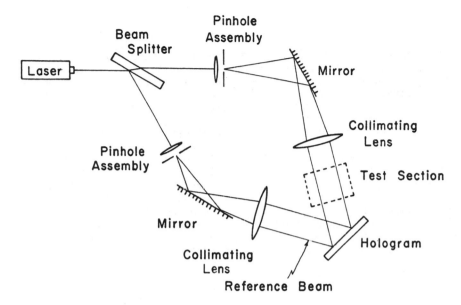

Fig. 1 - Schematic diagram of a holographic interferometer with plane-wave illumination

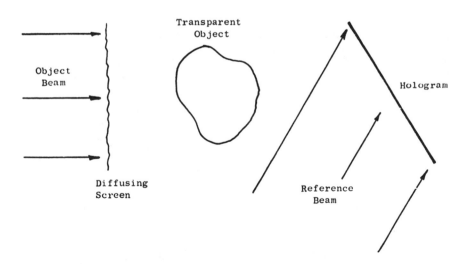

Fig. 2 - Schematic diagram of a holographic interferometer with diffuse illumination

those shown in Fig. 3(a) and (b) can be observed. These two fringe patterns
were photographed through the same hologram to depict how the interferogram
changes as viewing direction is varied. The fringe patterns corresponding to
all viewing directions compatible with the aperture of the hologram and
diffuser are recorded in a single double-exposure hologram. This represents
the same information which could be obtained from a multitude of Mach-Zehnder
interferograms, each recorded with a different orientation of the flame.

Analysis of interferograms with a single optical probing direction have
been discussed earlier in this session. Only radially symmetric flames, or
thin sheet flames, can be measured quantitatively. However, using multi-
directional interferometric data from a diffuse-illumination interferogram,
asymmetric flames can be studied. The inversion problem which must be solved
to analyze this data is also applicable to various emission and absorbtion
techniques discussed earlier in the worksbop. By counting fringes in an interferogram, values can be assigned to the topical pathlength, ϕ, which is
defined by

$$\phi = \int [n(x-y)-n_0]d\ell \qquad (1)$$

where n is the refractive index in the flame, n_0 is that of the ambient, and
$d\ell$ is a differential length of the optical ray through the flame. The data
consist of a set of measurements of the integrals of $[n(x,y)-n_0]$ along many
lines traversing the flame in various directions. This is indicated in
Fig. 4, where $f(x,y) \equiv n(x,y)-n_0$. The mathematics of the inversion of
line integrals was first worked out abstractly by the mathematician Randon in
1917. The appraoch I wish to outline here is due to Bracewell (Ref. 2), who
solved an analogous problem in radio astronomy.

For simplicity consider the line integral $\phi(y)$ obtained by counting
fringes while viewing the interferogram in a direction parallel to the x-
axis, as shown in Fig. 4. The two-dimensional Fourier transform of f(x,y)
is defined as

$$F(u,v) = \int\!\!\int_{-\infty}^{\infty} f(x,y)\, e^{-i2\pi(ux+vy)} dxdy. \qquad (2)$$

If F(u,v) is known, then f(x,y) can be determined by inverse transforming,

$$f(x,y) - \int\!\!\int_{-\infty}^{\infty} F(u,v)e^{i2\pi(ux + vy)} dudv. \qquad (3)$$

Now the one-dimensional Fourier transform of the data, $\phi(y)$ is

$$F_1\{\phi(y)\} = \int_{-\infty}^{\infty} \phi(y)e^{-i2\pi vy} dy, \qquad (4)$$

Fig. 3 - Two views of a single holographic interferogram produced with diffuse illumination

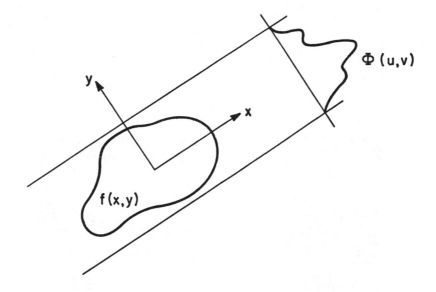

Fig. 4- For each viewing direction the line integrals along a parallel set of rays are measured

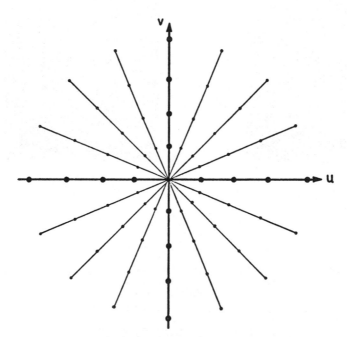

Fig. 5 - By using data from several viewing directions, the Fourier transform can be evaluated at points along a set of radial lines in the transform plane

but since

$$\phi(y) = \int_{-\infty}^{\infty} f(x,y)dx,$$

this becomes

$$F_1 \left\{\phi(y)\right\} = \int\!\!\int_{-\infty}^{\infty} f(x,y)e^{-i2\pi vy}dy. \qquad (5)$$

Thus

$$F(o,v) = F_1 \left\{\phi(y)\right\} \quad , \qquad (6)$$

that is the interferometric data can be used to determine the Fourier transform of $f(x,y)$ along the v-axis in the transform plane. For each viewing
direction, the transform can be evaluated along a radial line in the transform plane (Fig. 5). Bracewell refers to this as the Central Section Theorem.

Practical analysis of multi-directional interferometric data consists
of using a fast Fourier transform algorithm to transform the data for each
viewing direction. Next values of the transform are interpolated into a
rectangular grid. These values are then inverted by fast Fourier transforma-
tion to give values $f(x,y)$. Then using appropriate formulas, such as the
Gladstone-Dale law (Ref. 3), the temperature distribution in a cross-section
of the flame can be determined.

If the test section cannot be viewed over a complete 180° range of
viewing directions, the transform is known on only a portion of the (u,v)-
plane. The problem then becomes ill-conditioned, and other analysis schemes
must be used to make approximate measurements (Refs. 4 and 5).

I wish to acknowledge the support of this research by the Office of Naval
Research and the National Science Foundation.

REFERENCES

1. L. O. Heflinger, R. F. Wuerker, and R. E. Brooks, J. Appl. Phys. 37,
   642 (1966).

2. R. N. Bracewell, Aust. J. Phys. 9, 198 (1956).

3. W. Hauf and U. Grigull in Advances in Heat Transfer, edited by J. P.
   Hartnett and T. F. Irvine, Jr., Academic Press, New York (1970), pp. 133-366.

4. D. W. Sweeney and C. M. Vest, Appl. Opt. 12, 2649 (1973).

5. D. W. Sweeney and C. M. Vest, Int. J. Heat Mass Transfer 17, 1443 (1974).

EXPERIMENTAL CAPABILITIES OF HOLOGRAPHIC INTERFEROMETRY IN COMBUSTION

D. W. SWEENEY
Purdue University

In this note, we discuss two experiments which use holographic interferometry to quantitatively reconstruct refractive index distributions.
One of the experiments deals with three-dimensional, or multi-directional,
interferometry to measure temperature; this work was done at the University
of Michigan a few years ago. This experiment is of interest here because
it demonstrates the feasibility of using interferometric measurements,
which are normally considered to be line-average measurements, to obtain
point measurements in general asymmetrical temperature distributions. The
other experiment is presently being conducted at the University of California,
Lawrence Livermore Laboratory; it concerns the use of holographic interferometry to reconstruct electron number densities in laser-induced plasmas.
This experiment demonstrates the extremely good temporal and spatial resolution that can be obtained with state-of-the-art laser systems. These
experiments are not directly related to combustion phenomena, however we
hope to apply these procedures at Purdue University for combustion diagnostics
in the near future.

Fig. 1 shows a schematic of the experimental system used in the experiment conducted at the University of Michigan. The details of the experimental
system are discussed in the literature; (Refs. 1 and 2) it should be noted,
however, that the test cell is illuminated with light from many different
directions. Light from each direction produces a different interferogram.
This multi-directional illumination provides the information necessary to
invert the integral equation which characterizes three-dimensional interferometry (Ref. 3). The test-object in one of these experiments was a heated,
horizontal, rectangular plate submerged in water. Fig. 2 shows two interferograms obtained from two different directions of view. The interferometric
fringes are produced because the free convection process heats the water above
the plate and changes the optical pathlength of the probe light as it passes
through the test cell. Fig. 3 shows the results of collecting the multi-
directional fringe data and reconstructing the temperature in the developing
convective plume about the plate using a computer. The "finger-like" structure

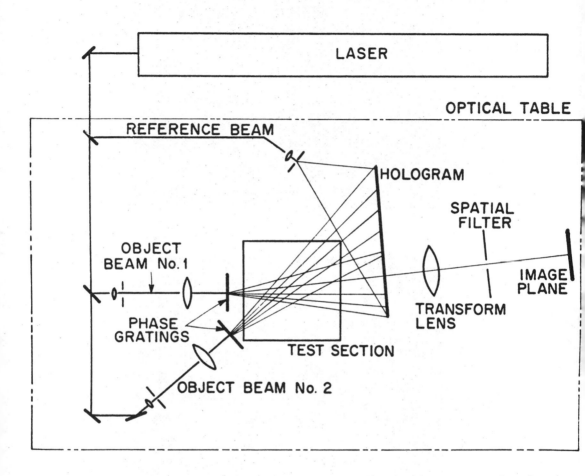

OPTICAL TABLE

REFERENCE BEAM

HOLOGRAM

SPATIAL
FILTER

OBJECT
BEAM No. 1

IMAGE
PLANE

PHASE
GRATINGS

TRANSFORM
LENS

TEST SECTION

OBJECT BEAM No. 2

Fig. 1 - Holographic interferometer in which the interferogram can be viewed
from a number of different directions.

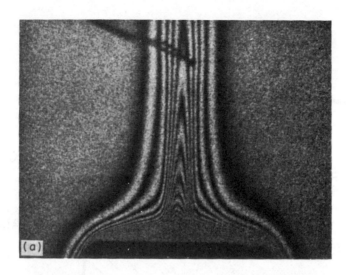

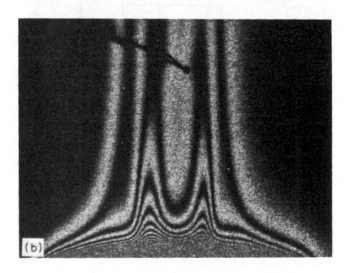

Fig. 2 – Two different views of a holographic interferogram of the plume above the heated rectangular plate. (a) View along the axis of the plate. (b) View along a direction 30° from the axis. The thermocouple junction is 1.05 cm above the plate.

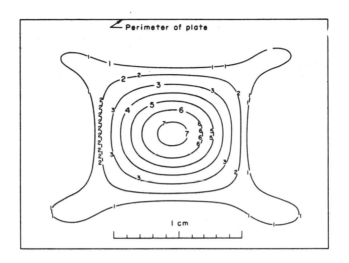

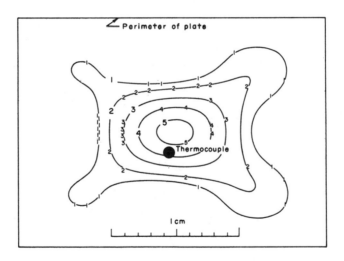

Fig. 3 – Isothermal contours in horizontal planes above the plate of aspect ratio 1.3. (a) o.5 cm above the plate. (b) 1.0 cm above the plate. Isotherms correspond to: 1(ΔT=.3°C), 2(ΔT=0.6°C), 3(ΔT=1.1°C), 4(ΔT=1.6°C), 5(ΔT+2.1°C), 6(ΔT=2.6°C), 7(ΔT=3.1°C).

of the temperature field is a result of the meeting of the convective plumes
from each edge of the plate. More detail concerning this work is available
in Refs. 1 and 2.

The experiments conducted at Lawrence Livermore Laboratory concern
reconstructing electron number densities in laser-induced plasmas using
holographic interferometry. In these experiments the refractive index (which
is the property directly measured interferometrically) is proportional to the
electron number density. Fig. 4 shows a schematic of the experimental system
used. A plasma is produced when a very intense light pulse from a mode
locked Nd-laser strikes a target. The duration of this heating pulse is
about 100 psec (10^{-10} seconds). Prior to striking the target, a portion of
the heating pulse is splitoff to be used as a probe beam for the holographic
interferometer. The probe beam is frequency tripled using nonlinear optical
techniques and manipulated so that it passes traversely through the plasma.
Fig. 5 shows an interferogram produced using this system. The fringe
positions on the interferogram were digitized and the electron number density
distribution was reconstructed. It is interesting to note that this system
has a time resolution of about 80 psec and a spatial resolution of about 10
μm. Again more detail concerning this work is available in Ref. 4.

Holographic interferometry offers two important advantages over
classical interferometry. First, because the holographic system compares
the wavefronts that have propagated through a test section at two different
times (rather than comparing the wavefronts that have propagated along two
different paths) the test-cell windows and associated optical components do
not need to be of high optical quality. Second, holographic interferometry
can provide multi-directional information which allows asymmetric refractive
index fields to be reconstructed. Holographic interferometry does have one
significant disadvantage for combustion diagnostics that is shared with classical
interferometry. The refractive index, which is the physical property measured
in any interferometer is a function of all the atomic and molecular species
present and their respective number densities. Since combustion processes
involve a number of species in unknown proportions, diagnostic information about
a single interesting physical property in combustion cannot be obtained interferometrically without auxilliary measurements. This problem could be overcome if an interferometric system could be designed for which the refractive
index would be a strong function of only one species concentration. D. Bershader
(Ref. 5) has suggested that this might be accomplished by adjusting the wavelength of the interferometer to nearly coincide with a resonant absorption
line of one species of interest.

Although these experiments using holographic interferometry are not directly
related to combustion, they do show a capability that can, we believe, be
applied to advantage for combustion diagnostics to provide new and useful
information.

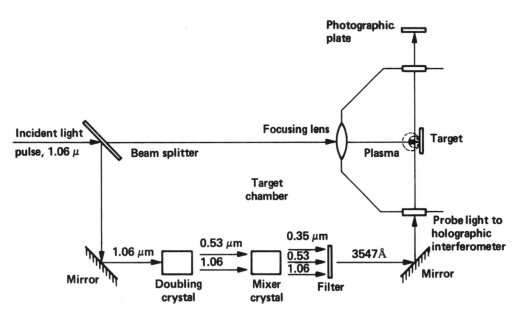

Fig. 4 - Schematic of experimental system used to produce laser-induced plasma. The probe beam is frequency tripled so that it will penetrate into the plasma without severe refraction.

**250 μm
wire**

Fig. 5 - Interferogram produced with experimental system shown in Fig. 4. The interferogram has a spatial resolution of about 10 m and a temporal resolution of about 80 psec. The maximum electron number density was calculated to be 2.8 X 10^{20} elec/cm^3.

REFERENCES

1. D. W. Sweeney and C. M. Vest, <u>Appl. Opt.</u>, <u>12</u>, 2649 (1973).

2. D. W. Sweeney and C. M. Vest, <u>Int. J. Heat Mass Transfer</u>, <u>17</u>, 1443 (1974).

3. See contributions by C. M. Vest elsewhere in these Proceedings.

4. D. T. Attwood, L. W. Coleman and D. W. Sweeney, <u>Appl. Phys. Letts.</u>, <u>26</u>, 616 (1974).

5. See contributions by D. Bershader elsewhere in these Proceedings.

DISCUSSION

EL WAKIL - You still have to measure concentrations in a flame, don't you?

SWEENEY - We measure the refractive index directly and we deduce whatever property we can derive from that information.

EL WAKIL - So you have the same problem as we have in the classical case.

SWEENEY - Yes.

BILGER, Univ. of Sydney - If you had a turbulent flame, then you could actually map out the contours of the refractive index of this flame with one pulse of a few nanoseconds or a few picoseconds. Am I right?

SWEENEY - Yes. Of course there is always the problem of resolution when we reconstruct in fine detail.

BILGER - How fine is that? Do you get down to a few tenths of microns?

SWEENEY - Theoretically you can get anything you like, but experimentally, using reasonable equipment, I would say about a millimeter, in a three-dimensional reconstruction.

EL WAKIL - Of all the holographic interferograms I have seen none of them really are as crisp and sharp as in the classical Mach Zehnder.

SWEENEY - It is true that many of the examples of holographic interferograms published in the literature are not as "crisp and sharp" as those obtained using a Mach-Zehnder. However, many of these holographic interferograms have been made in hostile environments through test-cell windows which are not of good optical quality. Some holographic interferograms are quite sharp; the interferograms of a bullet in flight made by R. Wuerker are a good example (page 425, Optical Holography, Academic Press, 1971). Also, to take advantage of the multi-directional capabilities of holography, phase objects are often back-illuminated with diffuse light. This introduces a source of noise called coherent speckle which can detract significantly from the "sharpness" of the interferometric fringes [cf. L. H. Tanner, J. Sci. Inst., 44, 1011 (1967)].

HOLOGRAPHIC TWO-WAVELENGTHS INTERFEROMETRY FOR MEASUREMENT OF COMBINED HEAT AND MASS TRANSFER

F. MAYINGER
W. PANKNIN

Institut für Verfahrenstechnik der T.U. Hannover

The two-wavelengths technique was already used a couple of years ago by Ross and El-Wakil (Refs. 1 and 2) for the study of a simulated drop of fuel, but then dropped because of different difficulties involved. However, this technique is so fascinating that we again tried to put the idea into practice by means of using holographic interferometry, instead of Mach-Zehnder interferometry. We thought that both some further theoretical work concerning the evaluation of the interferograms and some experimental improvements would make this measuring technique fit for use.

All interference methods, both the classical ones - as there are Michelson and Mach-Zehnder techniques - and holographic interference techniques first allow only the measurement of refraction index changes in the test section. Only if the reference field is known and if the refraction index was changed only by temperature or concentration or pressure, the interferograms can be evaluated without additional assumptions or measurements. In chemical engineering and combustion research there are however many processes where heat and mass transfer occur simultaneously. Since in all these cases the refraction index is influenced by temperature and concentration changes simultaneously, additional information is needed for the evaluation of the interferograms. The combination of interferometry with conventional measuring techniques using probes or thermocouples has severe drawbacks. The devices can disturb the process, are often not fast enough for the measurement of rapid events, or they are too big to allow examinations in thin boundary layers

It is therefore desirable to try to obtain the additional information also by using optical techniques.

Since the refraction index is a function of the wavelength, the idea of two-wavelengths technique is, to take two interferograms simultaneously in

order to determine both temperature and concentrations of two-component systems
from the difference of the interferograms. This principle was already successfully used in plasma physics to determine the electron and ion density of a
plasma (Ref. 3).

The substances which one usually encounters in combined heat and mass
transfer, however, have a much smaller dependance of the refraction index on
the wavelength of light as one finds when examining a plasma. Therefore the
optical set up, the measurement of the interference orders and the evaluation
of the interferograms requires much more accuracy than usually needed in one-
wavelength techniques. Holography provides much help to overcome the experimental difficulties and thus enables many more problems to be examined by
using this technique. First the evaluation equations and the necessary corrections of the interferograms shall be discussed. Then the experimental set
up will be explained and some applications will be given.

From the interferograms, which were made for example by using the double
exposure technique, one first gets the changes of the optical path along the
light beam (Ref. 1). S is the interference order in multiples of a wavelength
λ. If we assume a straight light path and a constant field during the exposure
of the comparison field, the equation can be integrated if a two dimensional
field (constant in light beam direction) is provided. We then obtain Eq. 2
which is the equation of ideal interferometry. If gases are to be examined
the Gladstone Dale equation gives the relation between refraction index and
density. Using in addition the ideal gas law we obtain Eq. 3, where N is the
molar refractivity. This molar refractivity is the sum of the refractivities
of the pure components a and b in the mixture. Eq. 4 gives the relation between
fringe shift S temperature T and concentration C. Since N is a function of
wavelength, two interferograms taken at different wavelengths λ_j and λ_k allow
to eliminate one of the unknowns T or C. From Eq. 5 temperature T can be obtained by evaluating both interferograms.

In Fig. 3 the principle of evaluation is demonstrated schematically. Two
interferograms with fringe shifts $s_j\lambda_j$ and $s_k\lambda_k$ are recorded.

The difference of the product $S\frac{\lambda}{N}$ is a value for $\frac{C}{T}$ The concentration C
can be found also by using the temperature and only one interferogram.

The picture already indicates that the measuring effect is very small.
It mainly depends on the optical properties of the system. In order to obtain
good measuring accuracies the deviations from ideal interferometry must therefore be taken into consideration. Three effects will be discussed.

First we consider the case that the boundary layer extends over the actual test length. Therefore the light beam receives an additional fringe shift
ΔS. This correction term can be calculated by representing the measured
fringe shift curve by a polynominal of e.g. 5th order and integrating Eq. 1
in Fig. 4. In two wavelength interferometry the differences of the two inter-
ferograms give the measuring value. The ratio of $\Delta S_j\lambda_j$ and $\Delta S_k\lambda_k$ depends on
whether the fringe shift is mainly due to concentration or temperature changes.

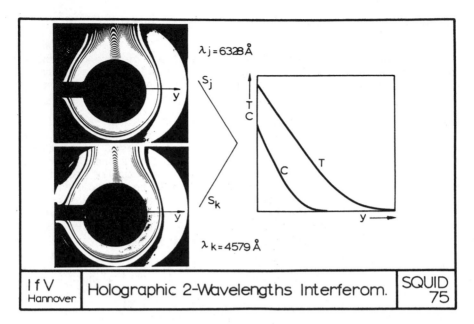

Fig. 1

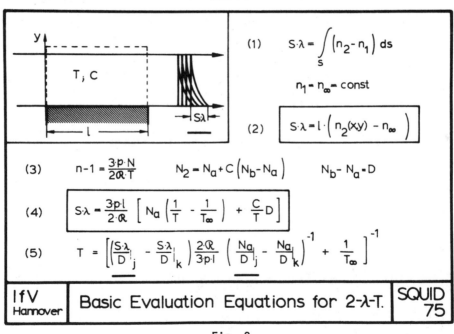

$$(1) \quad S \cdot \lambda = \int_S \left(n_2 - n_1\right) ds$$

$$n_1 = n_\infty = const$$

$$(2) \quad \boxed{S \cdot \lambda = l \cdot \left(n_2(x,y) - n_\infty\right)}$$

$$(3) \quad n - 1 = \frac{3 \cdot p \cdot N}{2 \cdot \mathfrak{R} \cdot T} \qquad N_2 = N_a + C\left(N_b - N_a\right) \qquad N_b - N_a = D$$

$$(4) \quad \boxed{S \lambda = \frac{3 \cdot p \cdot l}{2 \cdot \mathfrak{R}} \left[N_a \left(\frac{1}{T} - \frac{1}{T_\infty}\right) + \frac{C}{T} D \right]}$$

$$(5) \quad T = \left[\left(\frac{S \cdot \lambda}{D}\Big|_j - \frac{S \cdot \lambda}{D}\Big|_k\right) \frac{2 \cdot \mathfrak{R}}{3 \cdot p \cdot l} \left(\frac{N_a}{D}\Big|_j - \frac{N_a}{D}\Big|_k\right)^{-1} + \frac{1}{T_\infty}\right]^{-1}$$

Fig. 2

272

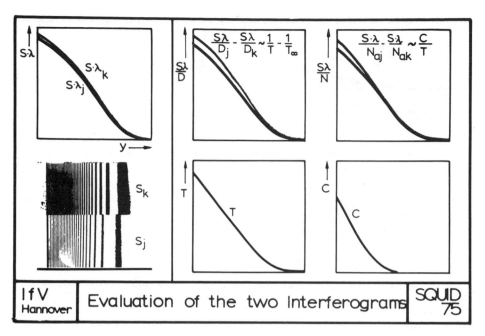

IfV Hannover Evaluation of the two Interferograms **SQUID 75**

Fig. 3

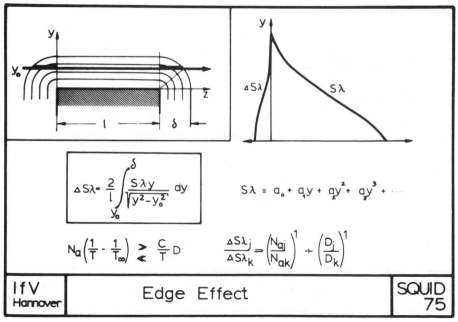

$$\Delta S\lambda = \frac{2}{l} \int_{y_0}^{\delta} \frac{S \cdot \lambda y}{\sqrt{y^2 - y_0^2}} \, dy$$

$$S\lambda = a_0 + a_1 y + a_2 y^2 + a_3 y^3 + \cdots$$

$$N_a \left(\frac{1}{T} - \frac{1}{T_\infty} \right) \gtrless \frac{C}{T} D$$

$$\frac{\Delta S\lambda_j}{\Delta S\lambda_k} = \left(\frac{N_{aj}}{N_{ak}} \right)^1 \div \left(\frac{D_j}{D_k} \right)^1$$

IfV Hannover Edge Effect **SQUID 75**

Fig. 4

273

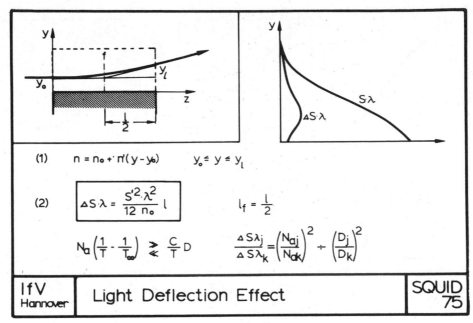

(1) $n = n_o + n'(y - y_o)$ $y_o \leq y \leq y_l$

(2) $\boxed{\Delta S \cdot \lambda = \dfrac{s'^2 \cdot \lambda^2}{12\, n_o}\, l}$ $l_f = \dfrac{l}{2}$

$$N_a \left(\frac{1}{T} - \frac{1}{T_\infty} \right) \gtrless \frac{C}{T} D \qquad \frac{\Delta S \lambda_j}{\Delta S \lambda_k} = \left(\frac{N_{aj}}{N_{ak}} \right)^2 \div \left(\frac{D_j}{D_k} \right)^2$$

| IfV Hannover | Light Deflection Effect | SQUID 75 |

Fig. 5

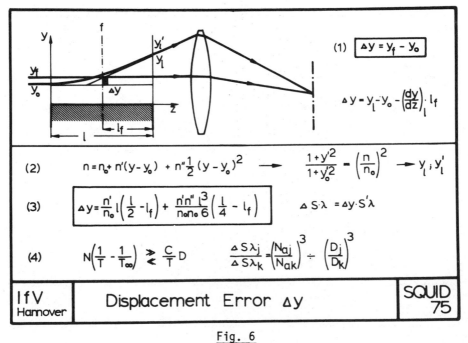

(1) $\boxed{\Delta y = y_f - y_o}$

$$\Delta y = y_l - y_o - \left(\frac{dy}{dz} \right)_l \cdot l_f$$

(2) $n = n_o + n'(y - y_o) + n'' \dfrac{1}{2}(y - y_o)^2 \quad \longrightarrow \quad \dfrac{1 + y'^2}{1 + y_o'^2} = \left(\dfrac{n}{n_o} \right)^2 \quad \longrightarrow \quad y_l \, ; \, y_l'$

(3) $\boxed{\Delta y = \dfrac{n'}{n_o} l \left(\dfrac{l}{2} - l_f \right) + \dfrac{n' n''}{n_o n_o} \dfrac{l^3}{6} \left(\dfrac{l}{4} - l_f \right)}$ $\Delta S \cdot \lambda = \Delta y \cdot S' \lambda$

(4) $N \left(\dfrac{1}{T} - \dfrac{1}{T_\infty} \right) \gtrless \dfrac{C}{T} D$ $\dfrac{\Delta S \lambda_j}{\Delta S \lambda_k} = \left(\dfrac{N_{aj}}{N_{ak}} \right)^3 \div \left(\dfrac{D_j}{D_k} \right)^3$

| IfV Hannover | Displacement Error Δy | SQUID 75 |

Fig. 6

274

This can be determined using an iterative way of solution, by neglecting first all the correction terms.

An additional fringe shift is due to the fact that the light beam is deflected towards the denser optical fluid. If we assume that within this region of deflection the refraction index can be represented by Eq. 1 the additional fringe shift is calculated by using Eq. 2. In this case we have, focussed onto the middle of the test section. The ratio of the correction terms is now proportional to

$$(\frac{Na_j}{Na_k})^2 \quad or \quad (\frac{D_j}{D_k})^2$$

Connected with the light deflection is a displacement λ_y of the interference lines. The beam which enters the test section at y_o is photographed at the position y_f. This is dependent on the focusing length l_f. If the refraction index is described by Eq. 2, the general equation for the light beam can be solved (Ref. 4, 5) and we get the exit values. The correction term 3 gives us the displacement error as a function of focusing length. Focusing onto the middle of the section gives the minimum displacement, which is a function of n' and n". The additional fringe shift is now proportional to

$$(\frac{Na_j}{Na_k})^3 \quad or \quad (\frac{D_j}{D_k})^3$$

A more refined analysis is given in (Ref. 6) where the third order terms are also applied to the calculation of fringe shift $\Delta S\lambda$ in Fig. 5. The final result then is that the overall correction term is a minimum for $l_F = 1/3$, with $\Delta S\lambda = n'^2 \, n" \, \ell^5 \, 1/30$. The distribution of this correction term is shown schematically in Fig. 7.

From the smooth temperature and concentration profiles a smooth fringe shift curve S results. The correction term is however proportional to the product of s' and s", and results in relatively strong local deviations of the measured values, which can lead to considerable errors in the final evaluation procedure. This was confirmed experimentally.

In Fig. 8 the measured fringe width b is plotted for the wall next boundary layer. The deviations from the theoretical straight curve lead to mistakes when the gradient near the wall is to be determined. Since both interferograms are affected, the resultant mistake can be kept small by careful interpolation between the fringes.

The experimental holographic set up is shown schematically in Fig. 9. It differs from conventional arrangements only by using two lasers simultaneously. The beams of a He-Ne-Laser and an Argon Laser are superposed in a beam splitter S. Then the beams are divided into reference wave RW and object wave OW, both with two wavelengths. The object beam passes through the test section TS and both waves interfere on the hologram H. When a double exposure is made, having a constant field first and introducing the heat and mass transfer before the second exposure, the hologram can be reconstructed afterwards. The two

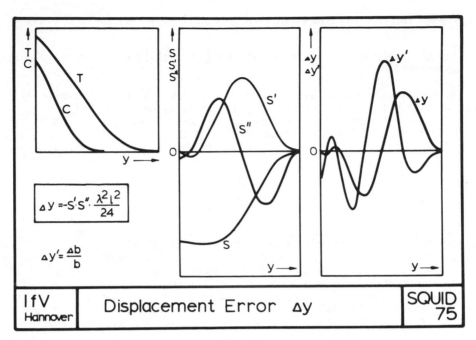

Fig. 7

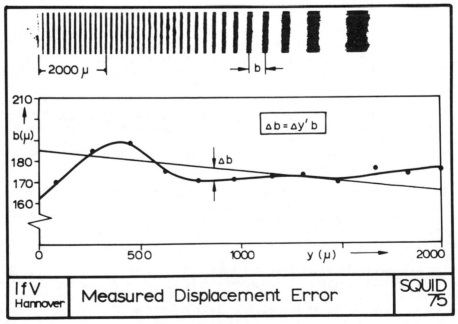

Fig. 8

276

resultant negatives. This is usually done by means of a microscope or photometer.

This procedure is easy, has however an important drawback. It is nearly
impossible to get an exact reference line for the two interferograms. This
problem was already encountered by Ross (Ref. 1). It can lead to a displacement of the two fringe shift curves to one another, resulting in enormous
mistakes.

Therefore a modified holographic set up was developed and tested.

The two interferograms are projected in this set up onto the hologram by
placing the camera lens between test section and hologram. The step of taking
sharp pictures by using the lens was thus combined with the recording of the
hologram. The hologram itself can now be used for measuring the fringe positions. This is done by illuminating it under a photometer alternately with
the two reference beams. This guarantees an exact correspondence of the fringe
shifts and improves considerably the obtainable accuracy.

It may be mentioned that both techniques result in interferograms which
are as crisp and sharp as in classical Mach Zehnder interferometry. All one
has to do is to carefully allign the optical set up and to eliminate interference patterns due to dust particles by using a pin hole in the focus of the
beam expanding lenses. Recent results show that interference lines with a
fringe width of less than 30μ can be recorded and evaluated (Ref. 7).

In Fig. 11 the resultant interferograms of a heat and mass transfer boundary layer are shown. The model consisted of a heated horizontal cylinder
covered by a thin layer of subliming naphtalene. Many of such interferograms
as well as interferograms obtained for heat and mass transfer on a vertical
plate were evaluated. One typical example is given in Fig. 12 where the resulting temperature and concentration profiles correspond exactly to the wall
values, which were determined by extra measurements. Experimentally determined
Sherwood and Nusselt Numbers are given in Fig. 13. The values are in good
agreement with theoretical investigations as e.g. by Wilcox (Ref. 8). The deviations are mainly due to the fact that for obtaining the Nu and Sh-numbers
the gradient of the temperature and concentration profiles had to be determined
near the wall, which can lead to some errors. The evaluation technique was not
the iterative way, proposed by Ross (Ref. 1) but the simultaneous solution of
the two interferograms.

When we got the invitation to this workshop we had only few time to apply
the two wavelength technique to the study of combustion phenomena. However the
results obtained within this short period may indicate how well this technique
works. The technique was used for the investigation of temperature and fuel
vapour concentration in a simulated drop of fuel, as earlier done by Ross.

Fig. 14 shows the dual interferograms of a burning n-Hexan cylinder. The
model consisted of a porous bronze cylinder of 9 mm ⌀ and 50 mm length, through
which the fuel evaporated. Since the light beam near the surface is deflected

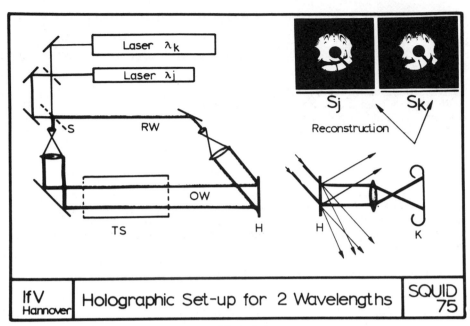

Fig. 9

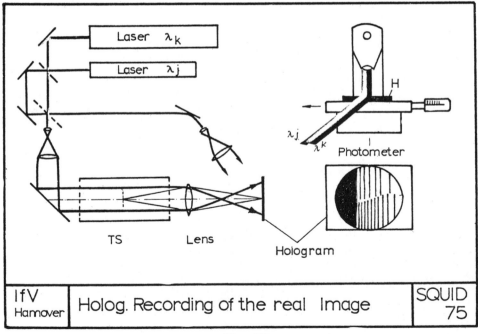

Fig. 10

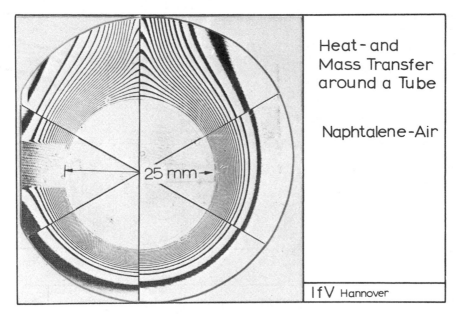

Heat- and
Mass Transfer
around a Tube

Naphtalene-Air

IfV Hannover

Fig. 11

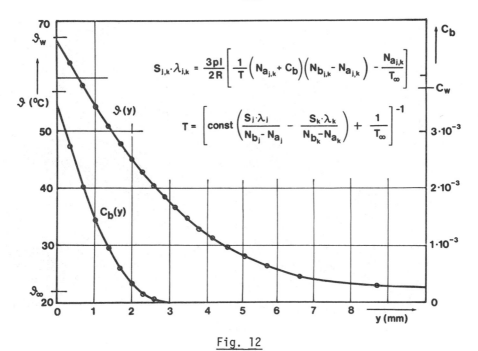

$$S_{j,k} \cdot \lambda_{j,k} = \frac{3pl}{2R}\left[\frac{1}{T}\left(N_{a_{j,k}} + C_b\right)\left(N_{b_{i,k}} - N_{a_{j,k}}\right) - \frac{N_{a_{j,k}}}{T_\infty}\right]$$

$$T = \left[const\left(\frac{S_j \cdot \lambda_i}{N_{b_j} - N_{a_j}} - \frac{S_k \cdot \lambda_k}{N_{b_k} - N_{a_k}}\right) + \frac{1}{T_\infty}\right]^{-1}$$

Fig. 12

279

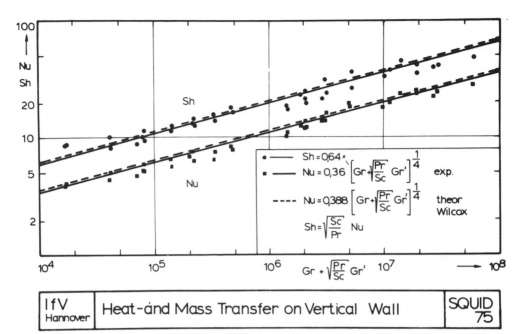

<u>Fig. 13</u>

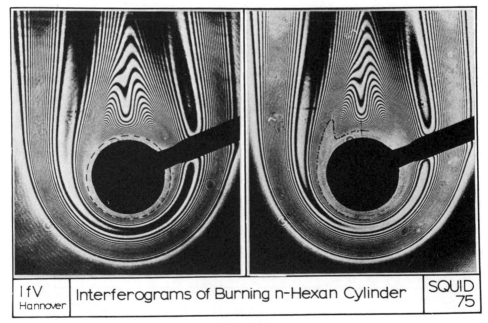

<u>Fig. 14</u>

into the surface (dashed circle) only outside of this region the boundary layer could be studied.

The fuel n-Hexan is not well suited for the two wavelength technique because of the very low dependance of the refraction index on the wavelength. Nevertheless the evaluation of the dual interferograms showed that the concentration of fuel vapour reached nearly zero at the distance 1.5 mm from the wall at the position 0^0 and about 3 mm from the wall at position 135^0. Therefore within this reaction zone the temperature was obtained by using one interferogram only. The fringe shift could be determined until near the wall where the fringe density is about 50 lines per mm which is an improvement compared to the results of Ross.

Additional thermocouple measurements confirmed the interferometric results. The region of fuel vapour can be estimated roughly by determining the fringe width curve. The beginning of the reaction zone is connected with the beginning of heat production and thus with a sudden change of the temperature and concentration gradient. This results in a sudden change of fringe density too, which can be seen in picture 14B indicated by the dotted curve.

First evaluations indicate that the fuel vapour concentration along this curve reaches values of about 0,2. The results obtained so far are in good agreement with measurements of Ross and the results reported in this workshop by El-Wakil who used gas chromatography as measuring technique.

The final picture may indicate some more applications of holography. A double exposure hologram is compared to a single exposure hologram. This was obtained by using a relatively long exposure time (1/50 sec). The interference lines which are seen in the right picture are lines due to the changing of the wavefront during the exposure time. The fringe density is proportional to wavefront gradient and the velocity of change. Further theoretical work in this very new field might reveal this to be a powerful technique for the study of flame front velocities and instationary reaction zones.

The use of two wavelength holography turned out to be a fine measuring technique for the study of simultaneous heat and mass transfer. The refined evaluation techniques and the better experimental arrangement allow the investigation even of mixtures whose optical properties do not vary much with the wavelength. This was confirmed by investigating air naphtalene and air-n-Hexan mixtures.

The development of tunable gas-lasers which allow a farther spread of wavelength, and the use of pulsed ruby lasers with frequency doubling promises an even larger field of applications of holographic two-wavelength interferometry. In this report only holographic two-wavelength technique was spoken about. As a participant of the workshop I think that all the other applications of holographic interferometry as well as holographic particle determination techniques were not discussed sufficiently enough.

We may therefore refer to some more of our own applications and experiences which were published in (Ref. 9, 10, 11).

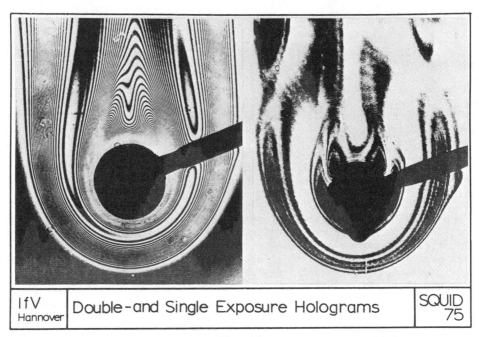

Double-and Single Exposure Holograms

Fig. 15

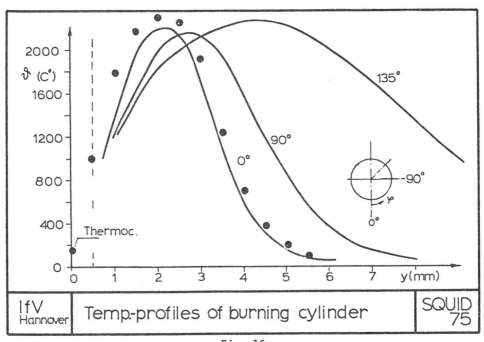

Temp.-profiles of burning cylinder

Fig. 16

References

1. P. A. Ross, Application of a two-wavelength interferometer to the study of a simulated drop of fuel. PhD thesis, University of Wisconsin (1960)

2. M. M. El-Wakil and P. A. Ross, A Two Wavelength Interferometric Technique for the Study of Vaporization and Combustion of Fuels, Liquid Rockets and Propellanta: Progress in Astronautics and Rocketry, Vol. II, Academic Press, New York (1960)

3. D. Meiners and C. O. Weiss, Interferometrische Messung der thermischen ZustandsgroBen von EdelgasstoBwellenplasmen. Z. Naturforschun, 28 a, 8, (1973)

4. U. Grigull, Einige optische Eigenschaften thermischer Grenzschichten, Intern J. Heat Mass Transfer 6, (1963)

5. W. Hauf and U. Grigull, Optical Methods in Heat Transfer Advances in Heat Transfer 6, (1970)

6. W. Panknin, Dissertation, Hannover

7. F. Mayinger and W. Panknin, Holography in Heat and Mass Transfer. Proceedings of the 5th Int. Heat Transfer Conf. Tokio (1974)

8. W. R. Wilcox, chem. Eng. Sci. 13,3,113 (1961)

9. W. Panknin, Einige Techniken und Anwendungen der holgraphischen Durchlichtinterferometrie CZ-Chemie Technik, 3 (1974)

10. W. Panknin, M. Jahn u. H. H. Reineke, Forced Convection Heat Transfer in the Transition form Laminar to Turbulent Flow in Closely Spaced Circular Tube Bundles, 5th Int. Heat Transfer Conf., Tokio (1974)

11. F. Mayinger, D. Nordmann u. W. Panknin, Chem. Ing. Tech. 46 (1974)

PART III
Probe Methods and Special Techniques

SESSION 8
Probe Measurements in
Laminar Flows
Chairman: R. M. FRISTROM

PROBE MEASUREMENTS IN
LAMINAR COMBUSTION SYSTEMS

R. M. FRISTROM
Johns Hopkins University

I. INTRODUCTION

One of the most fruitful methods of studying combustion processes has
been the use of measuring probes. In this discussion we will consider the
applications, problems and limitations of such studies. The introduction
of any probe, even an optical probe* always produces some disturbance and
it is a quantitative question whether the required information is compromised
beyond the point of usefulness.

The variables which are required to characterize a combustion system
are velocity, temperature and composition as a function of position and
time. If the system is steady state and possesses some symmetry, e.g.
Bunsen flames or flat flames, the required number of variables can be greatly
reduced. For example, one dimensional premixed or diffusion flames can be
realized in the laboratory. With this geometry and known initial flows, the
system can be completely determined by measuring s variables where s is the
number of species. This assumes conservation of mass and an equation of
state. In principle the requirement could be reduced to s-n by applying
conservation constraints to each atomic species individually, where n is the
number of atomic species involved in the incoming molecules. This is not
usually done because diffusion is so important in combustion that elaborate
calculations are required. Instead the conservation laws can be used to
check the quality of the data (Ref. 1, p. 88). Because this is an over-
determined system it is often possible to derive a variable which is difficult
to measure directly from the other variables. For example, if absolute
composition is known, local density can be calculated. Temperature can be
calculated from density and molecular weight using the equation of state.
Velocity can be calculated from the inlet mass flow and local density.
Similarly, missing concentrations can be deduced (See Table I).

---

*Optical beams of sufficient intensity can induce reaction, inhibit reaction,
 liberate heat and even levitate particles.

Table I - Measured and Calculated Variables

Each line of the table gives the symbol for the variables which must be measured directly by some technique. The others are then obtained by calculation.

Flame system	Independent variable z	Dependent variable						
		Distance z	Time t	Velocity v	Area ratio A	Density ρ	Temp. T	Concentration X_i, N_i
Flat [10]	z		Calc.	Calc.	A	Calc.	T	$X_i; i = 1, 2, \ldots, s - 1$
Flat	z		Calc.	v	Calc.	Calc.	T	$X_i; i = 1, 2, \ldots, s - 1$
Flat	z		Calc.	Calc.	A	Calc.	Calc.	$N_i; i = 1, 2, \ldots, s$
Spherical [11]	z		Calc.	Calc.	Calc.	Calc.	T	$X_i; i = 1, 2, \ldots, s - 1$
Conical [12]	z		Calc.	v	A	Calc.	Calc.	$X_i; i = 1, 2, \ldots, s - 1$
Expanding flame kernels	t	Calc.		Calc.	Calc.	Calc.	T	$X_i; i = 1, 2, \ldots, s - 1$
Theory	T	$\int \left(\dfrac{dT}{dz}\right) dz$	Calc.	Calc.	Const.	Calc.		$X_i; i = 1, 2, \ldots, s - 1$

In more complex geometries such as axially symmetric diffusion flames, a two dimensional manifold of variables must be measured and in the general case, a three dimensional manifold. If it is desired to derive rate of reaction or heat release information from the data, it is necessary to know not only the local intensive variables (temperature, velocity and composition), but also their first and second derivatives and the appropriate diffusion coefficients, thermal conductivities and coefficients of thermal diffusion. Determining rate of species production and heat release is difficult in most laboratory systems and virtually impossible in most practical systems. Therefore, the experimentalist must usually settle for more modest goals than complete analysis. Much useful information can be obtained from such measurements and we will now discuss some techniques which can be used for such measurements.

II.   VELOCITY PROBES

Local velocity must be known to derive rate processes. Several probing techniques have been used: Pitot probes- particle visualization, etc. (Refs. 1, 2, 3).

A. Pitot Tube - The pitot tube method of measuring velocity is standard in aerodynamics (Ref. 4). The principle is simple: if a tube connected to a pressure-measuring device is directed against a fluid flow, it will register a pressure which is proportional to the square root of the velocity. In flames these pressures are low, but measurable. (Ref. 2) (Fig. 1). These measurements are difficult to interpret because the probe must be small compared with the flame front thickness and boundary layer corrections become important. The measured pressure depends not only on velocity, but also Reynolds number which is in turn a complex function of temperature and probe diameter. These problems are discussed in Ref. 4.

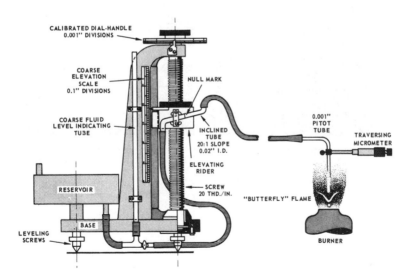

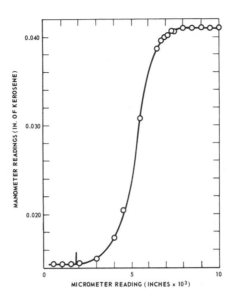

Fig. 1 - Apparatus for Pitot tube measurements in flames with typical profile (Ref. 2)

B. Flow Visualization with Particles - Another method of studying combustion aerodynamics if flow visualization with suspended microscopic dust particles. This is a standard aerodynamic technique (Refs. 1, 2, 7).

To be suitable for tracer studies a particle must be small, non-volatile, and non-reactive. Particle introduction disturbs a flame, the degree depends on the type, size, and number of particles. Particles can be visualized photographically using a timed, repetitive illumination. From such a picture velocity can be obtained by direct measurement (Fig. 2).

Common sources of error are accelerational lag, thermomechanical effect, and the requirement that the particle be very small compared with the flame thickness.

If a precision of three per cent is acceptable, then particle-tracer techniques can be used for quantitative studies. (Ref. 1).

With the advent of lasers, another particle method called laser Doppler velocimetry has been developed. It is based on the principle that the light scattered from particles will be shifted in frequency by the Doppler effect. By using a suitable detector and mixer for the scattered and unscattered

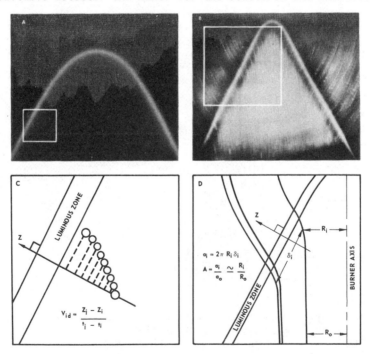

Fig. 2 - Measurement of velocity and area ratio in flames using particle track techniques (Ref. 7)

beam, a beat signal can be obtained which is proportional to the velocity of the particles. This very powerful technique is discussed in another session of this Workshop (See Ref. 3, for instance).

C. Other Methods of Measuring Gas Velocity - Many methods of measuring velocity are not applicable to combustion studies, because of the high temperatures or high spatial resolution required. The hot wire methods used to study boundary layers and turbulence can give serious errors because of the temperature gradients. An interesting variant is the pulsed hot wire of Westenberg and Walker (Ref. 5) which uses the heated wake of a pulsed hot wire as a tracer.

D. Avoidance of Aerodynamic Measurements - Aerodynamic measurements are among the most idfficult and least precise of combustion measurements. Therefore, it is desirable to avoid them or minimize the dependence on aerodynamic parameters. In flames velocity profiles can be calculated from area ratio measurements, and density determinations obtained from thermocouple or pneumatic probe traverses.

III. PROBE THERMOMETRY

Probes provide the most direct method of determining local temperature. Probe diameter should be small compared with the product and be rugged enough to stand the high-temperature corrosive flame environment. Thermocouples have found the widest usage in combustion studies.

A. Thermocouple Measurements - Thermocouple measurements make use of the thermoelectrical property of metals. This potential is reproducible and is a function of the materials chosen for wires. It is independent of the method of making the junction (wires may be welded, soldered, or simply twisted together) so long as good electrical contact is maintained, and pro- vided there is no appreciable temperature gradient across the joint. A large number of thermocouple pairs have been studied (Ref. 1, 6, 7) but only a few are suitable for flame use, notably the Pt, Pt-10% Rh and Ir, Ir-40% Rh couples (See Table II).

The advantages (See Table III) of thermocouple measurements are: (a) they can be made with high precision; (b) they are small (0.001 cm) and (c) they can withstand high temperature.

The principal source of error is radiation loss. Corrections can be made so that temperatures reliable to 10 and 20°K, positioned with a resolution of 50 microns, can be obtained (Refs. 1, 7, 8). This error can be eliminated by using the "null method" in which the thermocouple is heated electrically to balance the radiation loss (Ref. 7). Temperature derivatives are primarily limited by the size of the wire used and the disturbances of the vibration and catalysis. Temperature differences as small as 0.1°C can be reliably measured, with positional undertainty of 10^{-3}cm. Such measure-

TABLE II - Limits of Some High Temperature Thermocouples (Ref. 6)

Couple or Material	Upper Temp. °C	Prob. Error °C	Max. Output Milli Volts	Comments
W/Ta	3,000	+50	23	Inert or Reducing
W/W-50Mo	3,000	50	8	Atmosphere only
W/W-25Mo	3,200	50	5.4	"
Ta/Mo	2,600	50	19.5	"
W/Mo	2,600	50	8.0	"
W/Re	3,200	50	3.4	"
Ir-20Re/Re-30Ir	2,600	40	11	Air Compatible
IR/Re-30Ir	2,400	35	15	"
W/Ir	2,400	20	41	Inert or Reducing
Mo/Ir	2,400	35	33	Atmosphere only
W/Pt	2,000	35	30	Atmosphere only
W/Rh-40Ir	2,100	35	27	"
Rh/Rh-8Re	2,000	--	7.4	Air Compatible
Pt-20Rh/Pt-40Rh	1,900	10	5	"
Pt-6Rh/Pt-30Rh	1,850	10	13.5	"
Rh/Pt-8Re	1,850	--	18	"
Pt/Pt-10Rh	1,800	3	19	"
Pt/Rh	1,800	15	30	"
Ir/Ir-40Rh	---	15	--	"

TABLE III - Comparison of Methods to Determine Temperature Profiles in Flame Fronts

Technique	Upper Temp. Limit (°K)	Spatial Resolution (cm)	Precision (K or T/T)	Displacement (cm)	Corrections	Effect in Flames	Cost of Apparatus
Thermocouples	3000	10 dia. (10^{-2})	1	5 dia. $(5x10^{-3})$	Radiation	Aerodynamic Wake & Catalysis	Moderate, low
Resistance Thermometer	3000	10^{-1}	1	10^{-2}	Radiation	"	Moderate
Aerodynamic Measurements	3500	10^{-3}	3%	slight	Acceleration, lag & thermo-mechanical effects	Quenching	Moderate
Optical Pyrometry	None	$5x10^{-1}$	5	None	Non-equilibrium Radiation	Additives	Moderate
Spectroscopic Line Intensity	None	$5x10^{-1}$	5	None	"	None	High
Pneumatic Probe	2500	10 dia. 10^{-2}	2%	5 .dia. $5x10^{-3}$	Orifice Coefficients	Wake & Catalysis	Low
X-ray Absorption	None	$5x10^{-1}$	3%	None	Mol. Weight	None	High
Interferometer	None	$5x10^{-2}$	1%	None	Mol. Weight	None	High
Inclined Slit	None	$5x10^{-2}$	1%	None	Mol. Weight	None	Low

ments are satisfactory for determination of derivatives, since the errors cancel.

Thermocouple thermometry is described in the literature and is discussed in many courses on electrical measurements. The techniques of fabrication of small noble metal couples and coating them with silica are described in Refs. 1 and 7.

A thermometer immersed in a gas stream will record a temperature differing from the true stream temperature due to kinetic energy transfer by stagnation in high velocity streams, conduction and radiation losses, and vibrational effects. These problems can be classified into two groups: the effects of the probe on the flame, and direct errors.

The central problem is the probe effects on the combustion system. This can be reduced by reducing size. This approach is limited by practical problems of fabrication or the heat transfer difficulties. Disturbances can be classified as aerodynamics, thermal, and chemical, and are discussed in some detail with respect to sampling probes in Refs. 1 and 7. The significant differences between the actions of probe thermometers and sampling probes can be summarized as follows.

The principal chemical disturbance of probes is the promotion of catalytic reactions on the thermometer surface which gives spuriously high temperatures and hysteresis. This is serious with metal surfaces, but it can usually be reduced by coating with non-catalytic materials, such as silica (Refs. 1, 7, 8).

The principal aerodynamic effect is the velocity deficient wake behind the thermometer which to a first approximation can be visualized as a local propagation of the flame front in this region.

Errors due to stagnation kinetic energy are negligible for combustion systems where the velocities lie below Mach 0.1. Conduction losses are small in most cases since the support wires can usually be aligned along isothermals.

Radiation is a major source of error. It is proportional to the fourth power of the temperature to the emissivity and inversely proportional to diameter (Eq. 1). These parameters are often not well known. One correction is based on the Nusselt-Reynolds Number correlation for cylinders:

$$\Delta T_{rad} = \frac{1.25\varepsilon\sigma T^4}{\lambda} d^{3/4} \left(\frac{n}{\rho\nu}\right)^{1/4} \tag{1}$$

Based on his measurements for quartz-coated wires, Kaskan suggests an $\varepsilon = 0.22$ (Ref. 8).

In this equation ε is the emissivity of the wire; σ is the Stephan-Bolzmann constant; λ is the thermal conductivity of the gas; d is the wire

diameter; and η is the viscosity of the gas. In these cases the effective
constant for a given thermometer can be determined by putting it in a gas
stream at a known temperature and measuring the resulting temperature.

B. Pneumatic Probe Measurements of Temperature - If the pressure
drop across an orifice is sufficiently high (pressure ratio >2.5) a sonic
surface forms in the throat and flow depends only on the upstream pressure,
temperature, molecular weight, and specific heat, with a minor Reynolds
number correction for the effects of boundary layer. If two orifices are
in series, the ratio of the pressure to the upstream pressure is given by
Eq. 2 (Refs. 1, 7, and 9):

$$T_1 = T_2(P_1/P_2)^2 \text{ K (Reynolds Number)} \qquad (2)$$

This provides a desirable method of temperature measurement since
it provides a connection between composition and temperature studies.
Calibration is required for quantitative work. It is not always necessary
to calibrate at high temperature environment, since Reynolds corrections
can be evaluated by changing density through molecular weight. This is
important since it is difficult to provide calibration temperatures above
1500°K. The orifices must operate in the continuum flow regime and the
radical and concentrations should not be high, since they recombine before
entering the second orifice, changing the molecular weight and ratio of
specific heats. It is convenient to make the first orifice a quartz probe
of the type used in composition sampling studies, the second orifice is
not critical.

It is desirable to minimize the volume between the orifices to
minimize equilibration time. Pressures can be measured by diaphragm
gauges or mercury manometers (See Fig. 3). McLeod gauges are not satisfactory because flames contain condensible gases.

IV. CONCENTRATION PROBES

The composition of combustion gases can be determined by probe sampling
and subsequent analysis.

Sampling probes can be divided into two categories: (1) Isokinetic
probes, which remove a sample at stream velocity; and (2) Sonic probes,
which remove the sample at sonic velocity. In the absence of reaction,
isokinetic probes collect flux, while sonic sampling collects local concentration. If the sample contains reacting gases, the reliability of the
sample depends on the rapidity of quenching. In isokinetic sampling,
quenching times are controlled by the ratio between stream velocity, thermal
conductivity, and reaction rate, which depend upon the probe diameter,
reaction rate, effective thermal conductivity, and the rate at which subsonic
gas stream can be accelerated without disturbing the sampled region. For

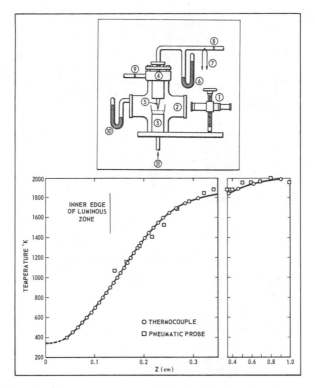

Fig. 3 - Temperature profiles using pneumatic
probe compared with thermocouple derived profile.
0.1 atm. CH_4 0.08- O_2- 0.92 flame (Ref. 7)

flames the required heat transfer rates are large so that isokinetic sampling
is used principally for slowly reacting systems, such as stack gases, or
very large systems, such as engines or furnaces. The principal advantage of
isokinetic sampling is that it samples flux, and the disturbance of two phase
flow is minimized. Thus, if one is interested in particulates, this type
of sampling is desirable.

    By contrast, sonic sampling radically disturbs the system in the region
of extraction, but offers the possibility of quenching rapid reactions.
Quenching time varies with orifice diameter. Quenching is accomplished by
adiabatic decompression, which simultaneously lowers pressure and temperature
of the sample. In most such systems the probe walls need not be cooled
because of the short residence time in the hot region of the probe.

    Samples can be taken in batches with sample bottles or introduced directly
into the analytical instrument through a continuous flow arrangement (Fig. 4).
Batch sampling allows analysis at leisure, but it is difficult to obtain
reliable analyses of absorbent species such as water. This can be minimized
by use of Teflon or polyethylene-lined sample bottles.

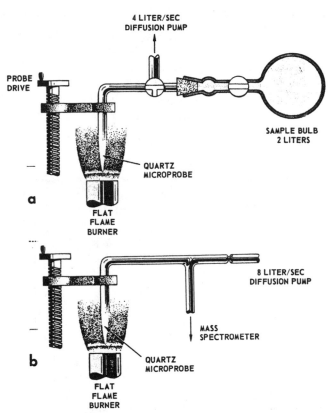

Fig. 4 - Batch and continuous flow
sampling of flames (Ref. 7)

Where absorption is a problem a continuous flow system is best. Absorbing
surfaces must ultimately come to equilibrium with the sample and the material
reaching the analytical instrument becomes identical with that entering the
probe. With Teflon lines only a few seconds were required to reach equilibrium
with a typical water ladened sample, while under comparable conditions glass
and metal systems required many minutes. One further precaution is necessary.
The system must be continuum flow throughout (i.e., tube diameters large
compared with the mean free path), and the pump must be isolated by a choking
orifice or by a capillary of sufficient length so that back diffusion from
the pump is negligible. This is necessary to avoid molecular separation which
occurs at low pressures. This would bias the sample and analysis. A typical
flow sampling system used in connection with a mass spectrometer is shown
in Figure 4.

The central problem in sampling combustion systems is to obtain a
representative sample and to interpret it either qualitatively or quantitatively in terms of the desired information. The withdrawal of sample should
either produce a quantitatively negligible disturbance of the system or produce

one which can be corrected. Quenching occurs through pressure and temperature drop due to expansion of the sample. The slowing of reaction is cumulative, and it can be seen intuitively that if the rate of pressure and temperature drop due to adiabatic expansion is rapid compared with the reaction rates, the sample composition will be quenched or "frozen." Bimolecular reactions as short as a few tens of microseconds should be frozen by probes. Water cooled probes at stream velocity can be unsatisfactory because of longer quench times and because flames are disturbed by bulky cooled surfaces. On the other hand, in engines where the scale is larger, such probes are very useful (Ref. 10). A recent bibliography of the field exists (Ref. 11).

A. Species in Combustion Systems - Combustion is usually associated with high temperatures, and steep temperature and concentration gradients. In such systems one finds not only reactants and products, but also intermediate and excited species such as vibrationally excited molecules, free radicals and atoms and ionized species (Table IV). Stable Species are those species which have lifetimes that are long compared with the sampling processes. The limiting time may range from a few milliseconds for fast flow sampling systems to hours or days for batch sampling. Most species with paired electron spins are stable but a few such molecules (e.g., O_3, H_2O_2, B_2H_6) are so reactive that they must be treated as transient species. Conversely, several radical species with unpaired spins are stable notably oxygen and the oxides of nitrogen and chlorine which can be treated experimentally as stable species.

Radicals and Atoms are important in combustion (Fig. 5) since the fuel and oxidizer do not react directly but are catalyzed through low activation energy paths involving radicals. A radical is a molecule (atoms are also considered molecules in this context) which has one or more unpaired electrons. It is not charged. In combustion, common examples are: $H\cdot$, O:, $OH\cdot$, and $CH_3\cdot\cdot$.

Because of their reactivity, particularly with walls, radicals are difficult to sample, but this can be accomplished in many cases (Refs. 23, 25, 26).

Ions are charged species which occur in low but non-equilibrium concentrations in combustion. As a result of chemi-ionization processes, in hydrocarbon flames the initial reaction is $O + CH \rightarrow CHO^+ + e^-$. Following this, other molecular ions are rapidly formed by ion-molecule reactions so that a great complexity of molecular ions are found in flames. (Refs. 12, 13) Relatively few molecules have stable levels for extra electrons; therefore, most of the observed ions are positive. Flames are neutral overall, and the major negatively charged species in flames is the electron. Special extraction techniques are required but since single charged particles can be detected, it is possible to measure the very low concentration of molecular ions in flames with satisfactory precision.

B. Data Interpretation - One is interested both in qualitative information, i.e. what species are present, and in quantitative analysis. Further, since combustion systems can have strong gradients, one is often interested in associating the analysis with a spatial position. Thus, the usual fruit of such studies is not a simple analysis, but a profile (see Fig. 5).

Often one wishes to deduce fluxes and rates of chemical reactions. This complex problem is discussed elsewhere (Ref. 1). Combustion systems contain

TABLE IV - Typical Species Distribution in a Premixed Laminar Flame

	Maximum Typical Concentration (mole fraction)	Examples
Stable Species	$10^{-1} - 10^{0}$	CH_4, O_2, H_2O
Atoms and Free Radicals	$10^{-1} - 10^{-2}$	$H\cdot$, $O:$, $OH\cdot$
Ions	$10^{-7} - 10^{-12}$	CHO^+, H_3O^+
Vibrational-Electronic	$10^{-5} -$	$HF*$

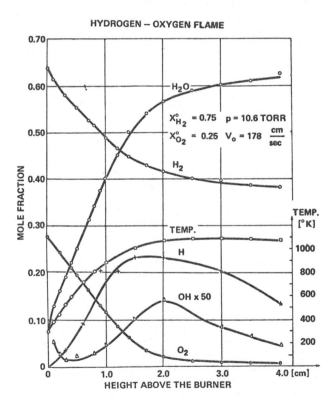

Fig. 5 - Composition profile of a low pressure hydrogen rich oxygen flame (Ref. 43)

298

steep gradients where substantial differences can occur between local concentration and local flux of a species (Fig. 6). Concentration is the amount of a species in a unit volume which is an inherently positive scalar quantity. Flux is the amount of material passing a unit area in a unit time which is a vector quantity and may be positive or negative. In the absence of concentration and temperature gradients these variables are numerically identical when expressed in dimensionless units (e.g., mole fraction and fractional molar flux).

In the simplest one-dimensional combustion system the reaction rate of a species is the spatial derivative of the flux vector (Fig. 6, Eq. 3).

$$R = d/dz(F) = d/dz[X(v + D/X \, dX/dz)] \qquad (3)*$$

To obtain rate data it is necessary to associate a composition with a position and temperature velocity as well as the first and second derivatives of the composition. In combustion systems where at atmospheric pressure the temperatures may range from 300 to 2000K and composition of a species passes from essentially zero to a maximum in a fraction of a milli-meter, this is difficult and often not possible.

C. Analytical Methods for Stable Species - Once a stable sample has been taken any convenient analytical technique can be used. The method of choice depends on the availability of equipment, and the complexity of the sample. The two most common methods have been mass spectrometry and gas chromatography, but spectroscopic methods such as IR and UV have also been used. These methods are discussed in standard texts.

Where the sample contains fewer than twelve species the method of choice is mass spectrometry because of its generality, sensitivity and rapidity. With more complex mixtures such as fuel rich combustion or polymer combustion, gas chromatography has the advantage of allowing the separation and analysis of complex mixtures. The combination of the two provides a very powerful tool for combustion studies. Spectroscopic methods are convenient for following certain species, such as CO which is difficult to determine in mass spectrometry or gas chromatography.

D. Unstable Species - Unstable species can be divided into two general categories; free radicals (i.e., unpaired electron species) and ions (i.e., charged species). Different experimental techniques are required for the two types. Unstable species are important in flame processes, but have not been studied as completely as stable species because of the difficulties involved. They are usually present only in low concentrations (10^{-2} - 10^{-8} mole fraction), and are too reactive for conventional sampling and analytical techniques.

---

*In this equation R is rate, moles/cm^3/sec; F is flux, moles/cm^2/sec; X is concentration, moles/cm^3; v is velocity cm/sec; z is distance (cm); D is the diffusion coefficient, cm^2/sec.

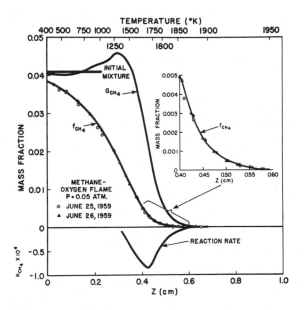

Fig. 6 - Concentration, flux and rate for
CH_4 in a 0.05 atm CH_4-0.08; O_2 - 0.92 flame (Ref. 7)

   1. Atoms and Free Radicals - Free radical species play an important role
in flame chemistry and these odd electron molecules enter into most flame
reactions. Most radicals are so reactive that they require special precautions
for sampling and analysis. This problem is not unique to flame studies.

   a. Calorimetric Methods - One classic method of determining atom concentrations is by calorimetry. Calorimetry has a number of advantages: (1)
the equipment is moderate in cost; (2) the method can be absolute; (3) good
spatial resolution can be attained using thermocouples or other probes.
There are certain serious disadvantages: (1) the method is not selective;
(2) the efficiencies of coatings, both catalytic and non-catalytic, are not
completely satisfactory; (3) calculation of the effective sampling region
for such a probe is difficult.

   In spite of these difficulties, these techniques in the form of a double
thermocouple have been used to study O atom concentrations (Ref. 14) and H
atoms (Ref. 15) and the method has been used by Rosner (Ref. 16) (Fig. 7)
in supersonic streams. This technique is satisfactory for simple chemistry.

   b. Emission Spectroscopy - Sugden and his co-workers have studied flame
radicals using the emission from traces of alkali metal salts as probes (Refs.
17 and 18). They have shown that the intensity of emission of the resonance
lines which are proportional to the concentration of free alkali metals can
be related to the concentrations of the radicals H and OH because of hydrides
and hydroxides existing in equilibrium with the radicals. This technique is
useful in regions where the metal-radical reactions are rapid compared with
the change in atom or radical concentrations.

   Another useful emission for radical studies in the "oxygen afterglow"
associated with the reaction $O + NO \rightarrow NO_2$. This emission is proportional
to the oxygen atom (and NO) concentration, since NO is regenerated rapidly
it can be considered to be constant. The emissivity provides a measure of O
atom concentration. This can provide a convenient measure of relative oxygen
atom concentration (Ref. 19).

   c. Exchange Methods - A number of elementary reactions are well enough
known that they can be used to estimate radical concentrations from isotopic
exchange rates. The most commonly used materials are deuterated compounds.
H and O concentrations can be inferred from the rates of reaction of D_2O and
N_2O (Refs. 1 and 20). It should be noted that a correction should be made
for the effect of deuterium substitution on the rate itself, since the rate
may be as much as 40% slower than the corresponding H reaction.

$$H + D_2O \rightarrow HD + OD \qquad\qquad (4)$$

$$N_2O + O \rightarrow 2NO \qquad\qquad (5)$$

   Since the concentrations of the deuterated compounds must be determined
by sampling and analysis (usually by mass spectrometry) some precautions
must be observed in avoiding wall exchange after sampling.

   d. Scavenger Probe Sampling - Radical concentrations can be determined
by combining microprobe sampling with chemical scavenging. This assumes that

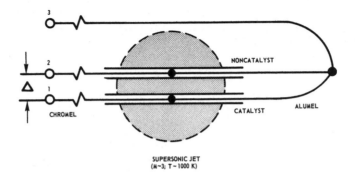

<u>Fig. 7</u> - Diagram of catalytic probe for determining atom concentrations (Refs. 1 and 7)

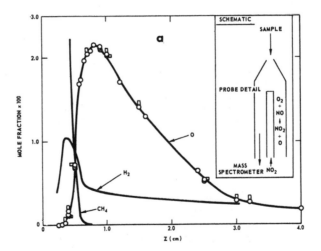

<u>Fig. 8</u> - O atom concentration in a methane-oxygen flame determined by scavenger probe techniques after Fristrom (Ref. 1)

after sampling by a microprobe radical concentrations are "frozen" sufficiently long for mixing with a reactant, a species which quantitatively produces an analyzable product. Two examples are the determination of oxygen atoms by the reaction $O + NO_2 \rightarrow NO + O_2$ and methyl by the reaction $CH_3 + I_2 \rightarrow CH_3I + I$.

The apparatus consists of a cooled quartz microprobe with provision for scavenger injection (Fig. 8).

   e. ESR Studies - Due to Seeman transitions in a magnetic field, many common radicals such as H, O, N, OH, halogen atoms, etc., can be detected with commercial spectrometers. This can be used for the measurement of absolute concentrations when calibrated against stable paramagnetic gases.

Electron spin resonance (ESR) has been utilized by allowing a flame to burn inside the resonant cavity of the spectrometer (Ref. 1). There are formidable problems of interpretation in this type of experiment. By combining probe sampling with ESR spectroscopy absolute atom concentration profiles were measured in flames with the apparatus shown in Figure 9. Gas samples withdrawn from the flame zone were pumped directly through the ESR detecting cavity. (Ref. 21) (Fig. 9).

   f. Molecular Beam Mass Spectrometry - For species which have a high surface reactivity collisionless flow inlet systems provide the only satisfactory inlet. Molecular beam inlet mass spectrometry was pioneered by Foner to establish the existence and identity of free radicals in flames and other reactive systems (Ref. 22). Two types of molecular flow inlet systems exist, the effusive and the supersonic. Effusive molecular beams are of low intensity and sample the boundary layer of a system. If wall processes are under study, are unimportant, or can be corrected for, this provides a satisfactory sampling system; otherwise, continuum sampling should be used. Continuum flow beams are intense but there are a number of problems. They are supersonic, the velocity distribution is narrow, and local temperature is low. (Fig. 10). Vibrationally and electronically excited states are frozen with the problems associated with cracking pattern changes.

Several problems are associated with molecular beam mass spectrometry of flames: (1) Mass separation by inlet flow; (2) change of cracking pattern with temperature due to changes in vibrational distributions; (3) polymer formation (Ref. 23).

For stable species microprobe sampling coupled with conventional analysis is usually quantitative except for strongly absorbed species. Molecular beam sampling is necessary for satisfactory sampling of such species. Free expansion produces separation due to Mach number focusing (Ref. 23). Interference of stable species with radicals can be reduced by lowering the electron beam energy below the threshold of ionization for stable species or by using magnetic separations. For trace molecular species the problem is more difficult. Calibration for expansion may be possible by combining information from a nonreactive trace molecule comparison with a knowledge of vibrational levels of the sample. Again if the species is a radical, problems can be reduced by lowering electron beam energy.

One of the major problems with molecular beam inlet mass spectrometry is that to form a satisfactory molecular beam with molecules which have made no wall collisions one must form a supersonic beam and skim out the center core. This can only be done by using a very wide angle sampling cone (>120°).

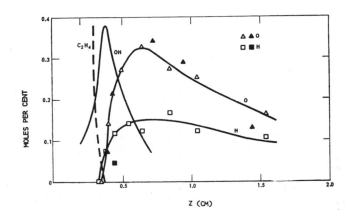

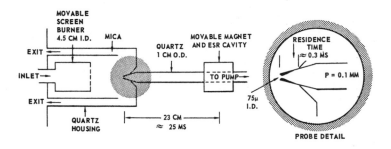

Fig. 9 - H and O atom profiles of ethylene-oxygen flames by probe sampling and ESR detection after Fristrom and Westenberg (Ref. 21)

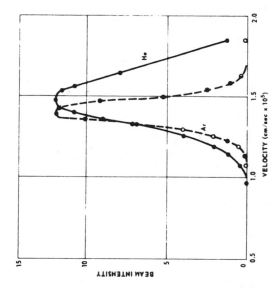

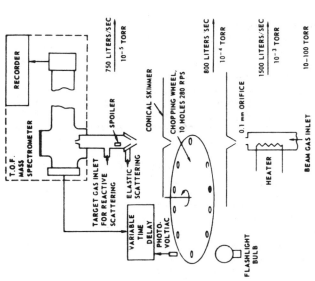

Fig. 10 - Velocity distribution of molecules in a supersonic molecular beam (Fristrom, p. 37 in Ref. 42)

Such a blunt probe has a strong perturbing effect on flames (Fig. 14).
The compromise which has usually been employed is about a 40° cone (Refs.
24 and 25). This does not visually disturb most flames and does allow beam
formation. Such a beam, however, contains many molecules which have made
wall collisions because of unfavorable aerodynamic configuration (Ref. 26).
This does not invalidate the analysis since the system is calibrated, however,
radicals which do not survive wall collisions may be lost. This problem
requires further study.

A mass spectrometer is not a primary analytical instrument, and for
precise work, standard samples must be used. Stable standards can be prepared,
but calibration can be a problem with strongly absorbed species such as water
and acids. The case of radical species is different and more difficult.
These species cannot be prepared as standard samples because of their reactivity.
Three techniques have been used: (1) Atoms can be prepared from their diatomic
parent by an electric discharge. Using a knowledge of the total pressure and
the cracking pattern of the parent species one can deduce the calibration
factor of the radical species, provided concentrations as high as a few percent
can be obtained; (2) A radical or atom can be titrated or scavenged in a flow
system and its concentration compared with that of a stable, known species;
and (3) One can look at an equilibrium system in which other species of the
equilibrium are known and deduce the sensitivity of the radical by difference
(Ref. 24). Since ion charges are known, ion sensitivities can be determined
directly provided the collection efficiency of the inlet system can be
determined.

2. Charged Species - The spatial distribution of charged species can be
measured by: (1) The Langmuir Probe, which measures dc resistance; (2) the
rf probe which measures energy dissipation in the microwave region; (3) the
photographic technique; and (4) the ion sampling mass spectrometer. The
first two techniques measure electron concentrations; the first and third
can measure either electrons or positive ions, but do not distinguish between
positive ions. The fourth technique allows the direct measurement of
individual positive ion concentrations. We will discuss the Langmuir Probe
and ion spectrometry. Discussions of the other two methods can be found
elsewhere (Ref. 1 and 7).

a. The Langmuir Probe - The Langmuir Probe was one of the earliest methods
for studying ion concentrations in flames. It is possible to measure ion or
electron concentration and effective electron temperature (Ref. 27). It consists of large area and small area electrodes (Fig. 11). At a given voltage,
current is limited by ions (or electrons) arrival at the small electrode. The
current is proportional to electrode area. If the small electrode is positive,
current is proportional to the electron concentration. If the small electrode
is negative the current is proportional to the positive ion current. The area
ratio between small and large electrodes must be very large to make the limiting
electrode positive, because of the high mobility of the electron. Complications
stem from the electrode size which affects the gradient and the plasma
potential which develops around an electrode immersed in a plasma. The
technique has been criticized because of the disturbance to the system being
studied; but with reasonable care, useful results can be obtained in systems
with spatial resolution which could be obtained by no other technique (Fig. 11).
The techniques are similar to polarography in electrolytes.

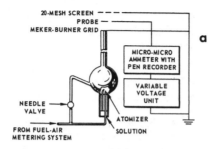

20-MESH SCREEN – – –
PROBE –––
MEKER-BURNER GRID

a

MICRO-MICRO
AMMETER WITH
PEN RECORDER

VARIABLE
VOLTAGE
UNIT

NEEDLE
VALVE

ATOMIZER

FROM FUEL-AIR
METERING SYSTEM

SOLUTION

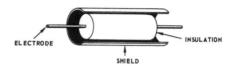

ELECTRODE

INSULATION

SHIELD

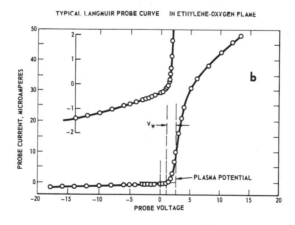

TYPICAL LANGMUIR PROBE CURVE   IN ETHYLENE-OXYGEN FLAME

V_w

PLASMA POTENTIAL

PROBE CURRENT, MICROAMPERES

PROBE VOLTAGE

b

Fig. 11 - The Langmuir probe technique for studying
ions and electrons in flames after Calcote (Ref. 13)

307

The energy from electric fields higher than a few megacycles is only
absorbed by free electrons because ionic particles are too massive to
respond. This method for studying electron concentrations has the advantage
of not disturbing the system. The disadvantages are low spatial resolution
and difficulties in determining exact path lengths and absorption coefficients.

b. Ion Mass Spectrometry - The best technique for identifying ions is
direct mass spectrometry. Reliable identifications can be made and quantitative
studies of ion concentration profiles are possible (Refs. 12 and 13).

The apparatus (Fig. 12) is similar to the conventional mass spectrometry
but no electron gun is used. A sampling orifice and a set of focusing electrodes are required. Considerable care must be devoted to the design of the
sampling inlet and pumping system. It is necessary to maintain low pressure
inside the spectrometer (mean free path large compared with the apparatus) to
avoid spurious ions.

V. APPLICATIONS

Probe sampling has been applied to a large number of combustion problems.
We will present several typical examples. Many more examples can be found in
the extensive combustion literature. Useful sources are the biannual International
Combustion Symposium Volumes, some fifteen of which have appeared in print (Ref.
28). The first ten volumes are indexed in Volume 10. Other sources are AGARD
publications, Old NACA reports, Combustion and Flame, Fuel, Fire Research
Abstracts and Reviews and other combustion journals.

A. Flame Sampling - Probing has been done extensively in the study of
laminar flames and the techniques are discussed in detail in Fristrom and
Westenberg (Ref. 1). There is a recent bibliography of the field (Ref. 11)
and there are several monographs (Refs. 20 and 29). A typical example of such
a study is given in Figure 5. Diffusion flames present a two or more dimensional
problem unless a symmetric system is analyzed. One such analysis is combustion
along the stagnation axis of a porous cylinder as in the example of Figure 13
(Ref. 30). Two dimensional diffusion flames have been studied qualitatively,
but we are unaware of any quantitative analyses.

B. Combustor Sampling - During the development of jet and rocket propulsion
following World War II, many combustion studies were made using probes. These
techniques are documented in Tine's survey (Ref. 10), the references previously
cited and a multitude of government reports such as the Ramjet Technology
Handbook (Ref. 31); the Princeton Series (Ref. 2); AGARD Publications (Ref. 32)
etc., many of which are still available. Two examples are illustrated in
Figure 14 (Ref. 33) using water cooled sonic probe and water cooled isokinetic
probes. Large water cooled probes are satisfactory for many combustor problems
because the rapid flow and high heat release make the disturbance offered
by the probe negligible. Problems connected with time variation in such samples
will be discussed by Billiger in the following paper in this symposium (Ref. 34).

C. Furnace Sampling - In the study of furnaces and low intensity combustors
sampling has also been done with probes of the water cooled variety both with
isokinetic sampling and sonic sampling. A discussion of furnace problems has
been given by Thring (Ref. 35). An example of multi-inlet probe used in furnace
studies is given in Figure 15.

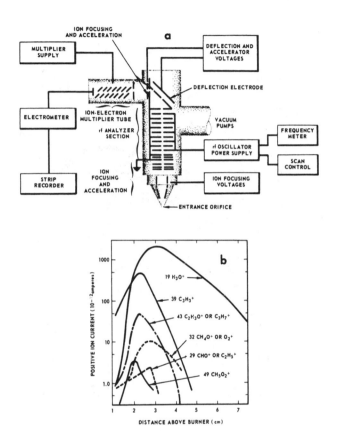

Fig. 12 - Determination of ion concentrations by mass spectrometry (after Calcote and King Ref. 13)

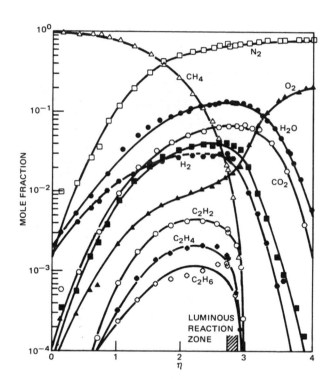

Fig. 13 - Composition profile along the stagnation axis of a cylindrical diffusion flame (Ref. 42)

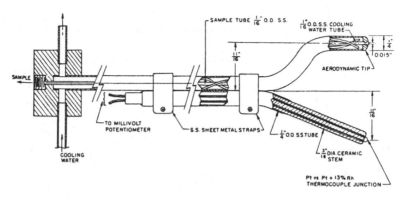

Fig. 14 - Probes for studying combustor performance
after Sawyer (Ref. 33)

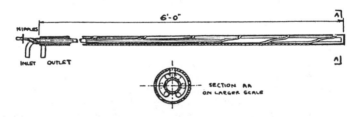

Fig. 15 - Multiple inlet water cooled probe for
furnace studies after Thring (Ref. 38)

    D. Rocket Sampling - High Pressure sampling presents many problems of
stress and high heat flux, but even in the case of a rocket chamber it has
been possible to sample using a supersonic inlet mass spectrometer (Ref. 36)
(Fig. 16).

    E. Supersonic Sampling - Sampling from a supersonic stream offers
special problems, because probes usually produce a bow shock which can alter
the sample. Special probes which swallow the shock have been used and
samples analyzed using gas chromatography (Ref. 37).

    F. Repetitive Phenomena - If a repetitive phenomena is reproducible
it is possible to follow both the time and space variation of the phenomena
by positioning the probe and varying the phase time of analysis. This has
been done in engines (Ref. 38) (Fig. 16) and in the study of spark ignition
(Ref. 39) (Fig. 17).

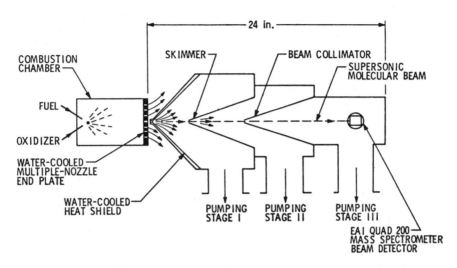

Fig. 16 - Molecular beam inlet sampling system for liquid
fuel rocket after Houseman and Young (Ref. 36)

   G. Condensed Phase Sampling - Since many combustion processes involve
condensed phase fuels, probing may be a useful technique for studying such
combustion processes. Several studies have addressed this problem, one
quenching the solid reaction by blowing out the flame with inert gas and
analyzing the solid by microtone sampling and Neutron activation analysis
(Ref. 40) (Fig. 18). The other used a low pressure liquid nitrogen probe
on a moving wire--analysis was by weight and wet chemistry. (Ref. 41
and 42) (Fig. 19).

                          VI. SUMMARY

   Probe sampling has been a versatile, useful tool in combustion problems.
It is a well established technique with an extensive literature. In the
future probing techniques particularly molecular beam inlet systems should
continue to be a valuable tool in combustion studies because of simplicity
and relatively low cost. They should be particularly useful when combined
with optical methods which can establish areas of applicability of probes.

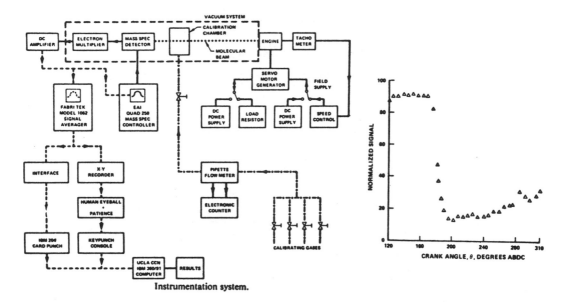

Instrumentation system.

Fig. 17a   Supersonic molecular beam inlet system for studying internal combustion engines (a) apparatus schematic (b) relative concentration of propane as a function of crank angle after Knuth (Ref. 38)

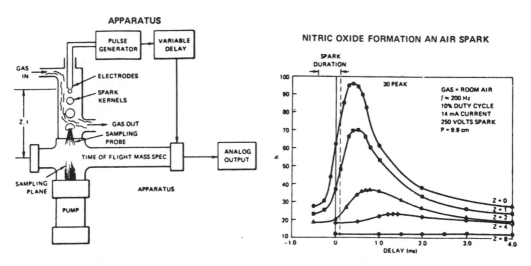

Fig. 17b   Spark ignition studies after Fristrom (Ref. 42)

313

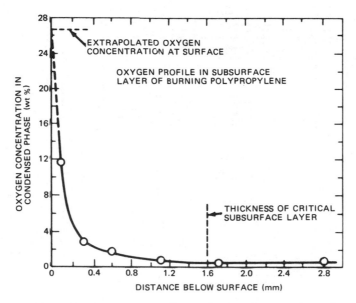

Fig. 18 - Microstructure of a polypropylene rod surface
burning in the candle mode after Stutz (Ref. 40)

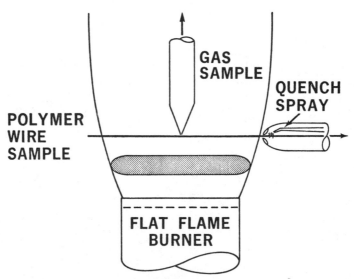

Fig. 19 - Apparatus for the study of the ignition of polymers
after Fristrom and Grunfelder (Ref. 44)

REFERENCES

1. R. M. Fristrom and A. A. Westenberg, Flame Structure, McGraw Hill (1965) p. 424.

2. B. Lewis, R. N. Pease, and H. S. Taylor, Physical Measurements in Gas Dynamics and Combustion, Princeton University Press, Princeton, NJ (1956).

3. N. Chigier, "Laser Doppler Velocimetry", This Symposium.

4. J. C. Quinn, "Laminar Flame Front Thickness", Harvard University, Combustion Aerodynamics Laboratory Report #5, May 1953.

5. A. A. Westenberg and R. E. Walker, "Absolute Low Speed Anemometer", Rev. Sci. Inst. 27 844 (1956).

6. V. Saners, "Thermocouple Measurements", Rev. Sci. Inst. 29 917 (1958).

7. R. Fristrom, "Experimental Techniques for the Study of Flame Structure", Bumblebee Report No. 300, Applied Physics Laboratory, The Johns Hopkins University, 8621 Georgia Avenue, Silver Spring, MD 20910 (1963) p. 187.

8. W. E. Kaskan, "The Dependence of Mass Burning Rate on Flame Temperature", Sixth Symposium on Combustion, Reinhold Publishing Co., New York (1957) p. 134.

9. D. W. Moore, "A Pneumatic Probe Method for Measuring High Temperature Gases", Aero. Eng. Rev. 7 #5 (1948).

10. G. Tine, Gas Sampling and Chemical Analysis in Combustion Processes, Agardograph Pergamon Press, New York (1961).

11. R. Fristrom, B. Kuvshinoff and M. Robison, "Bibliography of Flame Structure Studies", Fire Res. Abs. and Rev. 16 (1974).

12. J. Deckers and A. Van Tiggelen, "Ion Identification in Flames", Seventh Symposium on Combustion, Butterworths, London (1959) p. 254--.

13. H. Calcote, "Ion and Electron Profiles in Flames", Ninth Symposium on Combustion, William Wilkins and Co., Baltimore, MD (1963) p. 622.

14. A. Smeeton Leah and N. Carpenter, "The Estimation of Atomic Oxygen in Open Flames", Fourth Symposium on Combustion, William Wilkins and Co., Baltimore, MD (1953) p. 274.

15. L. Hart, C. Grunfelder and R. Fristrom, "The Point Source Using Upstream Sampling for Rate Constant Determination in Flame Gases", Combustion and Flame 23 (1974) p. 109.

16. D. E. Rosner, "The Theory of Differnetial Catalytic Probes for the Determination of Atom Concentrations in High Speed Non-Equilibrium Streams of Partially Dissociated Gases", A. R. S. 32 (1962) p. 1065.

17. M. Bulewizc, C. James and T. Sugden, "Photometric Investigation of Alkali Metals in Hydrogen Flame Gases in the Measurement of Atomic Concentrations", Proc. Roy. Soc. A 227 (1954) p. 312.

18. P. J. Padley and T. M. Sugden, "Chemiluminescence and Radical Recombination in Hydrogen Flames", Seventh Symposium on Combustion, Butterworths, London (1959) p. 235.

19. C. G. James and T. M. Sugden, "Use of NO-O_2 continuum in the Estimation of Relative Concentrations of O atoms in Flame Gases", Nature 175 (1955), p. 252.

20. C. Fenimore, The Chemistry of Premixed Flames, Pergamon Press, New York, (1964).

21. A. A. Westenberg and R. M. Fristrom, "H and O Atom Concentrations Measured by ESR in C(2) Hydrocarbon-Oxygen Flames", Tenth Symposium on Combustion, The Combustion Institute, Pittsburgh, PA (1965) p. 473.

22. S. Foner and R. Hudson, "The Detection of Atoms and Radicals in Flames by Mass Spectrometric Techniques", J. Chem. Phys. 21 (1954) p. 1374.

23. F. T. Greene (Editor), Molecular Beam Sampling Conference, Midwest Research Institute, 425 Volker Blvd., Kansan City, MO 64110 (1972).

24. J. Peters and G. Mahnen, "Reaction Mechanisms and Rate Constants of Elementary Steps in Methane-Oxygen Flames", Fourteenth Symposium on Combustion, The Combustion Institute (1972).

25. J. Biordi, C. Lazzara and J. Papp, "Molecular Beam Mass Spectrometry Applied to the Determination of the Kinetics of Reactions in Flames I Emperical Characterization of Flame Perturbation by Molecular Beam Sampling Probes", Combustion and Flame 23 (1974) p. 73.

26. J. Fenn, Private Communication.

27. H. Calcote and I. R. King, "Studies of Ionization in Flames by Means of Langmuir Probes", Fifth Symposium on Combustion, Reinhold Pub. Co., New York (1955) p. 423.

28. International Symposium on Combustion, Vols. 1-14, Biannual after Vol. 4, (1953), The Combustion Institute, Pittsburgh, PA.

29. A. Gaydon and H. Wolfhard, Flames, Chapman and Hall 3rd Ed. (1970) p. 392.

30. H. Tsiji and I. Yamaoka, "Structure of Counter Flow Diffusion Flames", Twelfth Symposium on Combustion, The Combustion Institute, Pittsburgh, PA, (1969) p. 997.

31. H. Kirk, "Facilities and Testing", Chapt. 13 of Ramjet Technology, available as TG 610-13, June 1968, Applied Physics Laboratory, The Johns Hopkins University, 8621 Georgia Ave., Silver Spring, MD 20910, also in Microfiche from NTIS as PB 179067 (June, 1968).

32. J. Surugue (Editor), Experimental Methods in Combustion Research, Agardograph Pergamon Press, New York (1961).

33. R. Sawyer, "Experimental Studies in a Model Gas Turbine", Emissions, (W. Cornelius and W. Agnew, Editors) Plenum Press, New York (1972) p. 243.

34. R. Billiger, "Probes in Turbulent Systems", This Symposium.

35. M. W. Thring, The Science of Flames and Furnaces, J. Wiley and Sons, Inc., New York (1952) p. 416.

36. J. Jouseman and W. Young, "Molecular Beam Sampling System for Rocket Combustion Chambers", p. 70 of Ref. 23.

37. R. Orth, F. Billig, and S. Grenleski, presented in the Symposium on Instrumentation for Air Breathing Propulsion Spet, 1972, to be published in Prog. Astronaut Aeronaut.

38. E. Knuth, "Direct Sampling Studies of Combustion Processes", a chapter in Engine Emissions Polutant Formation and Measurement, G. Springer and D. Patterson (Editors), Plenum Press, New York (1973).

39. R. Fristrom, "Flame Sampling for Mass Spectrometry", Int. J. for Mass Spec. and Ion Physics 16 (1975) p. 15.

40. D. Steutz, "Basic Principles in Polymer Combustion", Paper #3, "Flammability Characteristics of Polymeric Materials", Symposium Univ. of Utah, Flammability Research Center, Salt Lake City, Utah, June 1971.

41. R. Fristrom, "Chemistry, Combustion and Flammability", Journal on Fire and Flammability 5 (1974), p. 289.

42. R. Fristrom, "Fire and Flame Studies Utilizing Molecular Beam Sampling", p. 55 in Ref. 13.

43. K. Hoyermann, et al, Thirteenth Symposium on Combustion (1971).

NEW RESULTS OF STUDIES OF COMBUSTION FLOWS OBTAINED AT THE ONERA

ROLAND BORGHI

Office National d'Etudes
et de Recherches Aerospatiales (ONERA)

ABSTRACT

This paper is a brief summary of experimental studies now in progress at the ONERA. These experiments concern not only laminar flames but also turbulent ones. Conventional techniques of gas sampling have been extensively applied, in order to point out certain characteristics of chemical reactions in flames, and particularly of pollution reactions.

I

The first study has been performed in a well stirred reactor, as used many years ago by Longwell and Weiss (Ref. 1). Fig. 1 shows this reactor; only the envelope of the cylindrical reactor is seen; the walls inside are made of silica; the injection of premixed fuel and air is made by a central tube with many small holes. In this type of combustor, if the inlet mass flow rate varies, the mean residence time inside the reactor changes, and consequently different rates of combustion are obtained. Then, by concentration measurements for different species inside the reactor, the different rates of species generation or removal are very easy to calculate.

The probes were small (internal diameter: 1.2 mm). They were water cooled and made of stainless steel or silica. The samples were driven into a low pressure system, then pressurized and analysed after the test. Fig. 2 shows the concentration measurements of O_2, CO_2, C_3H_8 (the fuel) and CO, as a function of mean residence time (t_S). From the curves, the reaction rates have

<u>Fig. 1</u> - The cylindrical well-stirred reactor.

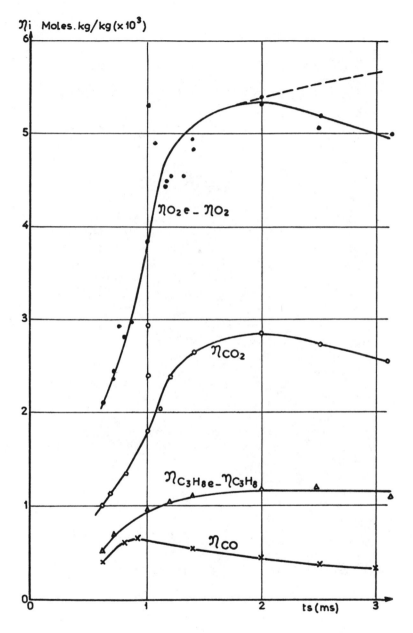

Fig. 2 - Sampling and analysis by gas chromatography.

320

been calculated. We have found that the rate of creation of CO_2 from CO follows the well known Hottel and Williams law (Ref. 2). An activation temperature for the global reaction (propane + oxygen→CO) was calculated. This value was lower than expected, approximately 7,500 - 8,000°K.

Fig. 3 illustrates the rate of production of the nitrogen oxides NO and NO_2 in the same combustor (chemiluminescence techniques). An interesting point is that NO_2 seems to be created before NO. The question, of course, is whether NO_2 is really present in the combustor or if it was created in the probe itself; after further experiments, we feel that NO_2, or at least most of it, was not created in the probe; it seems that NO_2 was produced without preliminary production of NO. However, if the mean residence time is sufficiently high, NO_2 makes room to NO in the reactor.

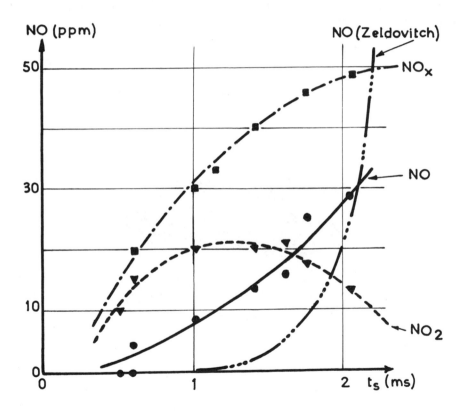

Fig. 3 - Chemluminescence measurements of nitrogen oxides in the combustor as a function of time t_s.

II

Another type of experimental research is the study of the development of a turbulent flame; a premixed air-methane flow is ignited by a pilot burner giving hot gases (2000°K). This type of flame is shown on Fig. 4; in this case the equivalence ratio of the fresh mixture was about 1. With the same kind of water cooled probes, we have obtained several concentration measurements; in this case the sampling was operated at atmospheric pressure, by pumping into a small cavity and the analysis was made by gas chromatography later on.

Fig. 5 shows the isoconcentration curves in the flame, for CH_4, CO and CO_2; the concentrations are mass fractions in percent of the mass of nitrogen (which is measured directly), and not of the total mass. It appears that CO is first produced by the combustion of CH_4, and later disappears, giving CO_2.

Fig. 6 shows the same concentrations in ppm of NO and NO_2. We note the same phenomenon as in the well stirred reactor: NO_2 is created before NO and is destroyed later. This particular aspect is not fully understood as yet.

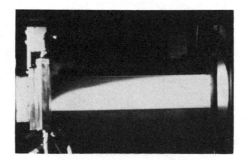

Fig. 4 - Turbulent premixed flame experiment.

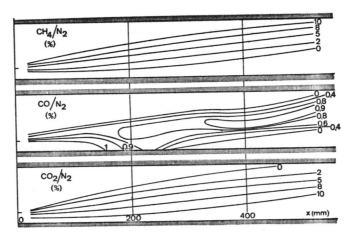

Fig. 5 - Methane-air combustion ($\varphi = 0.8$) - main components.

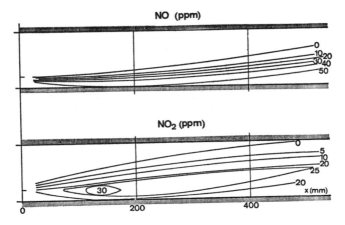

Fig. 6 - Methane-air combustion ($\varphi = 0.8$) - nitrogen oxides.

III

At the present time, a theoretical assessment of these experimental results is in progress; our results will be reported at the 2nd European Symposium of Combustion (Refs. 3 and 4).

REFERENCES

1. J. P. Longwell, M. A. Weiss, "High Temperature Reaction Rates in Hydro-
    carbons Combustion," IEC, Vol. 47 n°8 (1955).

2. H. C. Hottel, G. C. Williams, "Kinetic Studies in Stirred Reactors:
    Combustion of Carbon Monoxide and Propane," 10th Symposium (Int'l) of
    Combustion (1965).

3. J. Duterque, "Production du NO et NO_2 dans un foyer homogène," 2ème
    Symposium European de Combustion - Orléans (Sept. 1975).

4. P. Moreau, J. C. Bonniot, "Production des Oxydes d'Azote dans une Flamme
    Turbulente," 2ème Symposium European de Combustion - Orléans (Sept. 1975).

IN SITU OPTICAL VERSUS PROBE SAMPLING MEASUREMENT OF NO CONCENTRATION IN JET ENGINE EXHAUST

W. K. McGREGOR
ARO, Inc.

I wish to report briefly on a study that has been made recently by two of my colleagues and I at the Arnold Engineering Development Center (AEDC). Jim Few and Bob Bryson have been engaged in a program to investigate the source of a discrepancy found during the pollutant emissions measurement program on the GE-YJ-93 jet engine conducted at AEDC for the Department of Transportation's Climatic Impact Assessment Program (CIAP) in 1971. In that program we were asked to make measurements of the hydroxyl (OH) concentration in the exhaust of the engine. Since OH is a free radical and does not survive probe sampling, the measurement must be made in situ, requiring an optical technique. Previous work had been done in the laboratory in developing narrow line UV absorption techniques for both OH (Ref. 1) and nitric oxide (NO) (Ref. 2) concentration measurements. Since the same apparatus can be used for both OH an NO measurements, it was agreed that measurements on both species would be made. The full results of these measurements can be found in the Proceedings of the Second CIAP Conference (Ref. 3). This brief report is only concerned with the measurements of the NO concentration.

The measurement technique used was quite straightforward. A resonance lamp was placed on one side of the engine exhaust and a spectrometer on the other, as shown in Fig. 1. The source for OH resonance radiation utilized a gas flow of argon bubbled through water and excited in the dc gas discharge tube at about 5 torr pressure; for NO a 12:3:1 mixture of $Ar:N_2:O_2$ was found to give a good γ-band structure with no detectable impurity lines. The receiver was a 1/2-meter grating monochromator for the field measurements. The transmission measurement for NO was made at the peak of the second bandhead of the 0,0 γ-band.

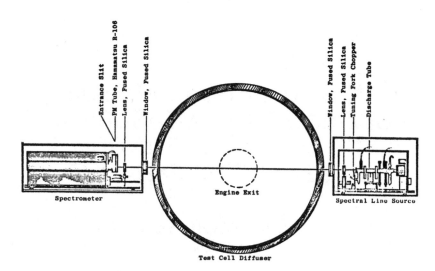

Fig. 1 - Diagram of installation of spectral absorption apparatus in propulsion development test cell (J-2) for OH and NO concentration measurements in exhaust of GE-YJ-93 engine.

Flight Condition		Absorption Measurement	Sampling Measurement
M	h(kft)		
1.4	35	75	56
2	55	165	70
		175	80
		278	110
2.6	66	323	100
		617	130

Military Power
Minimum Afterburner Power
Maximum Afterburner Power

Fig. 2 - Average NO concentration in the exhaust of a YJ93-GE-3 engine; comparison of values measured in situ by UV absorption and measured by sampling.

The engine was operated under simulated high altitude flight conditions over a range of power throttle settings, including afterburning.
A probe sampling system and a full set of gas analysis instrumentation
was used to make detailed maps of pollutant emission products, including
NO. A chemiluminescent analyzer, commercially made, was used to determine
the NO concentration in the sample. The absorption measurements were related to concentration of NO through a room temperature laboratory calibration after making corrections for Doppler broadening due to the much
higher temperature of the engine exhaust. The results of the probe sampling and in situ optical measurement are shown in Fig. 2. Note that the
average density of NO across the exhaust does not change very much with
engine power setting. However, because of the temperature increase, the
concentration as shown in the column at the right of the graph (in parts
per million, ppm) increases considerably during afterburner operation.
Note from the other column that the NO concentration as determined by the
sampling method is less than that determined by the optical method by factors of 2 to 5.

The above result received little attention because it was unsupported
at the time. A program was subsequently initiated at AEDC to confirm or
disprove the discrepancy between sampling and optical methods. A jet engine combustor system was assembled and an experiment configured as shown
in Fig. 3 so that both probe sampling and optical measurements could be
made in a combustor exhaust stream under controlled conditions. In addition,
absorption cells were placed in the sample transfer line near the probe and
also near the analyzer.

Work was also done in the laboratory on the absorption characteristics
of the NO γ-bands at elevated temperatures and at higher pressures than encountered in the high altitude engine test. From this work using a heated
absorption cell it became clear that a theoretical model was required to
extend the calibration beyond the limitations of the absorption cell ($\sim 1000°$
F). Such a model was developed and applied both to lower pressure situations where Doppler broadening dominates (Ref. 4) and to higher pressures
where collisional broadening is dominant. In fact, the model was used in
conjunction with the experiment to determine improved collisional broadening parameters (Ref. 5).

The result of a large number of measurements in the combustor exhaust
over a range of fuel-to-air ratios is given in Fig. 4. These results show
two important things:

1.   The in situ optical method gives NO concentrations much larger
than probe sampling with the discrepancy increasing with the fuel-to-air
ratio.

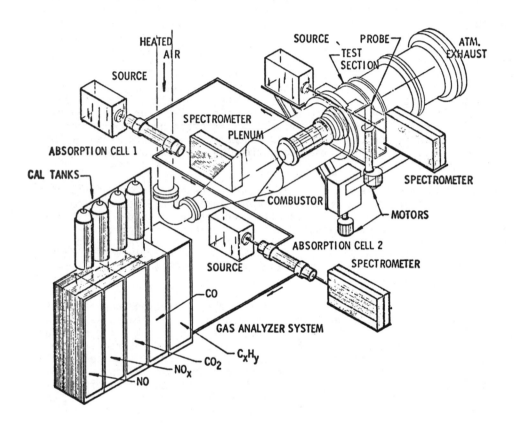

HEATED AIR

SOURCE

SOURCE

PROBE

ATM. EXHAUST

TEST SECTION

SPECTROMETER

PLENUM

ABSORPTION CELL 1

CAL TANKS

SPECTROMETER

COMBUSTOR

MOTORS

ABSORPTION CELL 2

SOURCE

SPECTROMETER

CO

GAS ANALYZER SYSTEM

C_xH_y

CO_2

NO_x

NO

Fig. 3 - Sketch of experimental apparatus used for evaluation of exhaust emissions measurements methods.

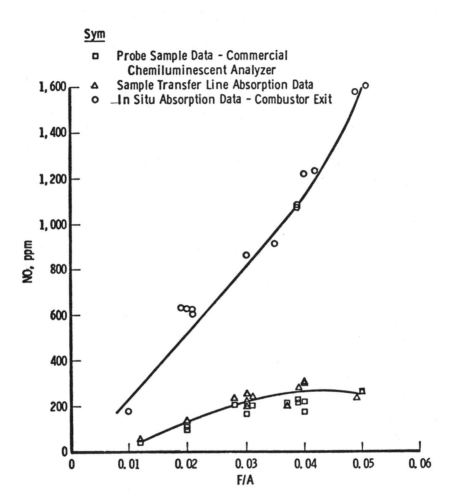

Fig. 4 - NO Concentration as a function of the fuel-to-air ratio for turbine engine combustor obtained by (1) gas analyzer measurement of probe sample, (2) resonance absorption measurement of probe sample, and (3) in situ resonance absorption at combustor exhaust plane.

2.   The values of NO concentrations within the sample transfer line as determined by the commercial chemiluminescent analyzer or the UV absorption through the absorption cell are substantially in agreement.

Thus, the results of the engine test (Ref. 3) have been substantiated. However, the source of the discrepancy is still at large.

REFERENCES

1.   Davis, M. G., McGregor, W. K., and Mason, A. A. "Determination of the Excitation Reaction of the OH Radical in H_2-O_2 Combustion." AFDC-TR-69-95 (AD695471), October, 1969.

2.   McGregor, W. K., Few, J. D., and Litton, C. D. "Resonance Line Absorption Method for Determination of Nitric Oxide Concentration." AEDC-TR-73-182 (AD771642), December, 1973.

3.   McGregor, W. K., Seiber, B. L., and Few, J. D. "Concentration of OH an NO in YJ93-GE-3 Engine Exhausts Measured in situ by Narrow-Line UV Absorption." Proceedings of the Second Conference on the Climatic Impact Assessment Program, Cambridge, Massachusetts, November, 1972.

4.   Davis, M. G., McGregor, W. K., and Few, J. D. "Spectral Simulation of Resonance Band Transmission Profiles for Species Concentration Measurements: NO γ-Bands as an Example." AEDC-TR-74-124, January, 1975.

5.   Davis, M. G., McGregor, W. K., and Few, J. D. "Transmission of Doppler Broadened Resonance Radiation Through Absorbing Media with Combined Doppler and Pressure Broadening (Nitric Oxide γ-Bands as an Example)." ETF-TR-75-115, June, 1975.

DISCISSION

DREWRY, ARL - Are you certain that you obtained an abrupt enough expansion in your probe?

MCGREGOR, ARO - The quick quench probe we used had - in principle - the very fast expansion necessary to "freeze" NO. It did not seem to make any difference in the end.

BILGER, University of Sidney - Did you check for NO_2?

MCGREGOR - Yes. Essentially we get a very small increase in total NO_x

concentration over the NO measurement in this case.

ROQUEMORE, AFAPL - I would like to mention briefly another program which involves in situ optical measurements of NO in a nonreactive turbojet engine exhaust. Basically, the program involves making NO measurements with two different types of optical instruments and comparing the results with probe measurements. The two optical methods are: (1) the UV absorption technique developed by McGregor and Few and (2) an IR absorption gas filter correlation technique developed by EPA for measuring smoke stack emissions. Measurements with both of these optical system in conjunction with probe measurements have been made at AFAPL in the low temperature (1200°F), non-reactive exhaust generated by a T56 jet engine combustor. The data from this program are presently being analyzed. The results will be reported in a session on aircraft emissions at the AIAA 14th Aerospace Sciences Meeting in Washington, DC in January 1976. [AIAA Papers 76-107 through -112].

MELLOR, Purdue University - I was a little confused by what you call a probe; is that a point probe or a rake? Is the optical method a line-average?

MCGREGOR - We have made detailed point samples and taken profiles across the stream by optical scanning. In this particular flow, we get essentially a uniform concentration profile except very near the edge of the flow. Then we integrate through the stream in three or four different ways.

FLAGAN, MIT - Did you correct for oxygen absorption?

MCGREGOR - We did not correct for oxygen absorption. What convinced us that there is no oxygen absorption is that we looked at the 1-1 or 2-2 bands of NO and there is essentially no absorption at low temperature. If oxygen had been there, it would have been present across the spectrum in the 1-1 band as well as in the 0-0 band we are using. We've also made measurements using a hydrogen lamp at 2400 angstroms and down at 2000 angstroms, with essentially zero absorption. You would expect the oxygen to absorb, or NO_2, or whatever else might be there to absorb it at those wavelengths.

GOULDIN, Cornell UNiversity - You're taking time and space averaged concentration and temperature measurements of a highly fluctuating flow. Could that be responsible for the difference?

MCGREGOR - That could account for a 20%, perhaps even a 50% difference. This is a combustion flow with a fairly completed combustion. I don't think there is any afterburning or any additional reactions. It is turbulent and there are corrections to be made, but not to the extent that would be needed to bring the probe sampling in agreement with the absorption measurements.

BIRKLAND, ARL - The data seem to indicate that this discrepancy is happening just at the tip of the probe, since the optical checks further down the sample line agree with the chemical analysis. Is anything happening in the decompression region right in front of the probe?

FRISTROM, APL/JHU - These are rather massive probes, and they are water cooled. I think it is conceivable that there is a disturbance associated with them. I am not sure about the stream velocities, but if you put a massive water cooled sink, you do get disturbances indeed. The other point that might be considered is that NO and NO_x are bad actors as far as pumping is concerned and one does have to be careful that the line is well flushed out. However, I presume he has taken care of this problem. My speculation would be that this is probably an interaction of the probe with the system.

PROBE MEASUREMENTS
IN TURBULENT COMBUSTION

R. W. BILGER
University of Sydney

SUMMARY

Measurements of composition, temperature, velocity and turbulence using hardware probes are reviewed. Probe interference effects can be made negligible in streaming flows but much more attention is needed to minimize this effect in recirculating and strongly swirling flows. The strong fluctuations in density present in turbulent combustion can lead to sampling errors. Velocities and species concentrations will actually be measured as the Favre averaged values. Measurements of NO and NO_2 are at best qualitative and a concerted effort is required to determine the phenomenology affecting these measurements.

I. INTRODUCTION

We shall be concerned here with the widely used technique for making composition, temperature, velocity and turbulence measurements in combustion flows which involve the introduction of a piece of hardware, a probe, into the flow.

The use of such a probe always raises two major questions:

    A. Does the probe interfere significantly with the combustion flow field?

    B. Does the measurement that we make or the sample that we collect truly represent what we expect or hope it does?

333

Such questions are complex enough in steady laminar flows but in turbulent flows they are made even less tractable by the complexities of the flow field and the effects that non-linearities can have on the averaging process. It is the hope, of course, that the new optical techniques will eliminate interferences of type A but it appears that they bring with them a whole lot of new problems of type B.

We are asked "to be as quantitative as possible in our performance assessments" so that the meeting can make a "comparison between the potentials of different measuring techniques." In large measure it is the uncertainties involved with probe measurements that have stimulated the development of the new optical techniques. This is not because we know that the error bars are plus or minus 40 percent and that this is not good enough, but rather that we are uncertain about how big the error bars should be. Error reduction and assessment in any measurement technique involves the following:

(a) Identification of all the phenomena that could significantly interfere with the measurement.
(b) Quantification of the effect of each phenomenon.
(c) Optimization of the variables of the measurement technique to reduce the total uncertainty in the measurement.
(d) Reconciliation of redundant measurements.

For probe measurements in turbulent combustion, we are getting a long but probably incomplete list under (a), have a lot of gaps in (b) and in consequence do a fairly rough job with (c). Most of the confidence we have in the data we obtain comes from mass and energy balances and other reconciliations (d).

Probe techniques are inherently simple and easy to use and certainly a lot less expensive than some of the newer methods. For these reasons they will and should continue to be used. Part of the role of the newer measurement techniques will be to assist in the better quantification of the errors associated with probe measurements. This review attempts to catalogue our present understanding of the phenomena involved and what can be said about the levels of accuracy.

II. PROBE INTERFERENCE

The presence of the probe can alter the flow field and the combustion patterns so that measurements are not being made of the combustion system it was intended to investigate. The disturbances can be largely fluid mechanic in nature or largely thermal or chemical but in general will be all three.

Fluid mechanic disturbances are likely to be most serious when probing recirculation zones and swirling flows. It is a well-known observation in fluid mechanics that the introduction of a probe radially into a vortex flow in a vessel can have a drastic effect on the flow. Fluid from the boundary layer on the walls flows radially inward along the probe and is deposited into the center of the vortex. The large radial gradients in static pressure are responsible and the probe provides a corridor of communication. What a good way of mixing fuel and air!

Strong dynamic pressure gradients such as are present in a jet can also have an effect. Any probe presents a blockage to the flow and the streamlines must bend to go around the probe. In fact they bend more easily into the regions with low dynamic pressure. This can give rise to a displacement effect on the measurement position.

That the probe can act as a flameholder must be an observation experienced by many people. In spray combustors probes can act as collectors, agglomerators, atomizers, vaporizers and generally as fuel redistributors. A large cooled probe can also be a significant heat sink.

In streaming flows, where no flow reversals occur and where distrubances from downstream have little effect on the flow upstream, probe disturbances to the measurements can be kept to a minimum. Such flows are mathematically described by parabolic equations. The flow disturbance is confined to the region up to about seven effective diameters upstream of the probe or its support. A common arrangement is in the form of a right-angled pitot tube. In this case the probe should project ahead of the support tube by at least seven support tube diameters. For static pressure measurements the distance should be even further. The business end of the probe should be as slender as possible. Position error effects due to stream-tube displacement in a dynamic pressure gradient will be of the order of 0.5 times the diameter of the equivalent square nosed cylinder.

In mathematically elliptic flows, such as recirculation regions, disturbances from the probe will always propagate upstream. Although the problem is obvious it has not been addressed in the literature to my knowledge, not even to the extent of saying that visual observations indicated no effect! Of course up to now we have been pleased to have any sort of measurements in such regions. But now the numerical modellers are saying: "Of course it is the experiments which are in error." I suspect that in many cases they are not too much in error but the experimenters must give us more evidence on which to make that judgement. Some measurements with and without the probe are needed: photographs; wall static pressure or temperature; even measurements from a second probe judiciously placed in the flow.

III. SAMPLING

A. Types of Probes in Use

Tiné (1961) gives us an excellent review of probe designs then in use. He also proposed a design for a water-cooled isokinetic probe of the pitot type. Fig. 1 from Kent and Bilger (1973) shows two physical realizations of this principle. They differ in the diameter of the sampling tube and the bluntness of the tip. They are cooled with hot water to ensure quenching of the sample without condensation of water vapor. The internal static tap is used to ensure isokinetic sampling (see below). Other realizations of the isokinetic principle have been used by Vranos et al. (1969) and Lenze (1970).

Many other probe designs in current use emphasize the use of sonic quenching rather than the principle of isokinetic sampling. Sawyer (1972) describes a probe having a hemispherical head with a small convergent-divergent orifice through which the sample is extracted. Schefer et al. (1973) discuss four different quartz and stainless steel probes all employing "aerodynamic quenching" with one of each type also water quenched. Lavoie and Schlader (1974) used a hemispherical head probe with a small sonic orifice. Orifice diameters range from 0.1 to 0.3 mm with outside probe diameter ranging from 4 to 10 mm for water cooled probes.

Mellor et al. (1972) describe a square-nosed probe with a converging sample orifice which has been used successfully in the sooty and droplet laden conditions inside a gas turbine combustor. The smoothly monotonic entry allows purging by blow back with high pressure nitrogen.

Some use is also made of quartz microprobes (Fristrom and Westenberg, 1965) in turbulent flows (Syred, 1975).

B. Aerodynamics of Sample Collection

Tiné (1961) gives a good discussion of the problems associated with sampling in non-homotropic gas flows, that is flows which are not uniform in density. In combustion density variations in the gas can be as high as a ratio of 10 or even 15 to one and these variations can occur over relatively small distances. For a probe sampling isokinetically, that is with sampling velocity equal to the mean stream velocity, there will be no segregation of light and heavy patches in the fluid. At non-isokinetic sampling conditions and for inhomogeneities small compared to the probe mouth there will be segregation effects similar to those found in sampling a particulate laden gas stream: low sampling velocity gives a bias toward the heavy components, high sampling velocity gives a bias toward the light components. Leeper (1954) has shown that the situation is further dependent on the acoustic impedance of the sample line. Where the impedance is not high and strongly resistive, oscillations will be driven in the probe mouth and sampling errors will occur even with a probe small compared with the scale of the inhomogeneities in the gas. Phase differences can be large and the bias can in fact go opposite to that expected by the simple rule of thumb quoted above, i.e., low sample velocity could give bias toward light components.

For near isokinetic sampling Kent and Bilger (1973) found that variations in sampling velocity equivalent to a variation in internal static pressure of one dynamic head gave rise to changes in H_2O concentrations of somewhat less

than one mole per cent (in about 20 mole per cent) and these results are in
agreement with those of Vranos et al. (1969) and Lenze (1970). Much more significant errors were found (Kent and Bilger, 1973; Lenze, 1970) to depend on
the shape of the probe head. The blunt nosed probe of Figure 1 was found to
give samples which correspond to a radial shift of about 3 mm in the direction
of low dynamic pressure as compared with the slender-nosed probe. The tests
were made in a turbulent jet diffusion flame of hydrogen in a co-flowing stream
of air. Lenze found similar results for turbulent jet diffusion flames of city
gas into still air where the dynamic head gradients are of opposite sign. In
each case the errors were small on the centerline and so the results cannot be
explained in terms of low or high effective sampling velocity; or in terms of
quench rate.

The phenomenology of sample collection with sonic orifice probes is not
well understood. According to the theory developed for isokinetic sampling
the bias toward the light fractions should be extremely large. However, this
will be offset by the effects of the blunt nose shape which in some cases may
dominate the situation giving samples biased toward the heavy fractions. The
sampling velocity also varies with the sound speed of the gas temporarily at
the choke plane in the orifice. For quartz microprobes this may be the dominant effect when the probe tip outside diameter is of the order of the Kolmogoroff microscale or smaller.

C.   Sample Averaging

Even if the sampling is unbiased in terms of collecting a representative
sample of the gas volumes sweeping past the probe the resulting composition
analyzed at room temperature will not represent the true time mean composition.
For isokinetic sample collection the composition analyzed should be the Favre
mean (Favre, 1969; Bilger, 1975):

$$Y_{i,s} = \overline{\rho\, U_s\, Y_i} \,/\, \overline{\rho\, U_s} \;=\; \overline{\rho\, Y_i} \,/\, \overline{\rho} \equiv \tilde{Y}_i \tag{1}$$

where Y_i is the mass fraction of species i and $Y_{i,s}$ is the sample value as
analyzed, ρ the density and U_s the sampling velocity; the superscripted tilde
denotes a Favre average defined as shown or more formally in terms of the
Favre probability density function (p.d.f.) (Bilger, 1975). The sampling
velocity U_s can be assumed constant provided the probe and sampling line have
a high and largely resistive acoustic impedance (see above).

For sonic orifice sampling and no aerodynamic bias (microprobes only?)
the composition as analyzed will be

$$Y_{i,s} = \overline{C_D\, \rho^*\, c^*\, Y_i} \,/\, \overline{C_D\, \rho^*\, c^*} \tag{2}$$

which in general falls somewhere between the true time mean and the Favre mean. Here c* and ρ* are the sound speed and density at the throat and C_D the discharge coefficient. Throat Reynolds numbers are generally of the order of 100 so Reynolds numbers effects on C_D can be appreciable.

Most theoretical methods will in fact predict the Favre mean (Bilger, 1975). The difference between the Favre mean and the true time mean can easily be as much as 40 per cent in some parts of the flow.

D. Quenching

Two basic methods are in use: the aerodynamic type and the convective type. Initial quenching rates claimed for aerodynamic quenching vary from about 10^8 °K/s for microprobes to 2 x 10^6 °K/s for the Sawyer (1972) convergent-divergent orifice probe. In order to sustain this quench it is necessary for supersonic flow or free molecule flow to be maintained downstream of the orifice. It is apparent that in many cases that this is not achieved and that the flow is shocked back up to stagnation temperature just downstream of the orifice. If the orifice itself is effectively cooled the quenching rates may still be very high.

Tiné (1961) discusses the calculation of quench rates for convective quenching. Bilger and Beck (1975) estimate an initial quench rate of 2 x 10^6 °K/s for the small probe shown in Fig. 1 with quenching to below 1000°K in 1 ms.

In turbulent combustion the quenching problem is probably not as severe as it is in laminar flames. For turbulent diffusion flames the rate of reaction is controlled by diffusion and not by kinetics and a slow quench will give an error equivalent to a corresponding position error in the streamwise direction; mean gradients of the composition in the streamwise direction are usually quite small. Slow quenching of course will allow recombination of high temperature dissociation products (e.g., H_2 and O_2) but the sample is normally dominated by material from regions relatively remote from the flame front. For turbulent premixed combustion the situation may be similar for this same reason.

Quenching rates may have a significant effect on trace species such as NO and NO_2 (see below). No attempt has been made to sample free radicals from turbulent flames.

E. NO_x Measurements

It is becoming increasingly evident that measurements of NO and NO_2 in combustion systems are very sensitive to the sampling and analysis technique. It is not long since Sawyer (1972) reported essentially no NO_2 in samples withdrawn from a gas turbine combustor whereas nowadays NO_2 often dominates the total NO_x sampled (Tuttle, 1975). Bilger and Beck (1975) and Lavoie and Schlader (1974) report totally different NO levels in hydrogen jet diffusion

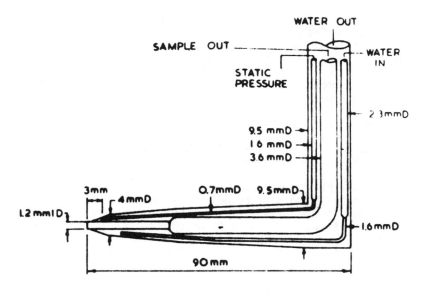

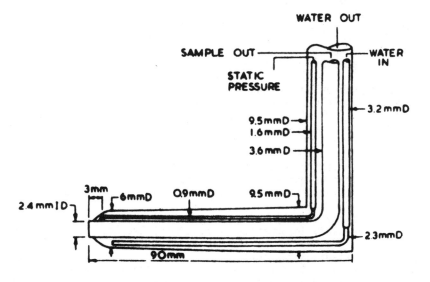

Figure 1. Isokinetic Sampling Probes Used by Kent and Bilger (1973).

flames into still air apparently without finding any NO_2. Schefer et al.
(1973) found strong effects of probe type on NO and NO_2 compositions measured
in a turbulent diffusion flame. Cernansky and Sawyer (1975) give a mechanism
for NO_2 formation in flames which would also give formation within the probe.
Tuttle et al. (1973) review a wide collection of data which shows influence
of probe material, sample line, soot filters, dryers for H_2O, "catalytic"
converters, and instrument type. England et al. (1973) report a strong influence of probe material on NO measured. Allen (1975) finds evidence of NO_2
formation in the probe during radical recombination reactions.

So far it has not been possible to reconcile all this evidence. It is
apparent that at least the following phenomena may be involved.

1. Aerodynamic biasing in sample collection
2. Thermal equilibration of NO/NO_2 which can be accelerated by
    a catalyst (probe, sample line, "catalytic" converter)
3. Reduction of NO_2 to NO by surfaces (probe, "catalytic" converter)
4. Reduction of NO (to N_2?) by surfaces (probe, "catalytic" converter)
5. Oxidation of NO to NO_2 by non-equilibrium species such as OH
    and HO_2 (probe)
6. Reduction of NO and NO_2 by gas phase reducers such as CO and H_2
    (probe, "catalytic" converter)
7. Oxidation of cyano and amino species (probe, "catalytic"
    converter)
8. Absorption and adsorption of NO_2 in sampling line and filters
    including carbon deposits
9. Absorption of NO_2 in water traps and dryers
10. Interference effects in instrument, e.g., C on NDUV measurements

Resolution of this problem is needed. Unfortunately laser Raman methods are
unlikely to be of help due to the low concentrations involved.

F.   Checks on Data

Due to the complexity of the phenomena involved it is not possible to
arrive at a quantitative estimate a priori for errors in concentration measurements. Estimates of errors can only be arrived at from redundant information.

The repeatability of the data gives only a minimum estimate of the error
levels, but is necessary information to determine the level of random error.

The most important checks that can be made are those involving atom
balances. In turbulent flows differential diffusion effects should be small
so that C/H ratios and N/O ratios should be preserved; in premixed flames
overall stoichiometry should be preserved also. It should be noted that such
atom ratio balances only give information on the accuracy of the analysis of
the sample withdrawn and not on whether the correct sampling has been made.
It is important to measure all species; significant information is lost in
dry-basis analyses. Fig. 2 shows N/O ratios obtained in the turbulent

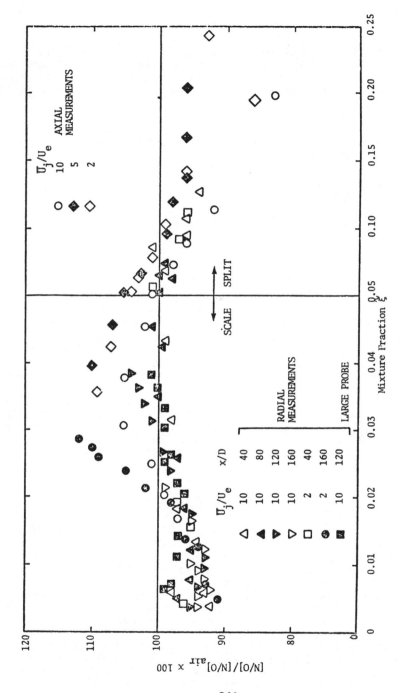

Figure 2. Nitrogen to Oxygen Ratio in Turbulent Diffusion Flames. From Kent and Bilger (1973).

341

hydrogen/air diffusion flame of Kent and Bilger (1973). Here $\tilde{\xi}$ is the Favre averaged mixture fraction (equal to the Favre averaged hydrogen element mass fraction) and is a measure of the stoichiometry. While some definite trend is evident the general level of accuracy is within ±5 per cent of the humid air value of 3.65.

In turbulent diffusion flames a check on sampling accuracy can be obtained by obtaining a mass flow balance for the fuel by integration across the flow assuming that the velocity is known.

$$\dot{m}_{fu} = \int_A \overline{\rho U \xi} \, dA \approx \int_A \overline{\rho} \, \tilde{U} \, \tilde{\xi} \, dA \tag{3}$$

$$\tilde{\xi} = \sum_i \mu_{fi} \, \tilde{Y}_i \tag{4}$$

where $\tilde{U}$ is the Favre average velocity, dA an element of cross-sectional area, and μ_{fi} is mass fraction of fuel atoms in compound i. Favre averaging is preferred since normally $U'' Y_i'' \ll \tilde{U} \tilde{Y}_i$ whereas $\overline{\rho' U'}$ can be a significant fraction of $\overline{\rho} \, \tilde{U}$ (double primes denote variation from the Favre mean). The most trouble is caused by the evaluation of $\overline{\rho}$ since $\overline{\rho} \neq \rho \, (\overline{\xi})$ or $\rho \, (\tilde{\xi})$ or $\rho \, (\tilde{Y}_1, \tilde{Y}_2, \ldots \ldots \tilde{Y}_i, \tilde{T})$ and some assumption for the p.d.f. of ξ is required. The situation is somewhat alleviated if velocities from pitot measurements are used since then only $(\rho)^{1/2}$ is involved (see later). Vranos et al. (1969) find integrated fuel flow to be accurate to within about 6 per cent of that metered into the combustor and Kent and Bilger (1973) obtain this accuracy for their higher velocity ratio flames; for lower velocity ratios they obtained errors as high as 20 per cent. In axisymmetric flow the integrals are particularly sensitive to the uncertainty in the mean density at the edges of the flame.

G. Accuracy Attainable

With careful technique in streaming (i.e., mathematically parabolic) flows using isokinetic sampling with slender nosed probes it is estimated that Favre averaged main species concentrations can at present be measured to an accuracy of within 10 per cent of their true values. No judgement can be made at this time on the accuracy of measurements made with sonic orifice probes. It is probable however that under streaming conditions quartz microprobes with proper quenching can achieve a similar level of accuracy for $\rho^* c^* Y_i / \rho^* c^*$, which usually lies between the Favre and time averages of Y_i.

At this time one can place little more than qualitative confidence in most measurements of NO and NO_2 in combustors. However with a relatively modest expenditure of research effort the phenomenology should be determinable and accurate measurements of at least total NO_x achievable.

For recirculating and strongly swirling flows the interference of the probe with the flow will always be questionable. However, with careful checks on the effect of the probe introduction on other measurements, disturbances can probably be kept down to an acceptable level. Close coordination with laser doppler velocimeter measurements will allow such checks and also element mass flow balances to be obtained.

IV. TEMPERATURE MEASUREMENTS

A. Thermocouples

In turbulent flows thermocouples need to be more rugged than those used in laminar flame studies since there is some tendency of fine thermocouples to follow the fluctuations in flame temperature which are often quite large compared with the mean temperature. Kent and Bilger (1973) used a Pt-5%Rh versus Pt-20%Rh thermocouple welded from 0.12 mm wire to give a 0.3 mm bead and coated with a non-catalytic coating of beryllium and yttrium oxide (Kent, 1970). With such large thermocouples radiation corrections become quite large ($\sim$ 400°K at peak temperatures) and it is necessary to have an accurate value for the emissivity (0.72 is given for this coating).

Other precautions include the use of a non-catalytic coating, particularly in premixed flames. In turbulent flames, there is no choice of side of entry for the thermocouple and the thermocouple will tend to stabilize the flame particularly as it passes from the burnt to the unburnt side. The tendency to stabilize the flame will be greatly reduced with a non-catalytic coating. The non-linearities found by Vlasov and Kokushin (1957) may have been due to this effect.

With luminous flames and hot-walled combustors radiation corrections become complicated by the need to account for back radiation which often cannot be accurately assessed. The use of radiation shields introduces further problems of disturbances to the flow that the thermocouple is exposed to including the possibility of density segregation during the effective non-isokinetic sampling. The suction pyrometer described by Beer and Chigier (1972) will also suffer from non-isokinetic sampling effects, but is known to give good performance in regions where density fluctuations are not large.

It is interesting to consider what sort of average the thermocouple will read. If it is assumed that the thermocouple is too massive to effectively follow the temperature fluctuations and there are no catalytic effects the average obtained will be:

$$T_{g,m} = \overline{\lambda_f \, Na \, T_g} \; / \; \overline{\lambda_f \, Na} \qquad\qquad (5)$$

where T_g is the gas temperature, λ_f the thermal conductivity at some appropriate (but varying) film temperature and composition, and Na is the Nusselt
number (also varying). This measured average temperature will probably differ
from the time mean temperature by more than the Favre average does but with
the opposite sign!

B.   Pneumatic Pyrometers

Pneumatic pyrometers for use in combustion flows have been described by
Beér and Chigier (1972) and Mestre and Benoit (1973). They employ two choked
orifices in series and obtain the temperature from the pressures measured in
the combustion chamber and the plenum between the two orifices. Once again
non-isokinetic sampling effects will be important although with a microprobe
design these could perhaps be overcome. Once again the average temperature
measured will not be the time average but a weighted average.

C.   Cooled Films

Cooled film probes have been used by Ahmed (1971) to study the structure
of turbulent premixed flames. Although these probes are essentially heat flux
transducers they are generally more sensitive to temperature than velocity
when used in flames and are accordingly considered here rather than in the
next section. They represent one of the few probes that can make effective
time resolved measurements in flames. An aspirated version is claimed to be
independent of velocity (Thermo Systems, Inc., 197-) but once again sample
segregation effects will be important. Although accuracy levels cannot be
expected to be high the information that can be gained from this system with
further development will be invaluable for the further understanding of turbulent flame structure.

V.   PITOT-STATIC MEASUREMENTS

The square nosed pitot tube is one of the most widely used instruments
for measuring total head in order to obtain velocity. Such a tube has a reading independent of yaw up to quite large angles and so the mean total pressure
measured $\overline{P_T}$ will be given by

$$\overline{P_T} = \overline{P} + \frac{1}{2} \rho \overline{\underset{\sim}{Q}^2}$$

where $\underset{\sim}{Q}$ is the velocity vector and $\overline{P}$ the static pressure. Since the density
fluctuations are high under turbulent combustion conditions Favre averaging is
best used. If the probe is aligned with the mean flow direction then

$$\overline{P_T} = \overline{p} + \frac{1}{2} \overline{\rho} \, (\tilde{U}^2 + \tilde{u''}^2 + \tilde{v''}^2 + \tilde{w''}^2)$$

For boundary layer flows where a static pressure measurement can be made in the non-turbulent fluid then the mean static pressure there, $\overline{p}_e$ is given by

$$\overline{p_e} = \overline{p} + \overline{\rho} \, \tilde{v''}^2 \qquad\qquad (6)$$

In many such flows $\tilde{v''}^2 \approx \tilde{w''}^2$ so that we have

$$\overline{P_T} - \overline{p_e} = \frac{1}{2} \overline{\rho} \, (\tilde{U}^2 + \tilde{u''}^2) \qquad\qquad (7)$$

The Favre mean velocity $\tilde{U}$ can be obtained if the mean density $\overline{\rho}$ and the turbulence level can be estimated. As mentioned above a proper evaluation of $\overline{\rho}$ requires knowledge of the p.d.f. of the composition in diffusion flames or temperature in premixed flames.

In three-dimensional flows pitch and yaw are also required. Beér and Chigier (1972) discuss the use of the five hole pitot and other instruments for making measurements under these conditions.

VI. TURBULENCE MEASUREMENTS

The advent of the laser-doppler velocimeter has made obsolete most of the earlier techniques for measuring velocity turbulence levels in flames. Günther (1970) gives a good review of these. They include the use of static pressure probes (see also Becker and Brown, 1969) employing equation (6) and microphone probes. Chigier and Strokin (1974) have used a gas tracer technique to obtain the turbulence level in a turbulent diffusion flame. It is doubtful if any of these techniques have a potential accuracy of better than ±20% and so will suffer in competition with the laser-doppler velocimeter which can obtain time resolved data as well as averages.

VII. CONCLUSIONS

In streaming flow probe measurements of velocity and main species mass fractions to within 10 per cent of their Favre average are achievable; and of temperature to within about 5 per cent of its time mean.

A concerted effort is required to understand the phenomenology affecting measurements of NO and NO_2 so that better measurements of these species can be made in turbulent combustion systems.

Much more attention is needed to determine and minimize the extent of disturbances caused by probes used in recirculation zones and swirling flows.

IX. ACKNOWLEDGEMENTS

This report was prepared while the author was a Visiting Professor at the Combustion Laboratory, School of Mechanical Engineering, Purdue University with support from the School of Aeronautical Engineering and the Environmental Protection Agency.

In Australia the work is supported by the Australian Research Grants Committee.

REFERENCES

1. A. A. Ahmed (1971): Application of the Cooled-Film to the Study of Pre-mixed Turbulent Flames, Ph.D. Thesis, McGill University, Montreal, Canada.

2. J. D. Allen (1975): "Probe Sampling of Oxides of Nitrogen from Flames" Combustion and Flame, 24, 133.

3. H. A. Becker and A. P. G. Brown (1969): "Velocity Fluctuations in Turbulent Jets and Flames," Twelfth Symposium (International) on Combustion, 1059.

4. J. M. Beér and N. A. Chigier (1972): Combustion Aerodynamics, Wiley.

5. R. W. Bilger (1975): A Note on Favre Averaging in Variable Density Flows, Project SQUID Technical Report UCSD-6-PU.

6. R. W. Bilger and R. E. Beck (1975): "Further Experiments on Turbulent Jet Diffusion Flames," Fifteenth Symposium (International) on Combustion, 541.

7. N. P. Cernansky and R. F. Sawyer (1975): "NO and NO_2 Formation in a Turbulent Hydrocarbon/Air Diffusion Flame," Fifteenth Symposium (International) on Combustion.

8. N. A. Chigier and V. Strokin (1974): "Mixing Processes in a Free Turbulent Diffusion Flame," Combustion Science and Technology, 9, 111.

9.   C. England, J. Houseman and D. P. Teixera (1973): "Sampling Nitric Oxide from Combustion Gases," Combustion and Flame, 20, 439.

10. A. Favre (1969): "Statistical Equations of Turbulent Gases," Problems of Hydrodynamics and Continuum Mechanics, Society for Industrial and Applied Mathematics, Philadelphia, PA, 231.

11. R. M. Fristrom and A. A. Westenberg (1965): Flame Structure, McGraw-Hill.

12. R. Günther (1970): J. Inst. Fuel, 43, 187.

13. J. H. Kent (1970): "A Non-Catalytic Coating for Platinum-Rhodium Thermocouples," Combustion and Flame, 14, 279.

14. J. H. Kent and R. W. Bilger (1973): "Measurement Techniques in Turbulent Diffusion Flames," First Australasian Conference on Heat and Mass Transfer, Monash University, Melbourne, Section 4.4, p. 39.

15. G. A. Lavoie and A. F. Schlader (1974): "A Scaling Study of NO Formation in Turbulent Diffusion Flames of Hydrogen Burning in Air," Combustion Science and Technology, 8, 215.

16. C. K. Leeper (1954): Sc. D. Thesis, Mass. Inst. of Technology, Cambridge, Mass.

17. B. Lenze (1970): Chemie-Ing.-Techn., 42, 287.

18. A. M. Mellor, R. D. Anderson, R. A. Altenkirch and J. H. Tuttle (1972): "Emissions From and Within an Allison J-33 Combustor," Combustion Science and Technology, 6, 169. Also Report CL-72-1, School of Mechanical Engineering, Purdue Univ., IN.

19. A. Mestre and A. Benoit (1973): "Combustion in Swirling Flow," Fourteenth Symposium (International) on Combustion, 719.

20   R. F. Sawyer (1972): "Experimental Studies of Chemical Processes in a Model Gas Turbine Combustor," Emissions From Continuous Combustion Systems, N. Cornelius and W. G. Agnew, Eds., Plenum Press, 243.

21. R. W. Schefer, R. D. Matthews, N. P. Cernansky and R. F. Sawyer (1973): Measurement of NO and NO_2 in Combustion Systems, Paper 73-31 presented at Western States Section/The Combustion Institute, Fall Meeting.

22. N. Syred (1975): Discussion to paper by R. W. Bilger and R. E. Beck, Fifteenth Symposium (International) on Combustion, 551.

23. Thermo Systems, Inc. (197-): Application Data for Chemical Combustion-Cooled Probes, Technical Bulletin TB3, Thermo Systems, Inc., Saint Paul, MN.

24. G. Tiné (1961): <u>Gas Sampling and Chemical Analysis in Combustion</u>
     <u>Processes</u>, AGARDograph 47, Pergamon Press.

25. J. H. Tuttle (1975): Personal Communication.

26. J. H. Tuttle, R. A. Shisler and A. M. Mellor (1973): <u>Nitrogen Dioxide</u>
     <u>Formation in Gas Turbine Engines: Measurements and Measurement Methods</u>,
     Report No. PURDU-CL-73-06, School of Mechanical Engineering, Purdue
     University, Indiana.

27. K. Vlasov and N. Kokushkin (1957): <u>Academy of Sciences USSR Division of</u>
     <u>Technical Sciences</u>, <u>8</u>, 137.

28. A. Vranos, J. E. Faueher and W. E. Curtis (1969): "Turbulent Mass Trans-
     port and Rates of Reaction in a Confined Hydrogen-Air Diffusion Flame,"
     <u>Twelfth Symposium (International) on Combustion</u>, 1051.

DISCUSSION

RHODES, ARO - Some hydrogen - air turbulent mixing experiments were done by Winterfeld and Schroter at DFVLR a couple of years ago. They compared the integral of their measured hydrogen concentration profiles to the total mass of hydrogen injected. When they sampled isokinetically, they lost 20 to 30% of the hydrogen. To get reasonable agreement, they went to probe pressure ratios 50% higher than isokinetic. Their advice is to pump as hard as you can.

BILGER - This has not been our experience in isokinetic sampling. But there might be many reasons for this loss: there is a significant dependence on the rate with which you pump the probes and on the shape of the head of the probe. If you have a fairly blunt nosed isokinetic sampling probe, you can get quite large errors just from the streamline shift due to the bluntness in this region of strong gradients. Lenze did too, but other people have different interpretations of the data.

You must establish carefully whether you are getting the right sort of velocity measurements and whether you are doing the integrations properly. You must recognize also that you have got Favre averages involved, and be sure that you are calculating this mean density correctly (which you can't do unless you assume some probability density function for the fluctuations in the turbulence).

HACKER, Univ. of Illinois, Chicago Center - I have a question related primarily to swirling flows. Provided you have located the probe correctly in the flow field - a major problem in itself - how do you then interpret measurements which are made within the wake created by the instrument?

CHIGIER, Univ. of Sheffield - One has to be extremely careful when making measurements in swirling flows. We have found that for instance, under flame conditions, the probe acts as a flame-holder: just by introducing the probe, you attach the flame to that position itself. It is an extremely difficult problem but the way we have tried to overcome this is to have very large flames and very small probes. Under these conditions, one is minimizing the effect.

Also we find that a very intense turbulent mixing is taking place and therefore, the type of probe interference which Bilger mentioned at the beginning of his paper might not apply to this very highly turbulent recirculation zone. Still measurements made under these highly turbulent conditions are delicate to interpret and we shall have to wait until we have an alternative technique to assess whether the streamlines patterns we obtain are true or not. Until then, all we can say is that a possibility of flame attachement exists. Also in these recirculation zones, there is an upstream and downstream stagnation point and these can be affected by the introduction of a probe.

GOULDIN, Cornell - It seems to me that there is another problem with sampling in these highly turbulent or swirling flows, beyond disturbing the streamlines. The instantaneous flow direction can be at a large angle to the probe axis and velocities are both positive and negative when you traverse a recirculation zone. The whole concept of isokinetic sampling in that kind of flow seems inapplicable to me.

BILGER - That's a very good point.

FONTIJN, AeroChem - You said that you could obtain a 10% accuracy on concentration. Was it for minor species or major species?

BILGER - It is for major species. For example the measurement of a mole fraction of 15% will produce Favre averages between 13.5% and 16.5%, if you do the measurements carefully.

BERSHADER, Univ. of Stanford - Fluid dynamicists are never happy unless they can put everything in non dimensional form. In your paper Eqn. 7), you mention that the turbulent component $\tilde{u}''^2$ has to be known in order to obtain an accurate mean velocity $\tilde{u}$. In gas sampling, you have similar errors on concentration measurements. I wonder whether you or others have done at least some qualitative analysis of these measurements, in terms of the size of your tube, or its radius of curvature, in relation to the scale of the turbulence or possibly other kinds of ratio. Has anyone derived a formula for the error in the concentration measurement as a function of the turbulence intensity or the spectrum of the turbulence?

BILGER - I think one could write those sorts of functional relationships, but you would end up with about ten parameters dependent obviously on Reynolds numbers, on scaling ratios, etc....It is not only the integral scale of the turbulence which is important, but also the Taylor microscale as far as the quartz microprobes are concerned. Certainly the turbulence intensity are important, even for the simple static probe. Brown and Becker wrote a very nice article in JFM last year just on the effects of the turbulence variables on static pressure measurements.

It would be a very rough problem to tackle analytically in any way. All we can do, I think, is to have confidence in checks like mass balance checks and to postulate that isokinetic measurements are reliable if they are set up properly and if your turbulence levels are not too high. As Gouldin said, once you get into the regime of 100% turbulence, isokinetic sampling doesn't make much sense. I would like to ask Chigier if he could say something about his experience with microprobes in turbulent flames. Have you checked mass balances and so on?

CHIGIER - Going from these large water - cooled probes to quartz microprobes, the enormous reduction in size was a great advantage. We haven't done anything like the checks that you have done, but I believe that Sawyer and others have made comparisons which show that if you use the quartz probe, you get a heating effect of the probe itself which under certain conditions can have advantages. For instance, when we have made measurements in oil sprays, having a hot probe had many advantages over a cold probe, because with a cold probe you get coking and deposition on the probe which clogs it up very quickly, whereas if you have the hot probe then you can carry on making measurements for a longer period of time. But I think with all these probing systems, we recognize that they have their limitations and until we manage to get an optical system to supersede them, the whole philosophy is to try and minimize them as much as possible.

BILGER - I wouldn't agree about the superseding. Optical techniques are going to be not only expensive but very demanding in human resources. You can't give them to a graduate student in the first year of his Ph.D. and expect him to make dependable measurements. We must instead compare the two kinds of measurement techniques, evaluate their respective merits and get better confidence in our probing ability.

HACKER - Have you made any quantitative estimates with respect to the effect of catalytic wall reactions on NO sampling procedures?

BILGER - Well, it is certainly one of the phenomena of concern to us (see part E of my paper). There are all sorts of other radical reactions, especially when you are sampling from very hot regions. But I haven't really made any computations.

HARVEY, NRL - The possibility of using the optical techniques to actually look at the flow around the probes is quite attractive. Also, if one makes both probe and optical measurements of the same flow and a difference is found, then a measure of the flow disturbance by the probe has been obtained. Perhaps holography might have some interesting role to play in this.

BILGER - Certainly it would be very nice to have some interferograms or holograms. They do time-freeze what happens when you sample in a stream. You should get some very nice pictures and be able to get a qualitative understanding of the flow.

MODELING OF
COMBUSTION PHENOMENA
WITH LARGE RATES OF HEAT RELEASE

P. T. HARSHA
R & D Associates

SUMMARY

The modeling of turbulent reacting flows is currently of considerable technical interest, but such flows present formidable difficulties to the analyst, particularly when problems involving the interaction of turbulence and chemical reactions are considered. Much of the modeling work in the literature is concerned with the closure of the moment equations for species mass fraction fluctuations in an essentially isothermal system. The question of the modeling of turbulent flows with large rates of heat release has received relatively less attention.

Work currently in progress at RDA has been concerned with the modeling of a complex turbulent flow with multiple phases and large rates of heat release, in a flowfield to which the classical boundary layer approximations do not apply. Our attempts to model this flow have involved an interplay of detailed moment equation derivations, closure modeling and the use of experimental data. In this note a summary of the modeling is described, in order to point out where currently available experimental information has entered the analysis, and to indicate the particular kinds of experimental data that are useful in the analysis of complex flows.

COMBUSTION MODELING

Conventional technqiues for writing the governing equations for a turbulent flow involve an assumption that the incompressible flow form of these equations may be used to describe a variable-density flow (e.g., Refs. 1,2). In effect, this assumption implies that the fluctuating density terms that arise in a derivation of the governing equations for a compressible flow (Ref. 3) are small. There is little direct experimental evidence for (or against) this assumption; however, in order to assess its validity, an indirect approach may be followed which will be briefly described in this comment.

Consider the unsteady-flow momentum equation

$$\frac{\partial \rho V_\alpha}{\partial t} + \frac{\partial (\rho V_\alpha V_\beta)}{\partial x_\beta} = - \frac{\partial P}{\partial x_\alpha} + \frac{\partial \tau_{\alpha\beta}}{\partial x_\beta} \tag{1}$$

where

$$\tau_{\alpha\beta} = \mu \left(\frac{\partial V_\alpha}{\partial x_\beta} + \frac{\partial V_\beta}{\partial x_\alpha} \right) - \frac{2}{3} \mu \frac{\partial V_\alpha}{\partial x_\beta} \delta_{\alpha\beta}$$

$$\alpha, \beta = 1, 2, 3$$

If, following the conventional Reynolds averaging procedure we define

$$V_\alpha = \bar{V}_\alpha + V'_\alpha \qquad\qquad \bar{V}'_\alpha = 0$$

$$\rho = \bar{\rho} + \rho' \qquad\qquad \bar{\rho}' = 0$$

and substitute these variables into the momentum equation, following Bray (Ref. 3), the resulting equation is

$$\bar{\rho} \frac{\partial \bar{V}_\alpha}{\partial t} + (\bar{\rho} \bar{V}_\beta + \overline{\rho' V'_\beta}) \frac{\partial \bar{V}_\alpha}{\partial x_\beta} = - \frac{\partial P}{\partial x_\alpha} + \frac{\partial \tau_{\alpha\beta}}{\partial x_\beta} - \frac{\partial}{\partial x_\beta} (\overline{\rho V'_\alpha V'_\beta} + \overline{\rho' V'_\alpha V'_\beta})$$

$$- \left(\frac{\partial (\overline{\rho' V'_\alpha})}{\partial t} + \bar{V}_\beta \frac{\partial (\overline{\rho' V'_\alpha})}{\partial x_\beta} \right) - \overline{\rho' V'_\beta} \frac{\partial \bar{V}_\beta}{\partial x_\beta} \tag{2}$$

which differs in arrangement from the form given by Bray because an error in Bray's definition of the turbulent shear stress has been corrected.

Closure of Eq. 2 requires modeling of the higher order correlations that appear in it. If we assume that the turbulent shear stress,

$$\tau_{T_{\alpha\beta}} = -(\overline{\rho V'_\alpha V'_\beta} + \overline{\rho' V'_\alpha V'_\beta})$$

is adequately modeled using, say, a two-equation model of turbulence, the remaining terms to be modeled involve the correlation of density and velocity fluctuations. Following Bray, write

$$-\overline{\rho' V'_\alpha} = \frac{1}{\bar{\rho}} \frac{\mu_T}{\sigma_\rho} \frac{\partial \bar{\rho}}{\partial x_\alpha}$$

where σ_ρ is a diffusion coefficient for density and μ_T represents a turbulent eddy viscosity obtained from the two-equation model of turbulence. The turbulent momentum equation then becomes

$$\rho \frac{\partial \bar{V}_\alpha}{\partial t} + \rho \bar{V}_\beta \frac{\partial \bar{V}_\alpha}{\partial x_\beta} = \frac{\partial \bar{P}}{\partial x_\alpha} + \frac{\partial_\tau T_{\alpha\beta}}{\partial x_\beta} + \frac{1}{\bar{\rho}} \frac{\mu_T}{\sigma_\rho} \frac{\partial \bar{\rho}}{\partial x_\beta} \frac{\partial \bar{V}_\alpha}{\partial x_\beta}$$

$$+ \frac{D}{Dt} \left(\frac{1}{\bar{\rho}} \frac{\bar{\mu}_T}{\sigma_\rho} \frac{\partial \bar{\rho}}{\partial x_\alpha} \right) + \frac{1}{\bar{\rho}} \bar{\mu}_T \frac{\partial \bar{\rho}}{\partial x_\alpha} \frac{\partial \bar{V}_\beta}{\partial x_\beta} \tag{3}$$

where it has also been assumed that the molecular shear stress may be neglected. But now the question arises: What is a proper value for the diffusion coefficient, σ_ρ?

The form of the model for the density-velocity correlation was chosen by Bray in analogy to established models for turbulent diffusion of a scalar. However, it has not been experimentally tested, so that empirical values of the diffusion coefficient have not been obtained; indeed, it is not clear whether the model itself is an appropriate one. Experimental data are clearly needed at this point, but in the absence of such data we must resort to an indirect approach, involving another kind of modeling of the turbulence equations.

For compressible flows, a more natural variable than the velocity, V_α, is the mass flux, ρV_α, and the set of governing equations for a turbulent flow may be rederived using this as the primary variable, (e.g., Favre, Ref. 4). Thus defining

$$\tilde{V}_\alpha = V_\alpha + V''_\alpha \qquad \overline{V''_\alpha} \neq 0$$

where

$$\tilde{V}_\alpha = \overline{\rho V_\alpha}/\bar{\rho}$$

the momentum equation becomes

$$\bar{\rho} \frac{\partial \tilde{V}_\alpha}{\partial t} - \bar{\rho} \tilde{V}_\beta \frac{\partial \tilde{V}_\alpha}{\partial x_\beta} = - \frac{\partial P}{\partial x_\alpha} - \frac{\partial}{\partial x_\beta} (\overline{\rho V_\alpha'' V_\beta''}) \tag{4}$$

neglecting the molecular shear stress. If it is now assumed that a two-equation model similar to that applied to the Reynolds-averaged equation may be used to obtain the shear stress $\tau_{T\alpha\beta} = -\overline{\rho V_\alpha'' V_\beta''}$, equation (4) is closed.

We can now view equations (3) and (4) as defining equations for the mean velocities $\tilde{V}_\alpha$ and $\tilde{V}_\alpha$. Stanford and Libby (Ref. 5) have obtained extensive data in a Helium-air pipe flow which has been used to provide a comparison between the Reynolds-averaged and mass-averaged forms of the various correlation terms that enter the equations of motion. From this data it can be observed that over the entire flow field the Reynolds-averaged value of the axial component of the mean velocity is essentially equal to the Favre-averaged axial mean velocity. Except near walls and at the end of the pipe this is also approximately true of the radial mean velocity component. Further, even at a point in the flow at which the density fluctuation intensity $(\overline{\rho'^2})^{1/2}/\bar{\rho}$ is 18%, the two forms of the axial shear stress term differ by no more than 3%. Thus, over much of the flow field the density fluctuation terms appearing in Equation (3) must be negligible (i.e., $\sigma_\rho \to \infty$) insofar as the prediction of the mean velocity field is concerned.

On the other hand, Stanford and Libby also report that near walls the "mass transport term" $\overline{u\rho'v'}$ can be significantly larger than the shear stress term $\overline{\rho u'v'}$; in these regions significant differences also exist between the two forms of the radial mean velocity component. This implies that in some circumstances the density fluctuation terms should be included at least in the modeling of the mean flow continuity equation, although little concrete evidence exists on which to base a determination of the appropriate diffusion coefficient.

It should also be noted that in all cases, in variable density, compressible flows, the form of the Favre-averaged equations is considerably simpler than that of their Reynolds-averaged counterparts. However, most of the available information on turbulent structure that has been used in the development of the newer models of turbulent flow relates to the Reynolds-averaged equations. In order to provide the necessary information for the development of sophisticated models of turbulent reacting flow, an urgent need exists for the measurement of Favre-averaged quantities in flows with large density variations.

REFERENCES

1. F. C. Lockwood and A. S. Naguib, Combustion and Flame, 24, 109-124, 1975.

2. R. P. Rhodes, P. T. Harsha and C. E. Peters, Acta Astronautica, 1, 443-470, 1974.

3. K. N. C. Bray, "Equations of Turbulent Combustion II, Boundary Layer Approximation," University of Southampton, AASU Report 331, 1973.

4. A. J. Favre, "The Equations of Compressible Turbulent Gases," Summary Report No. 1, Contract AF61(052)-772, Institut de Mechanique Statistique de la Turbulence, Universite d'Aix-Marseille, 1965.

5. R. A. Stanford and P. A. Libby, Physics of Fluids, 17, 7, 1353-1361, 1974.

DISCUSSION

BILGER - I think that you said that the Reynolds stresses were very much like the Favre stresses. That's not the way I read the paper: there was quite a difference between the two, a significant one, I think, for the density fluctuation terms.

HARSHA - That depends on how the shear stress is defined; there is some ambiguity. The apparent stresses in both the Favre-averaged and Reynolds-averaged momentum equations arise from the momentum flux term

$$\frac{\partial}{\partial x_\beta} \left(\rho V_\alpha V_\beta \right)$$

If this term is expanded in Reynolds-averaged form, the result is

$$\frac{\partial}{\partial x_\beta} \left(\overline{\rho} \overline{V}_\alpha \overline{V}_\beta + \overline{\rho V'_\alpha V'_\beta} + \overline{V}_\alpha \overline{\rho' V'_\beta} + \overline{V}_\beta \overline{\rho' V'_\alpha} + \overline{\rho' V'_\alpha V'_\beta} \right)$$

while the corresponding Favre averaged expression is

$$\frac{\partial}{\partial x_\beta} \left(\overline{\rho} \tilde{V}_\alpha \tilde{V}_\beta + \overline{\rho V''_\alpha V''_\beta} \right)$$

The ambiguity arises in the Reynolds-averaged definition: as Stanford and Libby (1974) point out, Morkovin has referred to the terms of the form

$$\overline{V}_\alpha \overline{\rho' V'_\beta} + \overline{V}_\beta \overline{\rho' V'_\alpha}$$

as "mass transport terms" implying that

$$\overline{\rho} \overline{V'_\alpha V'_\beta} + \overline{\rho' V'_\alpha V'_\beta}$$

represents the apparent stress. On the other hand, Stanford and Libby also show that from the definitions the expression

$$\overline{\rho V''_\alpha V''_\beta} = \overline{\rho} \overline{V'_\alpha V'_\beta} + \overline{\rho' V'_\alpha V'_\beta} + \overline{\rho' V'_\alpha} \overline{\rho' V'_\beta}/\overline{\rho} \tag{1}$$

connecting the two apparent stresses can be derived. Comparing this with the Favre-averaged and Reynolds-averaged expressions for the momentum divergence term, it is clear that the "mean flow" terms are different. Thus, it is not immediately certain which terms in the two forms of the equations of motion should be compared, nor is it clear how modeling assumptions may be applied.

Clearly the differences between the Favre-averaged and Reynolds-averaged equations extend to all of the terms, and ad hoc modeling is dangerous. However, it is worth noting that computations which use the Reynolds-averaged equations ignoring the density fluctuation terms are in reality solving the Favre-averaged equations. Further, many of the measurements on which modeling assumption depend are, to a greater or lesser extent, measurements of the Favre-averaged variables. There is a clear need to clarify both the definition of the apparent stresses and the currently available measurements, and for further study, including detailed measurements, of the Favre-averaged variables in flows with significant density variation.

BILGER - I think we both agree that what we really need is Favre-averaged quantities from the measurements.

TURBULENT MIXING

R. A. STREHLOW
University of Illinois

The comments that I am going to make have to do with the nature of the turbulent flow that we are studying and the types of turbulent flows that are involved in an engine environment. I think there are some very large qualitative as well as quantitative differences between these flows. The major thrust of this conference is the question of how to study turbulent mixing processes. If you look at turbulent mixing, there is completely nonreactive mixing; this is the type that Stanford and Libby[1] have studied in great detail, measuring concentration and velocity fluctuations and profiles. The next escalation in complication is hypergolic reactive mixing. This type has been studied in a shear layer. Batt, et al.[2] reported at the AIAA meeting in San Diego on turbulence on a study of the nitric oxide-oxygen reaction in a shear layer. This is a system where you don't need piloting because as soon as the gases contact each other they react rapidly at the ambient temperature. The third complication is, of course, the diffusion flame. In this case, the flame is piloted. You can have unreactive mixing at room temperature because if you don't have a pilot you don't have a diffusion flame. However, if you have a diffusion flame, you have to have a pilot region and the turbulent mixing process is completely different than either the nonreactive or the hypergolic mixing processes. Diffusion flames are really of two types. There are the ones that are held by a small premixed pilot and there are the types that are held by recirculation zone piloting.

There are differences in the way these different mixing processes behave and these differences are quite sizable. My feeling is that one should recognize these differences. We talk around them now, but nobody ever explicitly talks about the differences. One should really recognize that these three classes of mixing are markedly different and one should possibly try to do experiments which will allow one to see how these differences manifest themselves as one goes from one type to the other type.

The next escalation in complexity at this point is the question of spray droplet holding or spray droplet flames with recirculation zone holding where you have a highly mixed recirculation zone of large extent holding a flame zone. The next step is, of course, recirculation holding of a premixed flame. In any highly turbulent premixed flame there is yet another, different type of interaction between the turbulence and the chemical reaction. The new and powerful measurement techniques which are now available should be used to try and delineate these differences instead of just taking averages as though every turbulent reacting flow is the same. I might mention that I think that it would be very useful to use these new techniques to look at a shear layer flow for the three cases; unreactive mixing, hypergolic reactive mixing and diffusion flames. Some work of this type is going on right now. Unreactive shear layer mixing is being studied by Roshko and Libby's apparatus is evidently going to be put on Roshko's equipment so they can measure concentration and velocity fluctuations and profiles in an unreactive mixing shear layer. The hypergolic reactive mixing has been studied, but should be studied in more detail in the same type of a shear layer situation. Also, a simple shear layer diffusion flame should be studied to try to find qualitative as well as quantitative differences between mixing behaviors in these three simple cases.

REFERENCES

1. R. A. Stanford and P. A. Libby, Phys. Fluids 17, 1353-1361 (1974).

2. R. G. Batt, T. Kubota and J. Laufer, "Experimental Investigation of the Effect of Shear Flow Turbulence on a Chemical Reaction," AIAA Paper 70-721.

DISCUSSION

BILGER - I have just a brief comment on this. Our diffusion flame which uses hydrogen doesn't have a pilot. The lip thickness doesn't really give you a recirculation zone. It is only something like 0.003 inch. If there is any sort of a recirculation there, it's a very, very small laminar one. Hydrogen diffusivity is so high that it seems to be able to diffuse and cause the flame to hang on there to quite high velocities.

STREHLOW - What I am talking about here is not an external pilot, it is a little premixed zone that holds the flame right at the lip of the burner, just like a candle.

BILGER - I don't think all of the blue region is necessarily premixed. You can keep it down in some cases to something like one mm^3.

**Particulate Measurements
in Combustion**
Chairman: A. M. MELLOR

PARTICULATE MEASUREMENT
IN THE EXHAUST OF GAS TURBINE ENGINES

RICHARD ROBERTS

Pratt & Whitney Division, United Technologies Corporation

The primary interest of an aircraft gas turbine engine manufacturer with regard to exhaust particulate measurements is of course visible smoke. In addition to a limit on plume opacity, industrial gas turbines must meet a limit on total mass of particulates emitted in certain localities. It is my intent in this discussion to briefly review our experience in making particulate measurements, to indicate some of the problem areas, and to explain some shortcomings of the techniques currently in use.

The principal particulate measurement being made today is an implied dry particulate measurement which is related to visible exhaust smoke. Exhaust smoke is obviously an optical phenomenon, representing transmission loss due to light scattering by carbon particles in the exhaust stream. These carbon particles represent a very small fraction of the total fuel mass flow but, due to particle dimensions on the order of the wavelength of visible light, scatter light very effectively. A measurement system has been evolved over the years, wherein a small fraction of the engine exhaust stream is diverted and passed through a standardized filter at known sample volume, and the loss in reflectivity of the filter is measured. Details of the smoke measurement technique are described in SAE Aerospace Recommended Practice (ARP) No. 1179 (Ref. 1). This smoke measurement system has been incorporated intact into the EPA Aircraft Emission Standards (Ref. 2).

The present exhaust smoke measurement technique is not completely satisfactory, since it attempts to relate a stained piece of filter tape to what is seen by an observer watching the airplane take-off. Such factors as engine size, number and location on the aircraft, ambient light conditions (clear vs overcast), and position of the observer all impact apparent plume opacity. With variation in these factors, quite different observations would be recorded by a person standing on the ground, yet the filter would be identically stained in each case.

In addition to the fundamental problem of relating a stained filter to an optical phenomenon, there is a practical problem related to sample volume. Due to the very high airflow passed by gas turbine engines, only a fraction on the order of 1/100 or 1/1000 of the engine airflow is directed through the filter. A single or multiport sample probe is positioned in the engine exhaust stream to extract the sample flow. Successive samples drawn through the same probe in the same location at steady state power are generally repeatable. However, repositioning of the probe can produce quite different results. Strategic location of the probe sampling system in order to obtain a valid representation of overall plume opacity is a problem that is usually addressed by trial and error. It is a particular problem for common flow engines, where the core engine exhaust and fan streams are discharged through a single nozzle. Nevertheless, this is the accepted way to obtain smoke measurements and it is the technique employed at Pratt & Whitney Aircraft to certify production engine smoke levels.

The problem of representative smoke sampling is diminished by the fact that newer engine models are relatively low in smoke level. I think we are now in a position to develop combustors for new engines which provide a smoke level below the threshold of visibility. The current smoke level of the JT8D engine model is right on the threshold of visibility. It meets the EPA standard for that engine when measured in the prescribed manner. However, exhaust smoke can be seen under some conditions.

Smoke number determined from a stained filter tape is an implied solid particulate measurement which is related to exhaust plume visibility. Total mass of particulates, which includes solid particulates as well as condensed matter which is liquid at ambient temperature, has historically not been a subject of concern for gas turbine engines. However, some localities limit the quantity of total particulates emitted by ground based generating stations. The Los Angeles Air Pollution Control District (LAAPCD), for example, has limited total particulate emissions to less than 10 lbm/hr (Rule 67). This regulation was primarily directed at large, direct fired generating plants but also included industrial gas turbine engines.

The prescribed total particulate measurement technique involved withdrawal of a sample from the engine exhaust stream and passage through an impinger, wherein the particulate matter is collected in liquid water by sharply turning the sample gas flow. The impinger sample is then reduced to three components (hydrocarbon solvent soluble, water soluble and insoluble) by manipulation of the sample.

It was anticipated that industrial gas turbine engines, which burn relatively highly refined liquid fuel, would have no trouble complying with the total particulate standard. However, when measurements were made, it was found that the total particulate sample weight extrapolated to a value approximately 30% above the regulation. Examination of the possible sources, including solid carbon (smoke), lubricating oil loss, engine hardware erosion and fuel ash, could only account for about half of the observed value. The problem was traced to the water soluble particulate fraction, where hydrated sulfate precipitate was found. Even though the sulfur content of the fuel was very low, easily meeting the companion regulation for gaseous SO_2, SO_2 in the exhaust gas was being absorbed in the impinger water (and any condensed water), and was being oxidized and transformed to solids in the form of sulfates. The LAAPCD particulate separation procedure would not allow the sample to be heated high enough to drive the water out of the hydrated crystal. In fact, the presence of solid sulfate was due to the use of a water filled impinger and would not normally have been classified as a particulate. It is now recognized that very large errors in particulate measurement can arise due to this mechanism (Ref. 3). The solution in the case of industrial gas turbine engines was simply to run the engine qualification test on low sulfur fuel.

This incident aside, the quantitative measurement of total particulate mass flow rate shares the problem of inferring a total mass flux from a sample which is a very small fraction of the total engine flow. In the case of exhaust smoke, a great number of very small particulates aggregate on the filter tape. But there are also a smaller number of relatively heavy particles shed from the combustor, primarily deposited carbon which spalls off in an intermittent manner. These have minimal impact on the reflectivity of the stained filter and do not contribute to the observed plume opacity. However, should some of these larger particles be trapped in the particulate sample, they will dominate the weight of the sample. It is necessary to run a number of successive samples in order to minimize the effect of intermittent mass dominating particles. It is our experience that direct measurement of total particulate mass flow from a gas turbine engine is fraught with non-repeatability.

REFERENCES

1. "Aircraft Gas Turbine Exhaust Smoke Measurement," SAE Aerospace Recommended Practice 1179, May 1970.

2. Environmental Protection Agency: "Control of Air Pollution From Aircraft and Aircraft Engines; Emissions Standards and Test Procedures for Aircraft," Federal Register 38 (136) Part II: 19076 (July 17, 1973).

3. W. C. L. Hemeon, "A Critical Review of Regulations for the Control of Particulate Emissions," J. Air Poll. Control Association, 23:376 (1973).

OPTICAL METHODS FOR
IN SITU DETERMINATION OF PARTICLE
SIZE-CONCENTRATION-VELOCITY
DISTRIBUTIONS IN COMBUSTION FLOWS

S. A. SELF
Stanford University

SUMMARY

Several problem areas in particle distribution measurements have been reviewed: particle characteristics, scattering phenomena and size determination.

I. CHARACTERISTICS OF PARTICLES IN COMBUSTION FLOWS

Particles play an important role in various combustion flows. Their characteristics depend very much on the fuel, combustion conditions and location in the combustor.

Soot (carbon) particles occur to a greater or lesser extent with gaseous liquid and pulverized solid fuel; commonly, their thermal radiation is a major contribution to radiant heat transfer. (Beer and Siddall 1972; Thring and Lowes 1972) They are normally of very small size (0.01 - 0.1 μm) though a small proportion of larger particles occur by agglomeration. Typical mass concentrations are ~1 gm/m^3. Their refractive index has a large imaginary component and absorption dominates over scattering. (Millikan 1962; Erickson et al. 1964)

In atomized liquid fuel and pulverized solid fuel combustors, the particles, as injected, are normally quite large (10-100 μm), and their concentrations very high near the fuel nozzles. In this size range, scattering

dominates over absorption and is highly anisotropic, with most of the scattered power in the forward lobe. In liquid fuel combustors, the droplets
rapidly disperse, evaporate and burn, and the concentration of particulates,
other than small soot particles, may be quite low in downstream regions.

With pulverized solid fuel (e.g. coal or char) the situation is more
complicated. Depending on the flame temperature, the mineral content (up to
20% by mass of SiO_2, Al_2O_3, Fe_2O_3, etc.) may partly remain in the solid phase,
partly exist as liquid droplets and partly is vaporized, to condense out
again to droplets and particles of fly ash in the cooler downstream regions.
Fly ash, as collected, typically has a broad size distribution with a mass
mean average of ~ 10 μm, though there may be significant numbers at larger
sizes, and large concentrations of micron and sub-micron particles. It is
usually light in color and is a good dielectric scatterer (small imaginary
part to the refractive index).

Besides such naturally occurring particles, it is often necessary or convenient to add suitable scattering particles to combustion flows for the
purpose of laser anemometry (Self and Whitelaw 1975), since in gaseous flames,
and in downstream regions of liquid fuel combustors, there may be too few
particles to employ anemometry at an acceptable data rate. On the other hand,
in the injection regions of liquid fuel combustors, and throughout pulverized
solid fuel combustors, the "naturally" occurring particles may be used for
laser anemometry. This can pose certain problems. First, especially in large
systems, the scattering may be so great as to make laser anemometry difficult
or impossible. Second, the large particles may slip appreciably relative to the
gas; in fact, the slip largely controls the rate of burning, since it determines
the rate of supply of oxidant and the removal of combustion products from the
vicinity of the particle. Important information on the combustion process
could be obtained from the velocity-size correlation if these quantities could
be simultaneously determined on individual particles.

II. ELEMENTS OF SCATTERING THEORY

The scattering of plane electromagnetic waves by a conducting or dielec-
tric sphere, radius a, was first treated by Mie, and for $\alpha \equiv (2\pi a/\lambda) \gtrsim 1$ is
usually known as Mie scattering to distinguish it from Rayleigh scattering
which applies for small particles and molecules ($\alpha \ll 1$). The whole subject
has been thoroughly developed, at least for spheres, and detailed accounts are
given in the texts by Van de Hulst (1957) and Kerker (1969). The differential
scattering amplitude is a function of the ratio α, the (complex) refractive
index m relative to the medium, the scattering angle, and the polarization, and
has been calculated and tabulated in terms of these parameters.

In the Rayleigh limit the results are relatively simple; the total scattering cross section (integrated over scattering angles) increases with
particle size as a^6, and the dependence on the scattering angle goes as

$(1 + \cos^2\phi)$. Thus, for (optically) sub-microscopic particles, the forward and backscattered intensities are equal. However this is probably valid only for $\alpha \leq 0.3$, i.e., for particles less than $0.05\,\mu$m diameter. For α approaching unity and increasingly for $\alpha > 1$, the total scattering cross-section generally increases as a lower power of a , becoming equal to twice the cross sectional area in the geometrical optics limit $\alpha \gg 1$, but exhibits a marked and rapidly fluctuating dependence on α because of resonances of the wavelength with the particle dimensions. At the same time the dependence on scattering angle develops a distinct structure with a forward maximum, a secondary maximum in the backscatter direction and a number of lobes in between. For $\alpha > 1$, the forward lobe may contain 10 to 10^3 as much energy as the backward lobe, and accounts for the greater sensitivity of forward scattering laser anemometers. For dielectric particles the scattered amplitude generally increases with the refractive index (real part). For lossy dielectrics the absorption increases with the imaginary part of m, the scattering amplitude decreases and the rapidly fluctuating dependence on α and scattering angle are smoothed out. For a given mass concentration of particulate, the total scattering cross section has a maximum for α in the range 1-5, depending on the refractive index.

Liquid droplet spheres are seldom monodisperse and the spread of sizes in a polydispersion smooths out the sharp dependence on α and ϕ. Solid particulates are commonly of irregular shape as well as polydisperse, so that it is difficult to make any precise predictions of scattering amplitude in this case.

III. OPTICAL SCATTERING METHODS FOR SIZE DETERMINATION

Optical scattering forms the basis of many methods and instruments for particle sizing which rely on the theory of Mie scattering (Green and Lane 1967; Orr 1966; Van de Hulst 1957; Kerker 1969; Heller 1963; Hodgkinson 1963; Hodgkinson 1966; Dobbins and Jizmagion 1966; Shifrin and Perelman 1965). Many of the methods involve extinction (turbidity) or photometric scattering measurements involving a large number of particles simultaneously and it is difficult to unfold the particle concentration-size distribution except perhaps in the case of spheres of known refractive index. In contrast, and of more interest here, are counting methods which measure the scattered power of single particles observed one at a time, and can measure the flux and size independently. The detected optical pulses are sorted in a pulse height analyzer which outputs a histogram with perhaps a dozen channels in the size range 0.3-10 μm. Several instruments of this type are commercially available, but require a sample to be taken from the flow and fed to the instrument. They essentially measure the absolute scattered power and require calibration against particles of known size. The performance of several of these instruments has been compared (Cooke and Kerker 1975; Whitby and Vomela 1967; Liu et al. 1974).

The possibility of adapting this principle for the in situ sizing of particles in flow systems, through the use of laser beams is attractive. Preliminary work on focused laser particle counter/sizers for in situ measurements has been reported by Belden and Penny (1972) and by Andrews and Seifert (1971) using scattered light and by Faxvog (1974) who used an extinction method for small absorbing particles. These studies gave encouraging results for sizing by pulse height analysis of signals from individual particles, but revealed a certain problem related to the use of focused laser beams. Since the intensity distribution is a rapidly changing function of the position relative to the geometrical focus, the scattered intensity or extinction depends strongly on the exact particle trajectory. Large particles passing at some distance from the focus, may give scattered intensities or extinctions comparable with those from small particles passing through the focus. In this case monodisperse particles would be measured as a distribution which depends on the instrument geometry. This problem does not arise in commercial sampling counters because the illuminated volume is larger than the capillary flow tube and the intensity is uniform over the cross section of the tube.

A possible method (Faxvog 1975) of overcoming the problem of non-uniform illumination is to define a region of uniform intensity common to two crossed beams by using two detectors and coincidence techniques. In this connection, the use of focused ribbon beams, i.e., line foci rather than point foci, appears to offer significant advantages. Another method is to compare the relative intensities of scattering from a single beam at two angles, or with two polarizations at the same angle. Hodgkinson (1966) has shown that the relative intensity at two suitably small angles can be a good measure of particle diameter when this is larger than a few wavelengths. Moreover the measure is insensitive to refractive index (Meehan and Gyberg 1973) and even to particle shape. It should be possible to combine such methods of particle sizing and counting with simultaneous laser anemometry using the particle counting type of electronics.

An alternative approach to in situ particle sizing has been taken by Farmer (1972, 1974) by making use of the dependence of the shape of the rf signal on particle size in a differential scatter, real fringe laser anemometer. For a spherical particle of diameter much less than the fringe spacing that passes near the focus, the rf signal is fully modulated, but as the particle diameter increases the depth of the modulation decreases. Defining the signal visibility V as the ratio of the rf amplitude to the low frequency pedestal amplitude, Farmer shows that $V = 2|J_1 (2\pi a/\lambda_F)|/(2\pi a/\lambda_F)$, where J_1 is the Bessel function of the first kind and order, a is the particle radius and λ_F the fringe spacing. This leads to a method for sizing the individual particles in a fringe anemometer provided the signal/noise is high and the size distribution does not cover too large a range. Because J_1 is an oscillating function, there is ambiguity for $(2\pi a/\lambda_F) > 1.22\pi$ where J_1 has its first zero. By adjusting λ_F, it is suggested that sizes over a 10:1 range up to $2a \sim \lambda_F$ could be determined. It should be noted that this method is a geometric method, in the sense that it depends only on the signal shape and not on the absolute signal amplitude. Thus far it has been demonstrated to function only for relatively large particles in the diameter range 10-120 μm. For smaller particles it may be necessary to employ a method utilizing the absolute scattered power or better, the relative power scattered at two small angles, discussed above.

In any such scheme of in situ particle counting and sizing, possibly combined with laser anemometry, it will becnecessary to restrict the control volume, by controlling the dimensions of the illuminated and/or viewed volumes, so that particles are detected one at a time (low duty ratio). Given this condition, however, it should be possible to count and size particles at rates up to 10^6/sec with available electronic techniques.

IV.   CONCLUSION

Several instrumental schemes have been described for the in situ counting and sizing of particles by optical scattering from focused laser beams, which also lend themselves to the simultaneous determination of particle velocity. It is anticipated that one or more of these schemes will find valuable application for determining size-concentration distributions and also the correlation of size with velocity, for relatively large particles (> 1 μ) such as are present in atomized liquid and pulverized solid fuel combustors. The extension of such technqieus to small absorbing particles such as soot is more problematical.

REFERENCES

1.   D. G. Andrews and H. S. Seifert, 1971 "Investigation of Particle Size Determination from the Optical Response of a Laser Doppler Velocimeter," SUDAAR Report 435, Stanford University (Nov. 1971).

2.   J. M. Beer and R. G. Siddall, 1972 "Radiative Heat Transfer in Furnaces and Combustors," ASME Paper 72-WA/HT-29.

3.   L. H. Belden and C. M. Penney, 1972 "Optical Measurement of Particle Size Distribution and Concentration," General Electric Report 72 CRD 006.

4.   D. D. Cooke and M. Kerker, 1972 "Response Calculations for Light-Scattering Aerosol Particle Counters," Appl. Opt. $\underline{14}$, 734.

5.   R. A. Dobbins and G. S. Jizmagion, 1966 "Particle Size Measurement Based on the Use of Mean Scattering Cross Sections," J. Opt. Soc. Am. $\underline{56}$, 1351.

6.   W. D. Erickson, G. C. Williams and H. C. Hottel, 1964 "Light Scattering Measurements on Soot in a Benzene-Air Flame," Combust. Flame $\underline{8}$, 127.

7. W. M. Farmer, 1972 "Measurements of Particle Size, Number Density and Velocity Using a Laser Interferometer," Appl. Opt. 11, 2603.

8. W. M. Farmer, 1974 "Observations of Large Particles with a Laser Interferometer," Appl. Opt. 13, 610.

9. F. R. Faxvog, 1974 "Detection of Airborne Particles Using Optical Extinction Measurements," Appl. Opt. 13, 1913.

10. F. R. Faxvog, 1974 Private communication.

11. H. L. Green and W. R. Lane, 1967 Particulate Clouds, 2nd Ed., Van Nostrand.

12. W. Heller, 1963 "Theoretical and Experimental Investigations of the Light Scattering of Colloidal Spheres," in Interdisciplinary Conference on Electromagnetic Scattering, J. Kerker, Editor, p. 101, Pergamon Press.

13. J. R. Hodgkinson, 1963 "Light Scattering and Extinction by Irregular Particles Larger than the Wavelength," in Interdisciplinary Conference on Electromagnetic Scattering, M. Kerker, Editor, p. 87, Pergamon Press.

14. J. R. Hodgkinson, 1966 "Particle Sizing by Means of the Forward Scattering Lobe," Appl. Opt. 5, 839.

15. M. Kerker, 1969 "The Scattering of Light and Other Electromagnetic Radiation," Academic Press, New York.

16. B. Y. H. Liu, R. N. Berglund and J. K. Agarwal, 1974 "Experimental Studies of Optical Particle Counters," Atmos. Environ. 8, 717.

17. E. J. Meehan and A. E. Gyberg, 1973 "Particle-Size Determination by Low-Angle Light Scattering: Effect of Refractive Index," Appl. Opt. 12, 551.

18. R. C. Millikan, 1962 "Sizes, Optical Properties and Temperatures of Soot Particles," in Temperature - Its Measurement and Control in Science and Industry Vol. III, Part 2, p. 497. A. I. Dahl, Editor; Reinhold, New York.

19. C. Orr, 1966 "Particulate Technology," Macmillan, New York.

20. S. A. Self and J. H. Whitelaw, 1975 "Laser Anemometry for Combustion Research," in Special Issue of Combust. Sci. and Tech. on Reactive Turbulent Flows (to be published).

21. K. S. Shifrin and A. Y. Perelman, 1965 "Inversion of Light Scattering Data for the Determination of Spherical Particle Spectrum," in Electromagnetic Scattering, R. L. Rowell and R. S. Stein, Editors; Gordon and Breach.

22. M. W. Thring and T. M. Lowes, 1972 "Luminous Radiation from Industrial Flames," Combust. Sci. and Tech., 5, 251.

23. H. C. Van de Hulst, 1957 "Light Scattering by Small Particles," J. Wiley and Sons, New York.

24. K. T. Whitby and R. A. Vomela, 1967 "Response of Single Particle Optical Counters to Nonideal Particles," Environ. Sci. and Tech. 10, 801.

DISCUSSION

C.P. WANG, Aerospace Corp. - Since the particle or droplet size distributions, their burning rate, nucleation and condensation, are important to the combustion process, I would like to mention some optical techniques for the measurement of particulate number density and size distribution. For small particles (up to a few μm), similar to the Laser Doppler Velocimetry, homodyne detection can be used to measure the particle size distributions (Ref. 1,2). For larger particles (larger than a few μm) laser interferometer (Ref. 3), and optical extinction measurement (Ref. 4) can be employed, as was pointed out by Dr. Self.

References

1.   C.P. Wang, "Remote Sensing of Particulate Air Pollutant by Light Scattering," J. Am. Opt. Soc. 63, 507 (1973).

2.   W. Hinds and P.C. Reist, "Aerosol Measurement by Laser Doppler Spectroscopy. I. Theory and Experimental Results for Aerosols Homogeneous," Aerosol Science, 3, 515 (1972).

3.   W.M. Farmer, "Observation of Large Particles With a Laser Interferometer," Applied Optics, 13, 610 (1974).

4.   F.R. Faxvog, "Detection of Airborne Particles Using Optical Extinction Measurements", Applied Optics, 13, 1913 (1974).

LASER SCATTERING FROM MOVING, POLYDISPERSE PARTICLE SYSTEMS IN FLAMES

S. S. PENNER
J. M. BERNARD
T. JERSKEY
University of California, San Diego

Theoretical expressions are derived for the photocurrent power spectra observed with laser scattering from moving, polydisperse particle systems. The analysis refers to self-beating of scattered radiation (to which we refer as the homodyne spectrum) and beating between the laser beams scattered in the same direction from two, equally-intense incident laser beams which arrive at symmetrical angles with respect to the scattering direction (to which we refer as an example of heterodyne spectroscopy or, more briefly, as the heterodyne spectrum). We use a spherically-symmetric Gaussian distribution for the illumination intensity in the coherence volume (scattering region) and obtain a Voigt profile for the photocurrent power spectrum in the general case. We have used the theoretical relations derived for moving systems in order to estimate the maximum values of the particle diameters which are measurable for Stokes-Einstein diffusion when radiation is scattered from He-Ne or Ar-ion lasers.

Particulate growths in flames have been measured by using laser-radiation scattering as a diagnostic tool. The results obtained in three different experimental studies of progressively greater complexity and better definition of particulate formations are described. In the first series of investigations, the magnitudes of the scattered signals were employed as measures of total particle concentrations in the diagnostic region. The

This research was supported by the National Science Foundation under Grant ENG-7411494 to be presented at the <u>Fifth Colloquium on Gas Dynamics of Explosions and Reactive Systems</u>, Bourges, France, September 1975.

371

spacial resolution achieved was an appreciable fraction of the flame width above a Bunsen burner. In these investigations, the axial velocity components were mapped with a laser-Doppler velocimeter in C_2H_2-air diffusion flames and in partially-premixed flames of C_2H_2 with air burning in air. The results obtained for partially-premixed flames are consistent with the well-known fact that the gas speed is strongly accelerated at the flame front. In pure diffusion flames, the axial velocity rose uniformly along the centerline and the flame boundary may not close along the axis. Scattering-signal strength increased with downstream distance in all flames because of production of particles above the burner port. The relative amplitudes of the scattered laser signals, as well as the rates of rise with downstream distance, decreased as the amount of air premixed with C_2H_2 increased, thus showing that both the absolute magnitudes and the rates of growth of particulate formation are favored in the more fuel-rich systems. Radial distributions of particulate growth rates have also been defined through the magnitudes of scattered signal intensities.

In the second series of measurements, a flat-flame burner was employed and the radial dependence of the measured signal strengths was used as a measure of mean particle size on the assumption that the scattering centers are monodisperse. Work of this type has been performed by a number of authors using such light sources as mercury or tungsten arcs and suitable collimating systems in observations on flames formed in air by burning methane, ethane, ethylene, and propane. Our studies using the more convenient laser light source have generally confirmed the semi-quantitative conclusions reached earlier.

In our final series of measurements, we have employed the frequency distribution of measured power spectra for particle sizing. Our experimental results indicate that measurements of particle diameters in flames can be performed by using photocurrent power spectra, in accord with our theoretical analyses.

PARTICULATE MEASUREMENTS IN COMBUSTION

C. L. WALKER

Detroit Diesel Allison Division, General Motors Corporation

Briefly, our experience with probes is similar to what you have just heard from Mr. Roberts. Particulates in exhaust streams of gas turbine engines are small enough, generally less than one-half micron, that there is no effect of changing the sampling rate in an isokinetic probe. The only exception to this conclusion occurred several years ago when a particular burner configuration had a tendency to deposit carbon in the engine which would periodically break off and come out as large carbon particles. This condition presents no real problem in particle measurement because any engine operating in this mode quickly eliminates itself along with any smoke problem.

Our initial experience with holography for particle measurement was also related to smoke. Based on analysis of droplet burning if the droplets are all less than a certain size then smoke will be insignificant. The goal in this study was to be able to discern droplets down to at least as small as 50 micron. Fig. 1 shows one of the setups used to obtain holograms of the spray. One can see the hologram in the foreground and the fuel nozzle apparatus in the background. The most successful holograms were made with only a very thin slice of the spray pattern illuminated. This holographic work, done by Dave Monnier at Allison, is illustrated in Fig. 2. The holographic reconstructions are shown in the lower part of the figure at 40X and 150X magnification. There are no droplets as large as 50 micron diameter. Also because of the edge lighting, all visible particles are essentially in focus. The magnified reconstructions were provided by the Vidicon system shown in Fig. 3.

We are currently planning to apply holographic techniques to a spark ignition piston engine with a quartz cylinder wall shown in Fig. 4. The chief investigator on this program at Allison is Bob McClure. Prior to the workshop he did not have the holocamera operating but he has taken high speed schlieren movies which we will now show so that you can see some of the details of in-cyclinder air motion under running conditions. Because of the uncooled quartz cylinder wall we limit the running to about 15 seconds. At 5000 frames per

second, that's quite a lot of data. This entire sequence is computer controlled. The computer starts the dynamometer, turns on the ignition and fuel,
starts the camera, and shuts the engine off. The observations I want you to
make are the differences in apparent air motion in the cylinder from stroke to
stroke on successive cycles.

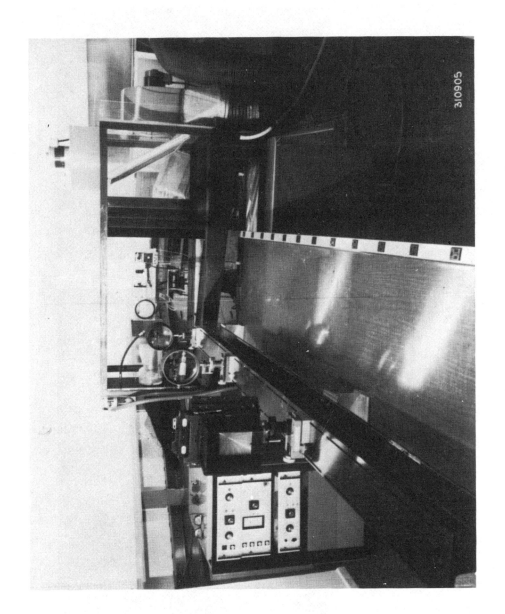

Fig. 1 – Setup for Fuel Spray Holograms

375

DEVELOPMENT OF A SPRAY
ANALYSIS SYSTEM THROUGH
THE USE OF HOLOGRAPHY

Fig. 2 - Development of Holographic Spray Analysis System

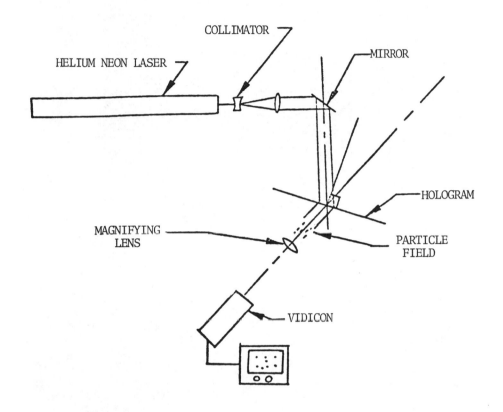

COLLIMATOR

MIRROR

HELIUM NEON LASER

HOLOGRAM

MAGNIFYING
LENS

PARTICLE
FIELD

VIDICON

Fig. 3 – Holographic Reconstruction System

377

Fig. 4 – Quartz Cylinder Spark Ignition Engine

BURNING SPRAY MEASUREMENTS

NORMAN A. CHIGIER
University of Sheffield

The measurements which we have made at Sheffield have been made in burning sprays of kerosene. This has been an extensive program which has taken place over a period of about six years and the method which we used for measurement is a shadowgraph photographic technique in which we measure the particles as they move inside the liquid spray. Measurements are made of the size distribution at the exit from the atomizer and then we make measurements of the change in size of particles as we proceed through the spray system. We have examined pressure jet atomizers which produce hollow cone sprays in the wake of a stabilizer disc and have also examined twin fluid air assist atomizers. The double flash technique which we use allows us also to measure the velocity of the particles as they are moving through the burning spray and we also measure the angle of flight. In addition to these photographic measurements in sprays we have measured temperature distributions by introducing very fine thermocouples. These are large sprays and we have also measured local air fuel ratios and we have come to the general conclusion that the behavior of these sprays flames is radically different to the behavior of a single droplet. Basically we find that the flame occurred at the outer periphery of the sprays and we do not have any evidence of single droplet burning i.e. with an envelope flame around individual droplets. In our opinion, this makes an enormous amount of difference to the prediction procedures for the formation of oxides of nitrogen and other species inside the flame. I would like to just show you very briefly some slides of our system and some of the principal conclusions. Fig. 1 shows one of the spray flames which we have in our laboratories. This is an air assist unconfined flame with an exit nozzle of about 1 mm. You can see the central spray structure and the flame front. Shadow photographs are taken with two high speed flashes with a flash duration of 1 microsecond and flash intervals of a few microseconds. A narrow depth of field is used which allows examination of a very thin slice of the spray itself. Fig. 2 shows a typical example of the spray, immediately after the

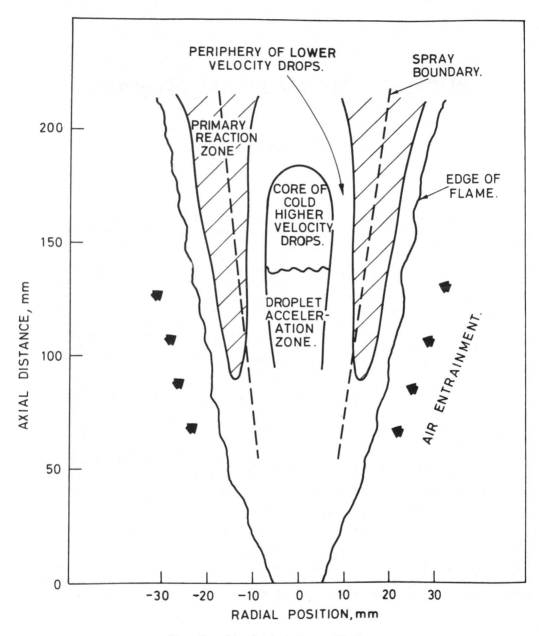

Fig. 1 - Air Assist Spray Flame

380

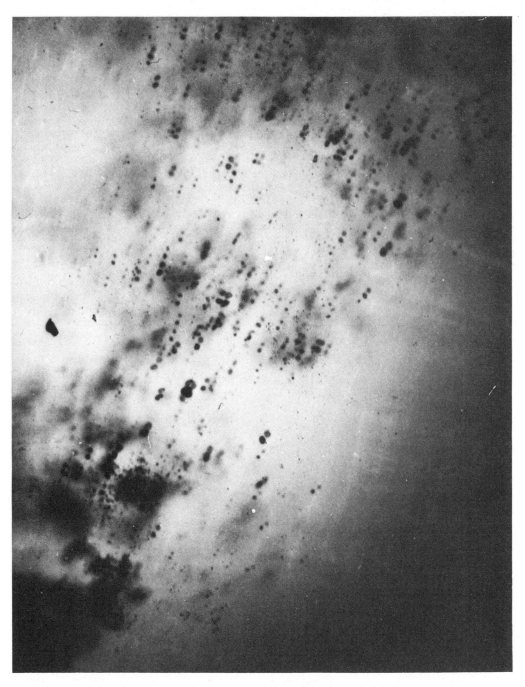

Fig. 2 - Double Image Flash Photograph of Liquid Spray

381

breakup region. We only examine droplets which are in focus and by calibration we can determine whether droplets are in focus or out of focus.
Fig. 2 shows examples of the double flash, a single particle then followed
by its image. From the direct measurement of the size of each particle we
can obtain a size distribution. By measuring the distance between the particles and dividing the distance by the measured time interval between sparks
we calculate the velocity of each particle and we also determine the direction of flight. Our studies on kerosene spray flames show that there is no
burning taking place in the central region of the spray. The particle size
disbributions are on the average about 70 microns with the larges particles
up to 300 or 400 microns. We find that there is a very rich mixture within
the spray because the droplets are vaporizing. Oxygen concentrations are
on the order of 1%. For these reasons, no burning takes place within the
spray and the flame is found at the outer periphery of the spray itself. We
have also carried out extensive studies both with air assist sprays and
pressure jet atomizers (Fig. 3). In both cases we found no presence of any
burning inside the spray itself. From these studies we have concluded that
it is not correct to translate measurements which are made on single droplets to the spray flame condition.

REFERENCES

1.   Chigier, N.A. and McCreath, C. G. Combustion of Droplets in Sprays,
      Acta Astronautica 1, 687-710, 1974.

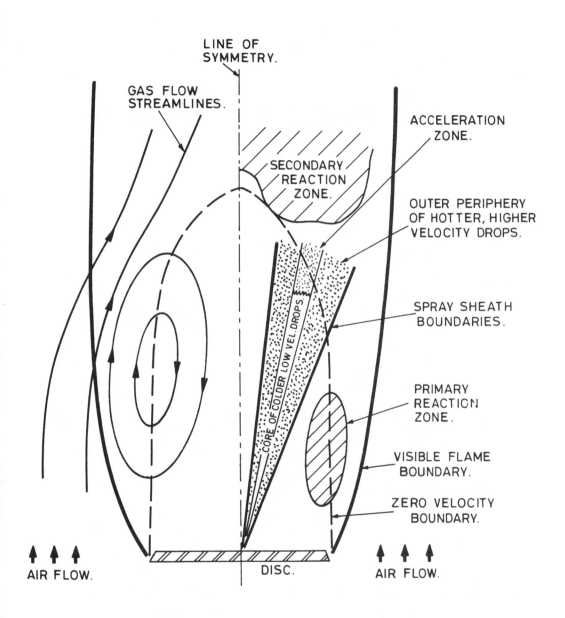

LINE OF SYMMETRY.

GAS FLOW STREAMLINES.

ACCELERATION ZONE.

SECONDARY REACTION ZONE.

OUTER PERIPHERY OF HOTTER, HIGHER VELOCITY DROPS.

CORE OF COLDER LOW VEL. DROPS.

SPRAY SHEATH BOUNDARIES.

PRIMARY REACTION ZONE.

VISIBLE FLAME BOUNDARY.

ZERO VELOCITY BOUNDARY.

AIR FLOW.

DISC.

AIR FLOW.

Fig. 3 - Pressure Jet Liquid Spray Flame in Wake of Stabilizer Disk

DROPLET EFFECTS IN COMBUSTION

R. C. FLAGAN

Massachusetts Institute of Technology

Recent experiments (Ref. 1 & 2) have been carried out at MIT to explore the effects of fuel atomizer characteristics on the fuel air mixing rates in combustors. The experiments were performed in a tubular combustor into which the primary air flow was introduced through stationary blade angle swirl vanes and the fuel was injected using either pressure type or air assist type atomizing nozzles. The rate of turbulent mixing was inferred from measurements of the mean molecular oxygen concentration as a function of distance from the atomizng nozzle during stoichiometric combustion in the manner of Flagan and Appleton (Ref. 3). Since the equilibrium O_2 concentration is about .2% for a uniform composition stoichiometric mixture, the oxygen concentration decreases along the burner length as fuel and air mix and burn, and providing a measure of the mixedness of the gases.

The air assist atomizer was used to inject both liquid kerosene and gaseous propane in the combustor. For both fuels the oxygen concentration decreased from the ambient level at the burner inlet to a few percent within the first two diameters along the burner length and decayed more slowly from that point to the end of the combustor. For both cases the rate of mixing in the jet dominated region of the combustor was very well correlated with the atomizer jet kinetic energy.

The pressure atomizer produced anomalous results. Although the jet kinetic energy was an order of magnitude less than the energy input of the air assist atomizer, the mixing rate in the first diameter of the burner was much more rapid for the pressure atomizer. The apparent mixing rate was much greater than the turbulence alone could produce. This can be attributed to the high concentration gradients surrounding the droplets which result in rapid mixing of the fuel vapor with air as the droplets burn.

Thus, it is apparent that droplet combustion can occur in at least two modes. For the air-assist atomizer, the droplets vaporize prior to burning so no droplet effects are observed. For the pressure atomizer the fuel vapor burns as the droplets evaporate giving an apparent enhancement of the mixing rate due to the very high concentration gradients which exist

for the entire droplet lifetime. Reducing the droplet evaporation time by
burning pentane rather than kerosene results in poorer mixing.

These observations do not answer the question of whether single droplet combustion occurs in practical systems. It does, however, indicate the
need to consider all possibilities since minor changes in combustor design
may drastically affect performance.

REFERENCES

1.   Komiyama, K., Ph.D. Thesis, Department of Mechanical Engineering,
      Massachusetts Institute of Technology, May 1975.

2.   Komiyama, K., Flagan, R. C., and Heywood, J. B., "The Effects of
      Droplets on Fuel-Air Mixing in a Burner," to be presented at the Second
      European Symposium on Combustion, Orleans, France, September 1975.

3.   Flagan, R. C., and Appleton, J. P., Combustion and Flame 23, pp. 249-
      267 (1974).

DISCUSSION
session on particulates

WOLFSON, AFOSR - I have two comments regarding the kind of research which is needed to improve airbreathing of ramjet systems. First, experiments with propane, methane or hydrogen as the reacting fuels are not what I would consider directly translatable to fuels such as kerosene or JP5. Particulate formation, spray droplet distribution, are not at all the same with simple fuels and complex ones. The other comment is in support of what Chigier just said. It is true that much work has been done on single droplet combustion, but it is not possible to translate the single droplet combustion data to a real situation of droplet formation and burning in a spray.

GLASSMAN, Princeton Univ. - I would like to take issue with these comments. We have made a film showing the burning of a single fuel droplet, and then, on the same film, what happens when the fuel is emulsified with 25% water: the droplet explodes, very wildly. As you can see we are interested in new kinds of fuel. Also, if these explosive characteristics are there, then droplets would explode outside the spray. In the film you can see little subdroplets subsequently exploding within the air. Therefore, let us not rule out single droplets as a fruitful area of research. To the contrary, let us always start with the simple experiment that will give us insight into the fundamental phenomenon. There are conditions where droplet burning happens even in a fuel spray. Also, if you take a particular impinging jet, you can produce droplets. On the other hand, if you have a film fluid swirl then maybe you will obtain a condition like a gaseous fuel jet. But if you take a twin fuel jet or emulsify the fuel, you will have droplets again. Therefore one cannot rule out this mode of combustion.

MELLOR - I have to disagree with Irv Glassman. We are doing Chigier's experiment in a confined duct at pressures and temperatures which are typical of inlet temperatures and conditions for gas turbine combustors. We tend to think that you can find all of the various extremes except droplet burning. We interpret our results not in terms of single droplet burning but rather single droplet evaporation.

One limiting case would be if the mixing was very rapid compared to the rate of addition of fuel to the vapor phase. In this case, the droplets have a significant effect on the flow and on the flame. The other case would be where the evaporation is very rapid compared with the mixing. In this case mixing would be the rate controlling step.

Chigier showed us today a spray flame which appeared to be controlled by mixing. We found some cases where evaporation could be a little bit more important. In any case, we found no evidence whatsoever of individual droplet burning for typical injectors.

The other point I wanted to make in answer to Wolfson's comments is that, while we are now using Jet A fuel, we will be going also to various other fuels because this is one good way to vary the rate of fuel vaporization.

S.S.PENNER, UCSD - I believe that Drs. Flagan and Glassman really put it properly: there are regimes where one extreme will prevail, others where another will. In liquid fuel rockets, droplet burning is extensively involved.

CHIGIER - Let me make it clear that I do not believe that it is possible to have single droplet burning in a practical system. We are getting more and more evidence from people who have been examining practical spray systems and they are not finding single droplet burning. When somebody produces a photograph of an envelope flame around a single droplet in a practical system, then we will accept that in that system it does exist. But we have not seen this kind of evidence in industrial furnaces, in open flames, in the laboratories at Sheffield, Purdue, in Japan, etc.

The important question, of course, is the location of the flame front in practical spray systems. If one finds an envelope flame around the droplet, one can accept all the detail studies which have been done for single droplets. If on the other hand, the flame is at the outer periphery of the spray, far removed from the drop surface, we have a totally different set of boundary conditions, involving turbulent heat and mass transfer from the flame, an entirely new ballpark!

FLAGAN, MIT - As a comment on the in situ soot measurements, there is a technique which has been developed by the carbon black industry in France, which we are just beginning to apply at MIT. We use a water cooled sampling probe. We inject into the very tip of the probe, along with the sample, water to flush the probe clear of soot deposition. We've only done preliminary measurements with this. So far, we have had no deposition and we have been able to get quantative mass loadings, not particle sizes but mass loadings within the primary zone of the combustor.

SELF, Stanford Univ. - The imaging type of measurements Chigier has described are very nice in that you get a picture, but there really is a serious data reduction problem. Maybe the graduate students that he had in England are different from the ones we have in the United States.

CHIGIER - Of course, the answer to your question is that there is no doubt that there is a difference between our students...(laughter). The first measurements which we made were done very laboriously: we took a ruler and we measured the size of each individual particle. This took many hours because thousands of droplets were photographed on each frame. We now have a Quantimet machine which allows us to analyse a photograph by electronic scanning. Within a very short period of time , one automatically gets the total size distribution. If one finds on the photograph some overlapping droplets, one just immediately cuts them out by circling them with a light pen. This is a rather expensive instrument (about $70,000), but there are many laboratories which have such image analyzing machines. It is thus possible to analyze data very much more quickly than we had to do with student labor.

PART IV
Forum
Chairman: S. S. PENNER

INTRODUCTION

PENNER, Univ. of California - This afternoon, we are going to evaluate the performance of the techniques we have discussed so far, from the point of view of their accuracy, spatial resolution, implementation time and cost.

We are interested, at this point, in a prototype combustor, and not in laminar or turbulent flames per se, or in field measurements. This focuses our field of debate in a way which is necessary if we want our recommendations to relate to the real engine world.

Implementation times will be evaluated in terms of the following broad categories:

> short: less than one year
> intermediate: from one to three years
> long: from three to five years.

These times assume that someone well trained gets adequate support to do the job. Now when I say well trained, I don't necessarily mean a diagnostics specialist. I mean that someone with appropriate training in spectroscopy, physical chemistry, mechanical engineering, etc., with good skill in experimental work, gets the job and support at a company, at a government laboratory or even at a University.

Costs for program implementation will be classified as follows:

> low: less than $50,000
> intermediate: from $50,000 to $150,000
> high: from $150,000 to $500,000
> out of signt: more than $500,000.

Let me repeat again that we are talking about prototype engine experiments. We are interested in point measurements, which are directly applicable to three-dimensional flow analysis.

To create a framework for the discussion, I have asked the chairmen of the following sections to write down in tabular form their estimate of the state of affairs and then to lead the discussion from the floor.

VELOCITY

C. P. WANG
Aerospace Corporation

Table I - An appraisal of velocity measurement techniques (C. P. Wang) [$u \equiv \bar{u} + u'$; the cartesian frame of reference of the vector (u,v,w) is arbitrary, so that the measurement capability shown here for u applies also to v and w.]

Measured Property	Accuracy	Resolution	Method	Implementation Time	Cost
$u,\ \bar{u}$	1%	1 mm^3	LV	Short	Low
$\overline{u'^2}^{\,\frac{1}{2}}$	1%	1 mm^3	LV	Short	Low
$\overline{u'v'}$	1%	1 mm^3	LV	Intermediate	High
$\overline{u'v'w'}$	1%	1 mm^3	LV	Long	Out of sight

WANG - We shall consider two kinds of laser velocimetry methods (LV): frequency tracking or single particle count. In both cases, there is no problem today with the measurement of u and $(u'^2)^{\frac{1}{2}}$ within 1% accuracy and 1 mm^3 spatial resolution. For correlated properties, I don't know if anybody has done that yet. I think the major difficulty is that, in the turbulence region, there is a beam propagation problem: usually the two beams cannot be made to cross. To measure a double correlation, one needs a three-beam crossing in order to measure in two directions simultaneously. Therefore, to obtain the same accuracy, implementation time and cost go up sharply. As for the triple correlation, they are difficult and expensive to measure and I am not even sure that it is necessary to perform such measurements.

ROBBEN, University of California, Berkeley - I would agree with your table. If one considers the man-years involved in measuring a triple correlation, the time-cost scale is indeed out of sight.

OWEN, UTRC - From my viewpoint, it is the mean velocity field $\bar{u}$ which I need most, as I mentioned yesterday. This is because we have a combined computational-experimental approach to these problems, and the ultimate aim is to simulate numerically the combustor flow field so as to cut down the design costs. We have to determine whether or not our analytical models are worthwhile. The first thing you have to do is to map out the mean flow field in these complex three-dimensional flows to the computations.

WANG - Mean velocity is no problem at all. We can do it with $5,000 or even less.

OWEN - We are doing it but we are not doing it with $5,000.

WANG - Or you can buy a commercial unit for $20,000.

OWEN - Let's make one thing perfectly clear now, as clear as we can. That is that experimental equipment may be commercially available but the experiments are not commercially available. You can buy a two-color dual LDV system for $77,000, for instance, but you are not under any circumstances going to take that device and use it in a combusting flow field by just taking it off the shelf. You have to know what you are doing and it's important that we make that clear.

PENNER - Consider a smart fellow, starting from scratch on a prototype engine. Would you consider Dr. Wang's estimate for $\bar{u}$ measurements (i.e., $50,000, less than one year) unrealistic?

OWEN - I would say it was a little optimistic. Besides, you can't generalize. In our case, we have a small burner, 5 inches in diameter. Therefore, problems of beam wander are small and we have no problems of intermittency due to infrequent beam crossing. In larger configurations, these problems are going to be there. Also, in our system, we are able to forward scatter, and have had no problem in getting good data rates. If we were in a geometry with no through optical access we would have to backscatter: the problem would become much more difficult.

PENNER - Of course, your burner is not a prototype combustor. Can one design a prototype combustor with such favorable instrumentation features?

OWEN - Unfortunately, no. Prototype combustors are not designed for the instrumentation which we want to use! Let me proceed with the properties on the table. When you time average the Navier-Stokes equations, you get this correlation $u'v'$. You can measure the time average of that correlation by taking a single beam system and rotating it 45° as you would do with a single inclined wire. For example, with the laser beams orientated at 45° to the axial plane, the quantities $\overline{V}$, $\overline{u'v'}$ and $\overline{u'^2 + v'^2}$ can be determined from vector addition of the mean velocities normal to the two fringe orientations and from the difference and sum of the two variances respectively.

WANG - But, in time correlations, you have to measure u and v simultaneously.

OWEN - I think that Dr. Self made the assertion that it was impossible to obtain real time information.

SELF, Stanford University - It's a question of seeding. You need practically a continuous signal to get u as a function of time. This is very difficult to accomplish in gases and especially, I would think, in prototype combustors.

OWEN - One of the difficulties was brought up earlier by Chigier and Asher, for the case of the mixing of two flows at different velocities. You have to make sure that the flux rate of the particles is the same in both streams. It is a difficult but very essential thing to do. Otherwise you could grossly bias your data in favor of the higher velocity stream.

In short, there are going to be problems in trying to get real-time velocities but then I don't think that we need that sort of information at this stage. The turbulence models and calculation procedures which are being used are a long way from solving the Navier-Stokes equations in a time-dependent way. All we need are those top three quantities on Table I, and they are no problem to get.

CHIGIER, University of Sheffield - Anybody using an LV instrument has got to understand what he is doing and it is not just a question of pressing a button. But a person who has a good education, is intelligent and knows how to make measurements, can learn certainly within one year how to use this instrument. I think we also have to bear in mind that the development rate of these instruments is extremely rapid and that LV is not only being developed for combustion purposes but for many other purposes. There are people not working in combustion, who are looking also at all these problems and who might be ahead of us in some aspects.

I would agree that the measurement of triple correlations is unnecessary at this time and since it would take a great deal of effort, I would discourage people from even trying. To measure double correlations in real time, one has to guarantee that if you have a set of fringes in the vertical plane and a set of fringes in the horizontal plane, a particle is crossing both of them at that same instant of time. This is a very difficult thing to organize and I don't know anybody who has succeeded in getting that correlation.

Now, an instantaneous measurement in any one direction is easy to make. And if you want to make it in another direction you just duplicate and if you

want to make it in the third direction you just triplicate your system. You
just multiply the cost. Not only are these things possible but they have
been done in wind tunnels. Measurements have been made of the instantaneous
velocity components in the various directions but not the correlations as far
as I know. And I would also agree it is not necessary at this time.

RHODES, ARO - We have made radial and axial velocity measurements on a
free jet in the last few weeks. It cost over $100,000 and it took about a year
and a half to set ourselves up. The high cost reflects the fact that it costs
about as much to build all the hardware you need to get the measurement system
into the test cell, as it does to buy the LDV system itself. Also, it must be
remembered that these figures do not include the cost of running the engine.

Also, we have made some measurements of the instantaneous Reynolds stress.
They are in reasonable agreement with the hot wire data taken at the same time
and they are both in reasonable agreement with the theory.

PENNER - We have not heard in this session from the probe people. I
assume from the discussions this morning that you do not feel comfortable in
measuring velocity vectors in multi-dimensional flow? (audible snicker from
the LV crowd.)

ECKBRETH, UTRC - Let us keep in mind that in prototype combustors such as
ours (Pratt & Whitney) there is virtually no optical access. I think that in
many cases, one must step back and perform research on rigs especially developed to allow these techniques to come in. For example, if you have an annular
combustor, chances are very good that you could incorporate a coaxial backscattering scheme, but not forward scattering.

PENNER - This would cost how much in your judgement?

ECKBRETH - A good experimental rig might cost up to $200,000.

WOLFSON, AFOSR - You are talking about extensive modifications of a prototype engine, optical access, more sophisticated types of structural arrangements and so forth... Judging from past experience, you can revise your figure
to half a million dollars.

PENNER - On the other hand, if you consider an already developed engine
that is simply modified for the purpose of learning how to use these techniques,
I suspect that the cost would be more in the $100,000 range than half a million.
Am I right in that?

ECKBRETH - It depends on how sophisticated a simulation you want.

OWEN - The experiments that we're going to make for EPA in this turbulent
diffusion flame are essentially designed for this particular laser velocimetry
system, including forward scatter. We are talking about $60,000 just to map
out the mean flow field and the root mean square of the axial velocity fluctuations in the initial mixing region immediately downstream of the injection zone.

WANG - Clearly the cost of a satisfactory measurement depends heavily on how easily you can gain access to the volume of interest. The cross section difference between forward scatter and backward scatter is of about 3 orders of magnitude. To compensate for this loss, backscatter equipment has to be far more expensive, for the same basic accuracy. So it is not so much the basic technique but whether you can put two windows on the combustor or just one, and how far inside the region you can probe.

PENNER - I think there is another way to summarize this. The college professors who have used LV find the technique readily usable, while the people in industry believe, with good justification, that it can only be done with great difficulty and, as Wolfson pointed out, at very high cost.

WOLFSON - I did not mean to infer that the cost is always large, but industrial programs, by their very nature, are done in big test cells and big facilities. You are not going to put a prototype combustor normally in a university environment.

PENNER - There is a possibility that we might want to do just that. Alternatively, it might be desirable to develop a central facility where the university investigators can have their proper impact and support for this kind of research.

CONCENTRATIONS

ARTHUR FONTIJN
AeroChem Research Laboratories

Table II - An appraisal of Concentration Measurement Techniques (A. Fontijn). X_{ijm} is the concentration of species
at a given excited state: electronic (i), vibrational (j)
and rotational (m). CARS = Coherent Anti Stokes Raman
Spectroscopy.

Measured Property	Accuracy	Spatial Resolution	Method	Implementation Time	Cost
X_{ijm}	5% (for major species)	mm^3	Raman	Intermediate	Low
	5% (for major species)	mm^3	CARS	Intermediate	Low
	7%	Poor	Absorption-emission	Intermediate	Low
	7%	mm^3	Fluorescence	Intermediate	Low
	7%	mm^3	Probes	Short	Low

FONTIJN - The concentration X_{ijm} of a species might have to be known in
some of its excited states. I might add that I am surprised that no one has
mentioned ionized species yet, since Langmuir probe techniques for concentration measurements in rocket exhausts already exist. In Table II, the achievable accuracies are for laboratory systems, as indicated by the preceding
speakers (partially in private conversations) for their various techniques.
The achievable accuracies in jet engines may be expected to be less.

I have put the Raman cost in the low range. That does not include the cost of installation in a combustor test cell. Spatial resolution can be very high. It has been quoted by Lederman as one mm^3. Accuracy is partly a matter of calibration; although it is theoretically possible from first principles, you cannot obtain good accuracy for absolute concentrations, you have to calibrate. The Raman accuracy, in principle, could be of the order of five percent if you are within a reasonable concentration range (major species). To be candid, I don't quite believe in any accuracy figure for any one method, unless I can compare my results with those of some other method. If you have two methods that are more or less independent and they are both accurate to five percent and agree, then I am willing to believe it!

CARS is an improvement as we have heard from Taran and Harvey, in terms of threshold concentrations. Now the cost of CARS is considerably higher than regular Raman, at least two times as high. The accuracy is comparable.

Absorption spectroscopy has to be supplemented by an "onion peeling" analytical treatment to obtain local concentrations. The accuracy is comparable to that of Raman but spatial resolution is poor. The threshold concentration is usually low. Therefore it is not necessarily an improvement over Raman when you are dealing with major species such as CO_2 and H_2O, but it is a factor of 100 better for lower concentrations (CO, NO,...). This is an oversimplification but I guess it is necessary at this point. The cost of this technique is certainly in the low category.

Now we get to the fluorescence technique, which I have discussed briefly earlier in the workshop. Like the Raman technique, it is a local not a space-integrated measurement method and therefore can have good spatial resolution. The sensitivity for those species for which fluorescence can be used can be much higher than the Raman technique. An extreme example is the work by Wang and Davis [Phys. Ref. Letters 32, 349 (1974)] who, utilizing a tunable laser source, measured OH concentrations at ground level in ambient air to a detection limit of about 5×10^6 cm^{-3}, corresponding to 0.2 parts per trillion. The other day, we measured (AℓO) in fluorescence in one of our flow reactors at room temperature, down to roughly 10^{10} cm^{-3} on the first attempt. The cost of laser-induced fluorescence I would put as comparable to the Raman technique.

Sampling probes have been tried out to a large degree; they may be subject to big errors by themselves. I wouldn't trust probe measurements especially in engines, because of the possibility of catalytic reactions while sampling. Under proper conditions, as discussed this morning by Fristrom, you could use probes. The accuracy then would depend on calibration.

If you have molecular species, calibration is relatively easy. If you have free radicals then calibration becomes quite difficult. Therefore, I think that the figures on Table II are the best achievable accuracy for molecules only. I don't think for instance, that you can achieve this accuracy for OH. You may be able to achieve it say for sodium atoms or some other atom where you have reliable f numbers, or where you can develop titration techniques to calibrate the system. But I am certain that for many free radicals you won't achieve this accuracy.

Probes have been used extensively - e.g. in laminar flames - so I would think that the implementation time is not that high. In fact, I would say that all the methods listed on this table are similar to the velocity measurements on that score: a lot of the effort is in adapting your engine to allow the measurements to be made.

PENNER - The costs you have listed in the table are the intrinsic equipment costs. This does not include salaries for the people working on it. Is this correct?

FONTIJN - Yes, the "cost" column should read low for equipment plus high for installation and labor costs.

PENNER - In short, the equipment cost in each case will be less than $50,000 and the actual implementation will bring it to something on the order of half a million or less.

LUDWIG, SAI - In the implementation time, one should consider that many of the cross sections of many of the species you are interested in are not known. Such are, for instance, fluorescence cross sections at the higher pressures met in combustion engines. Even for radicals at low pressures some of the absorption coefficients are not known.

TARAN, ONERA - Concerning CARS, I should like to make a few comments. Firstly, the accuracy is not that good since, with a given sample, one observes signal fluctuations on the order of 20% from shot to shot; the nature of these fluctuations is not fully understood, but we suspect changes in beam quality to be responsible for them (see Ref. 3 of our presentation). Secondly, this accuracy is somewhat independent of the pressure of the gas mixture. Thirdly, the concentration threshold may vary considerably from one method to the other; it is of about 10^{-3} in volumetric mixing ratio for CARS.

MCGREGOR, ARO - I think it's impossible to make the kind of generalization I have heard so far. The accuracy statements, the resolution statements, all depend on the particular experiment, on the burner you have, and on a number of other factors. Give me a specific case and I can tell you accurately the accuracy, the resolution and so forth.... but in this general way, no.

WRAY, Physical Sciences, Inc. - Also, accuracy is a function of money. To go from 5% to 2%, for instance, might triple the cost of the laser which you need!

MCGREGOR - You can also buy accuracy with longer implementation times.

SETCHELL, Sandia Labs - I would like to voice an opinion. The predicted capabilities of Raman spectroscopy have been demonstrated only for simple laboratory flames. We are just starting to learn of the limitations due to spectral interferences, turbulent fluctuations and unknown scattering properties of more complex molecules. Yet, you are quoting times and costs on your chart for actual prototype engines. I find those to be quite optimistic.

FONTIJN - The accuracies listed on Table II correspond to the best combustor conditions. In an environment full of particles, for instance, I anticipate the accuracy to become less. Also, please note that each of these techniques can measure only certain species and are limited in their range. To lift this limitation you may want to acquire a dye laser to scan the UV spectrum for instance: it would multiply the cost by 3 or 4.

In my opinion, simple fluctuations in the system are what prohibit you from measuring much more accurately, but I doubt whether on a jet engine anybody is interested in measuring properties more accurately. Even in a laboratory system, unless you do analytical chemical work, I don't think you could do much better.

PENNER - I find myself very disturbed about this whole discussion. The part of this activity which I know best is absorption spectroscopy. To hope for 7% accuracy with any kind of spatial resolution in a jet engine is a pipe dream. I can now understand why McGregor is worried about defining the system.

SELF, Stanford Univ. - I would just like to make the general comment that a lot of the experiences we are hearing about concern small scale experiments designed to suit the diagnostics. I've been struggling for several years in a really fierce hostile environment and we have to make such measurements on a shoestring. There is a big difference in time and money between measurements on a bench-top flow system and in a large industrial rig.

PENNER - That is the point we are trying to clarify. Some fantastic statements have been made regarding these techniques and they are probably correct. Now, if some people are willing to try these on some kind of jet engine prototype, the funding agencies won't be able to tell me five years from now that CARS, for instance, is a toy for academicians rather than something to try in a real engine. What is really limiting progress? I think what is limiting progress are the lack of adequate funding and the multiplicity of workshops (laughter).

WOLFSON, AFOSR - Let's really be practical about this. We are here to accomplish a purpose so let's not live in a dream world either. I know that the area is not sufficiently funded. No area is sufficiently funded today. But, on the other hand, let's not deceive ourselves either. It is important to first establish realistically that you could accomplish the job if money were available.

PENNER - Dr. Fontijn, who knows more about fluorescence than probably anybody around, says that he is now ready to try this in an engine with a reasonable expectation of success. Behind you is Dr. Taran, who says the same thing about CARS.

WOLFSON - But you have just told me that if you had all the money in the world you are not even going to get 7 percent accuracy...

FONTIJN - There are a number of things to say. I agree with Ludwig's comment that cross sections are often not known. Some laboratory measurements are needed there. Also, we were not talking about one technique to measure every species, as appears to be implied in some of the questions. Different species require different techniques. For the jet engine environment, you are not going to measure major combustion products via fluorescence because they absorb outside the near UV/visible and because of radiation trapping. You are not going to measure the minor products with laser Raman. Again to the point of the accuracy, the values mentioned on Table II are the best achievable from these techniques, i.e., in favorable conditions. But I don't think that in jet engine work, you are interested in these high accuracies. In fact, I would think that you would be very happy with a factor of two, wouldn't you? About the question of not being ready yet for real combustor measurements, I don't see why, given three years and the necessary funds, we couldn't apply to an engine what has been done with fluorescence in the lab. I would think that the same holds for CARS. Wouldn't you, Taran, agree?

TARAN - The present equipment is very sensitive to vibrations and temperature fluctuations. I would estimate a three to five years development time before making measurements on an engine.

PENNER - Is a three year estimate reasonable for it?

TARAN - Three to five years....

TEMPERATURE

SAMUEL LEDERMAN
Polytechnic Institute of New York

Table III - An appraisal of Temperature Measurement Techniques (S. Lederman. The distinction between T and T_{vib}, T_{rot} refers to the fact that the Raman and Absorption-emission techniques use the equilibrium Boltzmann ratio of two concentration measurements of specific vibrational Q branches or rotational lines.

Measured Property	Accuracy	Spatial Resolution	Method	Implementa- tion Time	Cost
T_{vib}, T_{rot}	5%	mm^3	Raman	Interm.	Interm.
T_{vib}, N_2	7%	mm^3	Rayleigh (Doppler shift)	Interm.	Interm.
T_{rib}, T_{rot}	7%	Poor	Absorption- emission	Interm.	Low
T	5%	mm^3	Probe	Short	Low

LEDERMAN - Raman measurement accuracy depends strongly on the signal-
to-noise ratio which in turn depends on a number of parameters, including -
inter alia - the laser power one can apply to the particles in the test
volume before causing breakdown. The best choice of Raman band depends on
the temperature range. At room or near room temperature one would try to
use the ratios of rotational lines as performed by Saltzman. He obtained
an accuracy greater than 5 percent. This choice would of course be dictated
by the very small Anti-Stokes vibrational signal at low temperatures. However at higher temperatures, including the combustion range, one can use
the vibrational Stokes to Anti-Stokes intensity ratio. With this method
one obtains a temperature accuracy of the order of 5% in the temperature
range from 500 to 2500°F. Lapp used intensity ratios of the ground state
and first upper state from the Stokes vibrational Q-branch band and obtained
the same order of accuracy for the temperature.

For spatial resolution, there is a trade-off with accuracy, since
smaller volumes include fewer scattering particles. Currently a 5% temperature accuracy corresponds approximately to a resolution of $1mm^3$. As for
the corresponding minimum species concentration, I would estimate an equivalent density of 0.1 atmosphere. I beg to differ with the previous speakers
regarding the cost which I consider intermediate, if not higher, exclusive
of the cost of engine modification and testing.

Probes are old methods which are available and proven. Their limitations in unsteady, three dimensional or destructive environments, have been
discussed earlier, but since they are immediately available, their short implementation time and low cost can be an asset in a number of practical situations.

BLOOM, PINY - Does the cost you mentioned include associated equipment,
such as on-line mini-computers or other equipment for the acquisition and
handling of data?

LEDERMAN - I only included in my estimates the minimal equipment necessary to measure the temperature (rotational or vibrational), a minicomputer
in this case. Of course, if one wants to automatize and obtain also concentrations, I would say that a data acquisition system would be more expensive, probably in the range from $100,000 to $150,000.

LAPP, GE - I agree with your basic comments. I too feel that it is unrealistic to expect to do anything sensible for less than roughly $50,000
in Raman Diagnostics for flames, considering the necessity for high-energy-
per-pulse lasers, signal detection apparatus that may have to work in the
tens-of-nanoseconds range, and associated data acquisition and processing
apparatus.

On the other hand, I don't think that anything like $150,000 are necessary for a computer-based system for density measurements, as long as we limit ourselves to major [and possibly intermediate (i.e. > 10 ppm)] concentration species and to simple data processing. Also it should be stressed that much of the apparatus required for this work can be used also in a wide variety of experiments in physical, chemical, and fluid mechanics research. Given this commonality, the investment in equipment is quite reasonable.

ECKBRETH, URTC - Costs do not even need be that high. There are commercially-available, flash lamp-pumped dye lasers retailing for between five and seven thousand dollars.* I am not going to guarantee their engineering reliability, but they are improving rapidly. They deliver one joule per pulse at a rate of a few pulses per second. If you buy that kind of laser, you could probably put a spectrometer together for a few thousand dollars, including the photomultipliers and narrow band interference filters. All in all, you could probably build a system for about $20,000 or $25,000.

LAPP - I agree that the cost of these new dye lasers is quite favorable and they may well be attractive in the 10^{-4} sec. to 10^{-2} sec range. But their reliability is not yet proven, and their inability to be Q-switched can limit their usefulness for shorter time interval applications. The more costly ruby and neodymium lasers have proven to be reliable. They can be Q-switched for luminous sources, as well as in the normal mode if desired.

I would also like to comment briefly, if it's appropriate, on fluorescence and Rayleigh scattering and on CARS as competitive methods.

Fluorescence is a very powerful tool in situations where we require a very high scattering cross section to ensure sufficient detection sensitivity. Even with quenching, fluorescence can still have an intensity high enough to be significantly more intense than Raman in many useful situations. However, in combustor flames, the hot gas is of somewhat unknown composition and temperature. Thus, we have quenching by a variety of species whose concentrations are often not well known, each of which has a different quenching cross-section. Furthermore, the cross-sections themselves are not known for all species of interest. This leads us to conclude that it is difficult to obtain quantitative data from fluorescence measurements. However, it should be stressed that much valuable qualitative data can be obtained, along with a limited amount of quantitative data.

---

*Phase-R Company, New Durham, N.A. See also Laser Focus 1975 Buyer's Guide, February 1975

Rayleigh scattering can be used to measure temperature sensitively, as has been suggested by Robben. But some problems still remain-notably in the area of the possible effect of different scattering cross sections for different gas components, producing alterations from a simple spectral profile. Species concentrations can only be determined in this fashion under limited conditions, because of the spectral overlap of the signatures.

CARS appears to have a tremendous potential value in these situations where it may be applied. It is relatively more delicate to instrument than some of the other techniques in combustion engineering environments. Furthermore, the effect of turbulence on CARS has not been clarified yet.

Perhaps we can summarize by saying that all of these optical techniques, as well as others not mentioned and the well-developed solid probes, have their particular virtues and shortcomings. Thus, we should look to combinations of these various techniques to aid our experimental programs in the ranges in which they are most appropriate. Vibrational Raman scattering, seems to be excellent in the high temperature range, for temperature measurements and for major species concentration measurements, especially in flows which exhibit strong temporal and spatial variations. Hopefully, the applicability of Raman scattering for such highly-stressed environments will compensate for the difficulty in dealing with lower concentrations.

PENNER, UCSD - Thank you very much. Can you expand on the possibility to make repetitive measurements, as might be desirable in turbulence work for instance? The 30 nanosecond pulse duration is for a single measurement. What is the repetition rate for such lasers?

LAPP - This depends on the type of laser chosen, and on the interaction of many experimental parameters. For highly luminous flames where you need a high energy per pulse (say 10J) to overcome background signals, the repetition rate might be as low as 1 pulse every few seconds. For many other situations, requiring only a few tenths of a Joule per pulse in the visible range, lasers can be found with a pulse rate of 10 pulses per second or more. To obtain a complete time history of the turbulence spectrum, we would require a laser repetition rate perhaps 3 orders of magnitude larger, which doesn't exist today with a pulse energy that is anywhere near adequate.

ROBBEN, UC Berkeley - Still you can double-pulse your laser, and obtain the full turbulence spectrum, by measuring the auto-correlation function. With two pulses, say 1 millisecond apart, and a repetition rate of 10 pulses per second, time averaging the auto-correlations will lead to the correct average value. A set of measurements at other pulse spacings will give to the complete auto-correlation function, and hence the turbulence spectrum by Fourier transforms.

THE SPRAY REGION AND
THE COMBUSTOR ENVIRONMENT

NORMAN A. CHIGIER
University of Sheffield

GLASSMAN, Princeton Univ. - The most difficult problem facing the turbine manufacturer is how well and how inexpensively he can design or develop his turbines. Therefore he doesn't necessarily have to understand the
combustor turbulence phenomena; he would like to know the temperature distribution of the gases coming out of the combustor can, their velocity and
also whether particles are present or not, because particles (solid and
liquid) give off hot spots on the turbine blades. These are the measurements we should concentrate on: the mean values of temperature, velocity
and particles concentrations. To measure these same properties at the
engine exhaust, one does not need some of the laser instrumentation which
was discussed at this Workshop. If I were running a big company, I would
insist upon the use of a good immersion probe, a sonic probe of the type
we used in ram rocket work for Project Squid 25 years ago. I would draw
out samples and analyze them. That would give me the information I need
to meet EPA pollution requirements.

If I wanted to do research in combustor design, then I could use the
sophisticated laser instrumentation effectively. If I have to design a
recirculation zone in a turbojet can or behind a bluff body in a ramjet,
I have emission and stabilization problems. Then I have to know about
turbulent reacting flows and to look into all the aspects which were discussed here. These problems are at a different level of sophistication than
the ones presently plaguing the gas turbine manufacturer.

BLOOM, PINY - It should be noted in this regard that the character of
the turbulence in the flow may be important for such things as heat-transfer
to turbine blades. Its relative importance depends upon the design situation, but increased operating temperatures are desired in a number of situations and the margin of safety of the blade material may become quite small.

CHIGIER, Univ. of Sheffield - When we talk about making measurements in real engines and under spray conditions, we have to bear in mind that in Europe and in particular in places like the International Flame Research Foundation, measurements have been made over a period of 20 years, on velocity, temperature, concentration of gases, concentrations of solid particles at various points. Also measured was the heat transfer from the flame to the surroundings, in large industrial furnaces under very dirty conditions, using heavy fuel oils and even using pulverized coal. A whole range of instruments was developed for making these measurements under such conditions. Measurements can be made inside the spray region by using stainless steel probes which are allowed to heat up, so that you will not get blockage of these probes. These measurement techniques have been established and I would say, that one could readily make these type of measurements inside an engine. It is about time that this was done and I think there are a number of people who are beginning to do this.

When we think of probe measurements, nobody really knows exactly what the accuracy is. We assume that it is not too bad and we talk about perhaps 5 or 10 percent if the disturbances are not very great. But in general, one is using large probes with volumes of the order of one cubic centimeter. They are cooled probes and they probably create disturbances in the flame. Nevertheless many measurements have been made and we are reaching the stage now where the advanced prediction procedures, such as those Spalding is making for flows with recirculation, reverse flow zones, etc..., are being directly compared with the measurements, and that's the state of the art today. In terms of the time required to do this in a practical engine, this can be done within one year at a relatively low cost because the cost of these probe instruments is really quite negligible.

For velocity, you use a Pitot tube which is water-cooled. When you come to temperature, you use suction pyrometers and some of us even believe that a bare thermocouple will also give you a reasonably good measurement. To measure your gas concentrations, use suction probes and gas phase chromatography. With these you can get a measure of many of the constituents. However, we measure nitrogen oxide distributions by the chemiluminescent technique. When it comes to particle size, we introduce large filter probes. We wait until the filter gets full, we pull it out, we measure the weight of particles at that point and the size distributions. It is a very tedious approach, but you can get the required information. When it comes to the question of heat transfer, we mustn't forget that for most practical systems, the whole purpose of the flame is to transfer heat and there are well established means of measuring the heat transfer and separating the measurement of the radiative and the convective heat transfer. One can use all of those techniques and they can be made in a short period of time at a relatively low instrumentation cost.

Now we come on to the question of <u>optical</u> techniques. There are a
number of us who have already established that you can use some of these
optical techniques in the type of laboratory which we have at Sheffield,
which deals with "large university flames" (laughter). Here you have small
university flames but there we have large ones: we have large laboratories
and large systems. In general, we look at open flames under atmospheric
conditions but I think that all those techniques which are being developed
can easily be adapted to the type of condition under which you want to make
the measurement. Now let us consider LDV. If you have a spray condition,
you immediately run into the problem that you have a spectrum of drop sizes.
We have established that you can measure the velocity of all size of
particles using a standard LDV system to get a mean velocity in any one direction. However, I think it is very important that we should be able to
simultaneously distinguish between the velocity, size and concentration of
the particles. There exist some ideas on how to do this, but they require
to be tested in the laboratory before they can be used in a real engine.
When it comes to concentration, I would say that the field of laser-Raman
spectroscopy is an extremely exciting one but it belongs at this stage in
the university and in the research laboratories and we are in no way near
to being ready to use this in a practical engine. I think if the rate of
development in laser Raman spectroscopy is as rapid as it was in laser velocimetry, then perhaps in three years time we would be ready to use it for
the purpose.

In terms of <u>particle size</u> distribution it is extremely important from
a practical point of view to know the size distribution of particles which
are entering the system and the size distribution of the particles which
are leaving it; there is a big need for a suitable instrument. That's why
I think so much emphasis needs to be placed upon the laser anomemeter because this offers a possibility of making such a measurement. The method
of putting in a probe and actually drawing it out at each point is such
a tedious one that one hesitates to recommend it. The optical method of
scannig right across the combustor provides an overall average of the size
distribution but that is not useful for us. We want to know particularly
the size and the number of the large droplets. Practical atomizers produce
these large droplets of about 100-200 microns, and they can pass through the
combustor and come out through the exit. That's where major problems arise,
so I would say that this is a region in which we need to concentrate our
efforts.

In your list of instruments you left out the <u>heat transfer</u> probes. These
can be used in order to separate the rate of heat transfer by radiation and
by convection. We can do this with relative ease by placing probes which can
be introduced into the system itself or at the walls. Correlations have been
established between the particle size distribution of soot particles and the
radiative heat transfer from flames.

In the general field of sprays, the velocity, temperature and concentration measurements have got to be made under dirty and difficult conditions. But they have been made in cement kilns and other environments which are much more difficult than those which we will have in our prototype engine.

PENNER, UCSD - What about holograms to characterize spray burning?

VEST, Univ. of Michigan - In my work in holography, I have not had really the experience of droplet or particle measurements under this kind of conditions. Still, I think that holographic techniques or parametric techniques may give a better feel for the particle size distributions than the methods discussed so far. We may be able to offer "instantaneous whole field" methods, the resolution of which will not be as good as that of these other methods, but which gives semi-quantitative and in some cases quantitative visualization of what's going on in the entire field. It's a complementary method and I hope that it will be improved further.

WOLFSON, AFOSR - Chigier brough up an interesting point. We are missing here a segment of the technological community which has addressed a number of these problems and have resolved some of these issues already: namely the chemical industry. They have been for years involved in things such as catalytic combustion, steel mill operations, for which they have developed pyrometers and other types of very unique instrumentation. Such people should be involved in our discussions.

BLOOM - Currently enhanced interests in detailed and sophisticated measurements of jet-engine processes appear to stem from an attentiveness to pollution effects, rather than the expectation of major improvements in such factors as performance, operating behavior and safety, which are the traditional concern in military applications. Although improvements in both performance and pollution control depend upon a steadily enhanced knowledge-base, it may be that, in the past, the pay offs of transfers of measurement technology from furnaces and chemical processes to engines were not judged to be high in military terms. However, the need to move our technical base forward for all purposes, certainly makes this type of technology transfer, where possible, most desirable.

PENNER - I might inject here a word of restraint regarding these wonderful things that are being done about combustors in the steel and chemical industries. I did spend quite a bit of time on these problems once, and I am still following the measurements they make and how they interpret them. My feeling is that most of the people here would accept neither the measurements nor their interpretation.

ROQUEMORE, AFAPL - I would like to comment on Dr. Chigier's statement concerning the immediate applications of laser Raman scattering (LARS) to practical combustion problems. I believe it would take 2 to 3 years for a laboratory with no experience in Raman spectroscopy to build a Raman system and develop the expertise required to apply LARS to combustion problems. There are several organizations which have developed a Raman capability which could be applied to practical combustion studies in the near future. AFAPL is just beginning a long term (5 years) program to determine the capabilities and limitations of the LARS technique as a combustion diagnostic tool. The program will consist of making Raman measurements of temperature and specie concentration profiles in a variety of different combustion environments. This will include measurements in the exhaust of an afterburning jet engine, at the exit of a ramjet and T56 combustor and in the controlled combustion environment of a combustion tunnel. The basic LARS system that will be used in these studies is described in a report by Leonard, (AFAPL-TR-74-100; available at DDC under A D number ADA003648). We have not performed sufficient work at this time to determine how applicable Raman will be for these combustion measurements. I would like to hear some comments from the Raman people about the prospects of making LARS measurements in practical combustion environments.

LAPP, GE - Raman temperature measurements have already been accomplished in many laboratories for flames and jet flows. For work in a combustor, there are no basically different problems in the physics of the measurement technique, but there are certainly profound difficulties involved in hardening the apparatus for the severe engineering environment. I feel that laser velocimetry is in a much more advanced state of development than Raman disgnostics at this time, as Chigier said, but I have no doubt that Don Leonard's apparatus will be effective in combustor measurements.

ROQUEMORE - An important feature of the AFAPL LARS system is that it is sufficiently hardened to operate in the field or test cell environment. It has functioned well in the close vicinity of jet engine exhaust where major species concentrations and temperature have been made. The basic equipment cost is about $80,000. However, the design, fabrication and tests bring the total cost to around $500,000.

REYNOLDS, Tech. Ops. - Holographic particle sizing equipment could be used to give you a size distribution as a function of the distance along the holographic path. Automatic read-out techniques are available but it makes the system more expensive. As we saw in the movie, holographic interferometers give us a lot of visual information. A lot of research has yet to be done to analyze all off this data quantitatively. But particle sizing could be done now because the technology has advanced that far.

RHODES, ARO - We have used laser holography in our engine test facility - in icing tests - to characterize particle sizes in sprays. It works well, if you do not use too much depth of field. If you have a deep spray you get a signal degradation. Again, I don't think we can overemphasize the problem of getting information out of the holograms. Right now we have spent a large amount of money to try to do a semi-automatic treatment and it is very difficult.

CHIGIER - Dr. Bloom raised the question of why a lot of the technology that was developed in furnaces was not applied to propulsion systems. For a long time, it was considered that the combustion efficiency was so high, of the order of 99%, that it was not worthwhile making any investigation of the combustion system. What has changed the situation is that now engines are required to satisfy environmental restrictions and also the increase in cost of fuel. The scene has changed and it now becomes necessary to know what's going on inside these chambers.

I am also very pleased with the comment in which Dr. Roquemore contradicted what I said about Raman spectroscopy. It shows that if you take people who are competent and you give them the support, you can drastically shorten the time it takes to apply to a very large system, an instrument which works well on a small flame. It has been shown that instruments which are optically sensitive can be made less sensitive by bolting them down and holding them rigidly.

For particle sizing, we use photography and we run into this problem of whether droplet images are in focus or out of focus. The advantage of holography is that you can take one hologram of a whole section of the spray and then you can play it back and move in and out of focus along the width of the spray. I think that these techniques are well developed and they are not particularly expensive. The analysis of these photographs has in the past been long and tedious, but it is now becoming much more manageable. Also, you can cut down a lot of effort if you count only those particles above 100 microns or so in diameter, for instance.

PENNER - Again we find that knowledgeable people are quite willing and ready to apply the scattering techniques to the spray combustion region.

TURBULENCE

S. N. B. MURTHY
Purdue University

MURTHY - I should begin by referring to a workshop held here at Purdue about a year ago on Turbulent Mixing in Non-Reactive and Reactive Flows. One of the important aspects of that workshop was the fluid dynamics of turbulent combustion. Three things became clear then: (i) Very useful computational schemes are coming into existence for reactive system calculations when turbulence is present. (ii) Extremely interesting advances are occurring in the understanding of turbulence structure in non-reactive systems with obvious overtones for reactive systems, for example, through the establishment of connections between turbulence strain and vorticity on the one hand, and turbulence strain and product formation through chemical action on the other. (iii) Finally, measurements in important flows are still lagging behind developments in theory.

In this workshop, some statements were made which should be kept in mind:

(a) Basic research in classical turbulence is unlikely to find direct application in combustion.

(b) At the same time, there were fundamental developments in turbulence which had important implications for combustion modeling, for example the coherent structure associated with large, strong eddies.

(c) In regard to the interaction of combustion with turbulence, one can only generate evidence through pressure velocity correlations and higher order correlation or through Mach number/compressibility effects.

(d) The coupling between combustion noise and turbulence is still ambiguous.

(e) The situation in regard to turbulent flows is such that, in order to avoid matching experiments to a theory that is essentially a postulate,

412

one has to accept considerable redundancy in measurements.

(f) Turbulence models were shown to be of practical value in view of
the great advances in computational techniques, for example the semi-
elliptic flows such as recirculation regions.

Now, at a certain level of sophistication, one can ask the question:
what measurements are required to establish the mutual interaction of combustion and turbulence?

Obviously, this question may only be answered in selected flow situations.
Some flow configurations of interest, in increasing order of complexity, are
as follows: a jet plume, a chemical laser cavity, a dual chamber burner with
and without flame stabilization, a dump combustor and a reciprocating engine
chamber. What I am trying to emphasize here is that the environment essentially determines the characteristic length and time scales.

In determining the type of measurements, it is necessary to bear in mind
that we are dealing with eddy sizes from 100 microns to 100 mm., a temperature
range of 20°C to 1500°C with fluctuations as large as 100°C, certain species
in ppm and certain products in percent ranges. In addition, we should note
the range of Mach numbers involved. If we take in account the various scales
of significance, the integral scale, the Taylor microscale, the Kolmogoroff
scale (length and time) and the Batchelor scale, we may find that in order to
obtain a sufficiently detailed picture of the turbulence, one may need 10^{10} -
10^{20} bits of information at 0.1% accuracy. This should be compared with the
practicable analog capability of 10^8 bits at 0.1% accuracy or with the digital
capability of building up a probability density function with 10^5 - 10^7 signals at signal frequencies of 30 kHz to 100 kHz.

Let me now divide my remarks into two parts: (1) questions pertaining to
turbulence modeling and (2) general advances in turbulence with chemistry.

If we consider the best of the models, we have basic equations for momentum, turbulent kinetic energy, dissipation and Reynolds shear stress, with
models for scales and dissipation. If chemistry is included, we include the
species balance equation and the reaction equations. Several things can now
be stated about the status of experiments:

(a) Experiments have been of little help to date in distinguishing
between gradient diffusion and closure models. Transport processes in
general may be looked upon as (i) wave-like in small time and diffusion-
like in large time, (ii) analogical to radiative transfer, or (iii) de-
scribable with closure schemes. Measurements are not available in the
detail necessary to distinguish between those.

(b) Active scalars have been sought to be replaced by passive scalars
in various models under various approximations. Measurements are required which can bring some order to those approximations.

(c) Reported measurement schemes seem to indicate some assumptions relating to (i) the number of measurements and the turbulence intensity and (ii) the number of measured quantities and the assumption of isotropy.

(d) Measurement techniques are not unambiguous about the measurements, whether they are Favre-averaged quantities or density-independent quantities.

(e) The type and extent of the measurements relating to pressure fluctuations do not permit a clear relation between either the pressure fluctuations and the eddy size or the magnitude, convection and decay of the fluctuations and the wall shear stress, for instance.

(f) Finally, although direct modeling of pdf or intermittency is being suggested, the extent of the measurements available pertaining to energy spectra, auto-correlations, probability distributions, two-point correlations and joint density functions only permit modifications to traditional modeling in special regions.

More specifically, in regard to velocity fluctuations, it is clear that while convective and production terms can be measured, the diffusion can be only partially measured and the dissipation and redistribution cannot be measured at all.

Similarly, in the case of scalar quantities, while the flux and the level of fluctuations can only be partially measured, the situation is very poor in regard to diffusion and redistribution.

Turning now to the general advances in turbulence with chemistry, experimentors have established some clarity in regard to (i) the postulated double structured nature of turbulence vs. coherent structure and (ii) the occurrence of significant events with a fairly sharp and definable statistical mean period. Detailed measurements are of course yet to be devised especially in the presence of scalar quantities.

Meanwhile, there is a growing feeling, essentially based on observations, that (i) there may exist a defect in the notion of a random velocity field superimposed on a mean flow, (ii) there may be possibilities of developing an interaction model between a time-dependent, quasi-ordered mode and a small-scale random mode and (iii) it is necessary to recognize that turbulence production at a fixed point has to be related to the large scale structure. These offer tremendous possibilities if adequate and systematic measurements can be made.

I would end by referring to the problem of engulfment at an interface which arises so often in combustion systems. We can formulate three questions as follows:

(a) What experiment and measurement will distinguish between (i) the random engulfment of large volumes of fluid by the large scale structure which will eventually become mixed by molecular processes through small scale eddies and (ii) the engulfment as a result of vorticity diffusion by viscosity and its augmentation by straining and microscale convolutions of the interface and as being independent of large scale motions that convolute the surface?

(b) What is the relation between the dynamics of a deforming, advancing turbulent entrainment interface and the dynamics of a turbulent flame front?

(c) What is the measurement that is required to establish a basic relation between turbulence intensity and scale and flame structure and propagation?

OWEN, UTRC - I think we are equipped today to make three-dimensional velocity measurements in turbulent flows with LDV, as it was discussed earlier. For temperature, Rayleigh measurements are good, but I don't think that they will work if you have particulates, as you will have in combustors. The weaker Raman measurements, as discussed by Lapp, should give us eventually specie concentrations primarily, but also temperature. For this we need, say, 5 years and $500,000. There is now a laser which would enable us to do this; it sells for almost $75,000.

ROQUEMORE, AFAPL - How are you going to measure correlations?

OWEN - By closely spaced double pulses. You can then build up the autocorrelation function over a period of time, then you can transform it into the spectrum at various scales. You can produce these double pulses from 100 microseconds to 10 millisecond intervals. You can reduce the interval time to as low as 1 microsecond, I think.

WANG, Aerospace Corp. - A few years ago we began an ambitious program of using Raman scattering technique to measure density fluctuations in a turbulent wind tunnel. We have built a 120 watt CW Argon Ion laser.* We can make continuous measurements or, if we wish, we can use a cavity dump technique and pulse this laser to several kilowatts at megahertz frequency. Unfortunately our financial support was terminated, and this high-power ion-laser is instead being used in a laser isotope separation program. If the original program were restored, I feel that we could achieve still better frequency response in our turbulence measurements.

RHODES, ARO - How do you differentiate between the randomness of the natural event and that of the laser pulse?

OWEN - I average out both over a period of time.

MURTHY - It looks as though the feeling exists that it is possible to make

---

*C. P. Wang and S. C. Lin, "Performance of a large-bore high-power argon ion laser," J. Appl. Phys. $\underline{44}$, 4681 (1973).

concentration and temperature measurements in turbulent-flows with some further development. What is still in question is the kind of flow configurations where one can make such measurements so that the modelers may be able to adopt at least a building-block approach with some confidence.

PART V

Review and Suggested Experiments

R. GOULARD
A. M. MELLOR
R. W. BILGER

INTRODUCTION

In this note, we shall consider the current measurement needs of the jet engine combustion engineers and scientists and we shall try to assess the potential of the probe and optical technologies to meet them.

In the first part, the available instrumentation will be evaluated and compared for the following criteria: non-interfering access to the flow, specificity, accuracy, sensitivity, space and time resolution, and cost effectiveness. The flow properties of interest are: velocity, pressure, temperature and the components of the flow (N_2, O_2, CO_2, H_2O, CO, NO, NO_2, OH, THC's and smoke).

In the second part, the different needs of the development engineer, the combustor researcher and the fundamental combustion scientist will be established separately. This distinction is essential, as each group has different objectives and constraints (see Fig. 1). The set of measurement criteria listed above are likely to be emphasized in quite a different manner by the test engineer facing a complex and specified combustor assembly, and by the chemist elucidating the kinetics of a particular reaction in an apparatus designed for measurement convenience. For example, those measurements generally available to the development engineer and/or designer might be termed performance-type (radial exhaust plane temperatures and average concentrations of pollutants), and change as a result of variations and interaction in the combustor operating conditions (inlet pressure, temperature, air flow rate, and overall equivalence ratio), combustor geometry, and fuel type. Dependence on such measurements (and upon intuition regarding the processes occurring within the combustor) can lead to totally unexpected results, particularly for changes in liner geometry (see for example Gradon and Miller, 1968, and Mosier and Roberts, 1974a). Diagnostic measurements (i.e. directly inside the combustor), not really of secondary importance, but not generally obtained except in water tunnel testing, would then include flow visualization in the diffuser and within and around the combustor. Also in this second category would be axial temperature and concentration profiles at selected azimuthal planes. These diagnostic measurements show directly the cause of design changes on exhaust plane parameters. Desirable levels of measurement criteria will be proposed for all three situations. Several experiments are suggested.

419

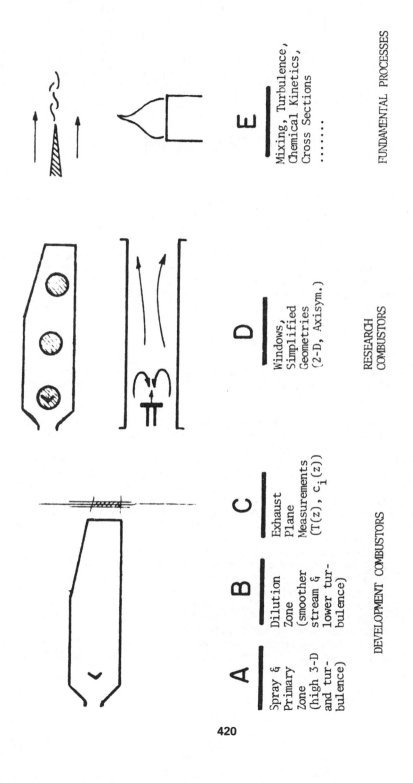

A — Spray & Primary Zone (high 3-D and turbulence)

B — Dilution Zone (smoother stream & lower turbulence)

C — Exhaust Plane Measurements $(T(z), c_i(z))$

DEVELOPMENT COMBUSTORS

D — Windows, Simplified Geometries (2-D, Axisym.)

RESEARCH COMBUSTORS

E — Mixing, Turbulence, Chemical Kinetics, Cross Sections ....

FUNDAMENTAL PROCESSES

Fig. 1 - Areas of Interest in Combustor Measurements

I. MEASUREMENT TECHNIQUES

In this part, we shall first define and discuss the criteria desirable for good measurements, especially in the context of combustion
gases. We shall then review and evaluate some of the current measurement techniques and their future prospects.

A. PERFORMANCE CRITERIA

One could list many requirements for effective measurements (Lapp
1973, Ludwig 1974, Hartley 1974, Parts 1974). For our purpose, they
might be grouped into the five categories mentioned in the introduction:
non-interfering access to the flow, specificity, accuracy, sensitivity,
space and time resolution, and cost effectiveness.

1. Non-interfering Access to the Flow

A characteristic aspect of a combustor is its relatively small en-
closed geometry where complex and potentially material-destructive phenomena take place. Hence, the least flow disturbance might lead to a
significant change in the energy transfer pattern, possibly destructive
to the combustor liner. Therefore, the instrumentalist is faced with
the challenge of devising a probe which does not melt and which disturbs
neither the flow nor the walls of the combustion chamber. This, of
course, is hardest to achieve inside a production combustor, but relatively straightforward in a "research" combustor or in a simplified
combustion setup.

The situation is summarized on the attached sketch (Fig. 2), where
the increasingly indirect methods to assess the flow properties at some
point A inside the combustor are listed from left to right. Clearly,

421

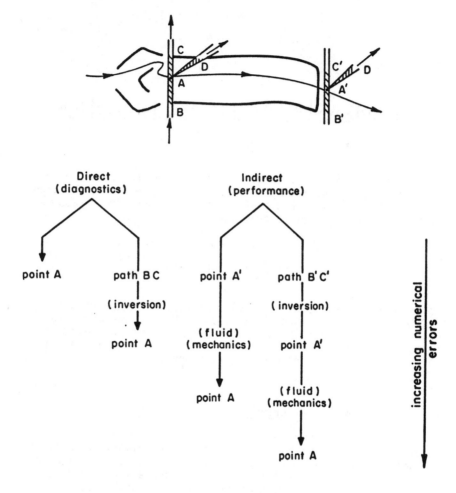

Fig. 2 - Direct and Indirect Measurements in a Combustor (Point A)

"path" measurements (BC, B'C') necessitate a further error-generating mathematical treatment, as compared to "point" measurements (A, A'). This drawback is often compensated by their better sensitivity: the absorption cross sections are orders of magnitude larger than scattering cross-sections (see Table 1 on page 13).

Similarly, the near-impossibility of probing inside a combustor makes necessary a fluid dynamics treatment of the streamline A A', in order to induce the properties of A from those measured at A'. There again, the accuracy obtainable in "open" conditions (A') might compensate for the errors of the fluid dynamics model, although it is still a highly qualitative procedure at this stage (Mosier and Roberts 1974b).

Hence the need to compare overall accuracies rather than purely instrumental ones. It is difficult to rate these analytical procedures in a general way. The fluid dynamics models are rudimentary at best and attempts at sophistication have yet to be correlated satisfactorily with combustor experiments (see Table 2, and also Roberts W*). Similarly, the inversion procedures used to convert path measurements to point by point data tend to be ill-behaved (Wang, 1970), whether they are based on the assumption of axisymmetry ["onion peeling" (Neer 1974, McGregor 1972)], or on advanced knowledge of some other profiles [frequency scanning (Chao and Goulard 1974)]. On the other hand, there exists a record of reasonable success (about 5% accuracy) in some specific cases of jet flow (Herget 1968, Blair 1974), especially where large scale turbulence is absent.

Another difficult aspect to quantify is the disruptive effect of solid probes inserted in the flow field. Not only do the provisions for cooling the immersed probes make them bulkier, but there are regions in the flow where the rate of chemical reactions could be strongly affected by the changes brought to the flow field by the probe or even a wire (Fristrom W). It is estimated later in Part II, that an obstacle area of 1% of the total combustor cross section would not induce any appreciable loss of accuracy. A more serious concern is due to losses in component concentration which might take place in sampling lines. In such a case (Roquemore 1975), even qualitative explanations are missing.

---

*Throughout this report, the symbol W will refer to the proceedings of the recent Squid Workshop (May 1975) on "Combustion Measurements in Jet Propulsion Systems" (to appear).

2. Specificity

Three possible causes can reduce the experimentalist's ability to isolate a property of interest in some small volume in space: the existence
of background signals from other parts of the flow, that of signals from
other sources in the small observation volume itself, and the dependence
of the signal strength on other properties than the one of interest.

     a.   background signals are dominant in combustion applications.
Setchell (1974) shows for instance that the luminosity of a flame contri-
butes from 1% to 10% to the Raman signal between 1600K to 2200K. Also
the combustor boundaries produce a strong black body signal in optical
measurements. These effects are often eliminated by measuring them independently. Those which display continuous spectra are often assessed by
measuring the background intensity outside the spectral line or band of
interest.

     b.   other components may contribute signals which overlap with
those of interest to the investigator. It is again relatively simple to
eliminate continuous spectral sources, such as particulates and soot. It
is more delicate to untangle two overlapping bands (e.g. H_2O and CO at
4.7 μm), but this is done to a good degree of accuracy, by the use of
spectral correlation plus masking filters (Burch 1974). Actually a common
technique for enhancing the signal from a given component is to use it as
well in a discharge tube as the source of radiation, thus correlating
emission and absorption implicitly (Jachimowski 1975, Neer 1974, Davis,
McGregor and Few 1975).

Some other techniques accomplish this separation readily: Raman
scattering shifts the signal with the laser frequency chosen for the experiment; it is usually possible to move the spectra of interest into a spectral range (e.g. visible) where the background of IR signals from the flame
is eliminated. However, this still leaves the possibility of an overlap
among Raman bands themselves, especially at the temperatures characteristic
of combustors. Leonard (1974) found that the Raman signal from the head of
the vibrational band of CO lies very close to that of N_2: with the thermal
broadening of such bands, overlap occurs and small amounts of CO might be
difficult to detect against the dominant concentration of N_2. Similarly,
the fluorescence generated by unburnt hydrocarbons (THC) tends to drown
the weak NO signals. Extensive time resolution or alternate approaches
(e.g. polarization) may become necessary to separate such signals.

A promising approach would be an absorption technique using a tunable
laser*, the narrow lines of which ($2cm^{-1}$) can select the center of a "clean"

---

*still limited, unfortunately, to rather low power levels such as 10^{-3}
watts cw (see Wang 1975).

line out of a whole band set (Hinkley 1971, Sulzman 1973) for most combustion products, including SO_2, OH, NH_3, etc.

   c.   An important consideration in measurement techniques is whether the acquired signal is <u>dependent on properties other than the particular one of interest</u>. Typically a pitot tube measurement will yield the velocity only if one knows the density. Also optical absorption and scattering are strongly influenced by pressure-and-temperature-controlled broadening mechanisms. Therefore an independent measurement of temperature is often necessary. Conversely, the standard optical method of temperature measurement is to measure the ratio of the populations of two excited states. Since it is a ratio, calibration is automatic and since any component can be used, the strong signals from N_2 can be used. Therefore, temperature turns out to be easier to measure accurately than the component gases concentrations themselves, especially in the low ppm range.

   If concentration measurements are needed but temperature data are not accessible, it is often possible to use those spectral intervals of the rotation-vibration bands which are relatively insensitive to temperature. An optimal choice of lines, with concomittant weighing, can bring this dependence to an arbitrarily small level (Chen 1975).

3.   Accuracy and Sensitivity

   These two aspects are linked closely. <u>Accuracy</u> is quantitatively* defined by the signal-to-noise ratio $\frac{S}{N}$ and <u>sensitivity</u> is defined by the threshold signal (in watts or photons) which can be measured. Clearly this threshold is a function of the noise level and those two quantities will be discussed together.

   To appreciate the noise problem in optical measurements it is perhaps simplest to write the fundamental equations describing these processes (see also Fig. 3):

$$\frac{E_m - E_s}{E_s} = - YN_i Q_{\nu i} e \qquad\qquad \text{Absorption} \qquad\qquad (1)$$

---

*In practice, one refers often to the inverse of $\frac{S}{N}$ in percentage points: a 1% accuracy corresponds to a signal to noise ratio of 100.

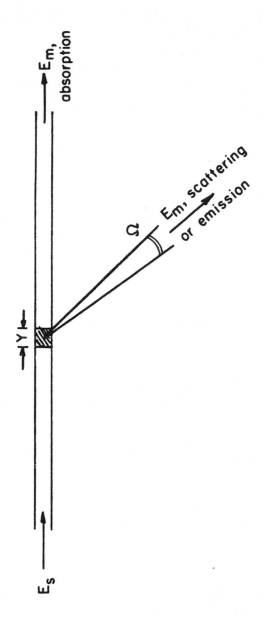

Fig. 3 - Optical Measurement Geometry

$$\frac{E_m}{E_s} = -YN_i Q_{\nu i}(\Omega e) \qquad\qquad \text{Scattering} \qquad\qquad (2)$$

$$\frac{E_m}{B_\nu \Delta\nu} = YN_i Q_{\nu i} e \qquad\qquad \text{Emission} \qquad\qquad (3)$$

where E_m and E_s are the energies measured by the sensor and emitted by
the source respectively. Y is the length of the beam section of interest,
N_i is the component i population per unit volume, $Q_{\nu i}$ is the cross section
of the component i at the frequency ν, e is the optical efficiency of the
system and Ω is the solid angle subtended by the collecting optics of the
sensor. $B_\nu \Delta\nu$ is the Plank function integrated over the frequency interval
of the sensor. For convenience, all these processes have been written in
the optically thin limit, the only one of interest most of the time. Typical values for Ω and e are 4.10^{-2} and 10^{-1} respectively.

In those equations, N_i is the unknown quantity. It will be known
accurately if all the other terms are known accurately too, and it will be
capable of sensitive measurements if low threshold levels can be measured
accurately for E_m-E_s (absorption) or for E_m (scattering). Note in passing
the weakness of the absorption approach (which requires an analytical inversion process if the flow is not uniform over all the segments Y along
the path) and that of the scattering process [whose cross-sections are
orders of magnitude less than absorption cross sections* (see Table 1)].
The emission process is often used for heat transfer measurements (Ludwig
1973, Beer 1974), but infrequently for diagnostics if a source can be
found so that $E_s \gg B_\nu \Delta\nu$ (usually available with current lasers). Note,
however, that the emission term (Fig. 3) may have to be entered as a correction factor on the right side Eq. 1 at least in high temperature flames
(Setchell 1974, Lewis 1975).

One might classify the various noise contributions by their origin
in the measurement system:

---

*Note also that in some cases, usually for radicals and low densities,
fluorescence also has the advantages of both scattering (point measurement)
and absorption (large cross sections), provided collision recombinations
(quenching) does not make the interpretation of N_i a difficult matter of
coupled kinetics.

TABLE 1

ORDER OF MAGNITUDE OF TYPICAL
PHOTON-PARTICLE INTERACTIONS

Particle	Interaction	Cross Section
10 μm diam.	Mie	10^{-7} cm^2/sr
0.1 μm diam.	Mie	10^{-13} cm^2/sr
Classical dipole	Oscil.	10^{-10} cm^2
"	Oscil. (collision broadening)	10^{-12} cm^2
Molecules	Absorption	10^{-17} -10^{-20} cm^2
"	Fluorescence	10^{-18} -10^{-21} cm^2/sr
"	Rayleigh	10^{-28} cm^2/sr
"	Raman (rot.)	10^{-30} cm^2/sr
"	Raman (vib.)	10^{-31} cm^2/sr

a.　Most sources of light, whether they are continuous spectra (Xe discharge), band (NO-RF or discharge) or even lasers, tend to be noisy. Typically the value of E_s oscillates randomly within $\pm$ 1% of its value for NO lamps (McGregor). Hence the absorptance $A \equiv YN_iQ_\nu$ cannot be determined (Eq. 1) within an accuracy better than 1%. This yields in turn a dynamic range of 50 corresponding to an absorptance interval 1% < A < 50%, since the low A's (< 1%) fall within the noise of the source, and the high A's (> 50%) approach saturation [i.e. the correct exponential expression (non-thin gas) shows that N_i varies very rapidly for small increments of A as A approaches unity]. Indeed, the results produced by this technique for NO (Davis 1974) correspond exactly to that small range of concentrations around the value $N_i = \frac{.5}{YQ_\nu} \backsim 2.5 \times 10^{16}$ cm^{-3} (i.e. $\sim 10^3$ ppm) [for the NO γ band (0,0), Y = 30 cm]. In principle, a simple way to increase the dynamic range of a source is to observe a set of bands of increasingly small absorption coefficients, Q_ν, thus covering a large range of N_i.

Source noise varies from technique to technique. It can be averaged out if suitable exposure time is available. But short time response is necessary in many situations and more stable sources, especially lasers, are desirable. Continuous lasers tend to be stable - for a price (Forsyth 1970).

b.　Absorption phenomena concern a relatively large number of photon-particle encounters (say, 10^{-10} watts or 10^{10} photons/sec.). Therefore no accuracy problem arises from the averaging of these inter-action events when defining the volumetric properties of matter itself.* In scattering measurements, however, it can be shown (Saltzman 1971, Lapp 1973) that the Raman process involves so few photons per unit time that the sensor located at a given scattering angle has only a statistical chance of receiving exactly the signal E_s from which N_i is derived in Eq. 2. A signal-to-noise ratio $\underline{S}$ is derived from Poisson statistics; it is equal to the square root $\overline{N}$ of the photon counts at the receiver: 10^2 photons yield 10% accuracy on N_i, 10^4 photons yield 1% accuracy, etc.... These very weak signal statistics are the key limitation of the Raman processes.

---

*The problem of linear beam propagation across turbulent flows is not critical for single path measurements but becomes substantial in ima-ging techniques (El Wakyl, W) and in point techniques (Self, W). Also, the loss of accuracy encoutered in test cell measurements, as compared to quiet laboratory conditions, seems to be due in great part to the degradation of performance of the optical system itself.

c.   Dispersive optics have been discussed extensively (Lapp
1973). There does not appear to be an accuracy limitation, given standard monochromator equipment, polarization filters, etc... Non disper-
sive optics (correlation method), interferometers (Fabry Perot) are
highly selective spectral selectors also (Burch 1974, Ludwig 1971, Gregory 1974).

d.   Detector noise and dynamic range are often the key limitation to a measurement. There exist textbooks (Platt 1968, Smith 1957,
Stewart 1970), handbooks (RCA-PT61) on this subject. A number of recent
reports have special sections to this problem (in particular: Lapp 1973,
Leonard 1974, Kildal 1971). There is a wide range of optical detectors
with different noise characteristics. It is customary to separate the
detector ability to convert an incoming photon into an electronic signal
(sensitivity, quantum efficiency,...), from its generation of additional
"noisy" signals (dark current) in the process of transforming this original electronic signal into a measurable current (noise equivalent power:
NEP).

In the UV and visible range, photomultipliers yield quantum efficiencies of the order of 10% to 25%. Dark currents in photomultipliers are
mostly due to thermal noise. Since efficient cooling, by liquid N_2 (77K)
for instance, is readily available, it is often possible to eliminate
dark currents altogether and to be able to measure minuscule currents corresponding to a few dozen photons only. This situation fits well weak
scattering signals, such as Raman, where photon flux levels are very small.
Salzman (1971) has shown that even with an uncooled photomultiplier (S-20),
an average instrument noise of 8 photons/sec. can be achieved, the Poisson
statistics of which correspond to an uncertainty of about 3 photons. Hence
the only accuracy limitations are due to the Raman statistics of volumetric
scattering (paragraph b.) and to quantum efficiency. The dynamic range of
photomultipliers is of about 4 or 5 orders of magnitude.

In absorption measurements, a double-path technique can be used. The
difference between the two signals is measured by a difference amplifier
It yields the number of photons absorbed, a much larger one ($\sim 10^{10}$ sec^{-1})
than in the scattering case. Thus it is not necessary to eliminate the
thermal noise by cooling, since its current can be kept well below (NEP =
10^7 photons sec^{-1}) the current generated by the photon flux E_m. This holds
true even for very low absorptances A ($\sim 10^{-5}$) and therefore for very low
concentrations $N_i \equiv A/Q_\nu Y$. In the case calculated by Sulzmann for tunable
diode lasers (1973), the signal-to-noise ratio corresponds to 10% for the
lowest concentrations calculated (several ppb-cm in some cases). It could
still be improved by cooling the photomultiplier and it increases in any
case for larger concentrations.

In the <u>infra red range</u>, photovoltaic detectors are dark-current-limi-
ted by shot noise (Ludwig 1973, Pratt 1968, Kildal 1971). In this case,
the noise equivalent power (NEP) is a function of the detector areas, the
bandwidth and the response factor of the detector. Accuracies and sensitivities are found to be of the same general order of magnitude as those obtained by photodiodes in the UV visible case.

  e. In summary, for the <u>whole measurement system</u>, accuracy is
expressed by the combined signal-to-noise ratio $\underline{\frac{S}{N}}$, where N corresponds to
all the noise contributions of the system to N the measured quantity
E_m. It is often stated in an overall way as:

$$\frac{S}{N} = \frac{\eta\, E_m}{(\eta\, E_m + \eta\, E_b + E_d)^{1/2}} \tag{4}$$

where η is the quantum efficiency, E_b the background photon flux and E_d
is the dark current equivalent noise (shot or thermal). For a known system (i.e. known η, E_b, E_d) it is then possible to assess the feasibility
of measuring a certain component population N_i by checking from Eqs. 1-3
if the outcoming flux E_m will produce a satisfactory accuracy $\frac{S}{N}$ in Eq. 4.

  Similarly, the sensitivity (i.e. the measurement threshold) is given
by that value of N_i which corresponds through E_m and Eqs. 1-3 and 4, to
the lowest tolerable value of $\frac{S}{N}$ in Eq. 4.

4. Space and Time Resolution

  From the viewpoint of the user, it is clear that the size and duration of the physical event which he wishes to analyze will put an upper
limit to the volume and time duration available for the measurement. In
the Part II of this report, these requirements will be expressed for combustor work. In part IB the capability of each measurement method will
be assessed. In this part we shall define and briefly discuss the criteria involved.

  <u>Space</u> resolution is a simple concept: it is the size, say in cm^3,
of the control volume on which the measurement is made. When one is concerned with flows, however, one cannot depend on a simply defined volume,
such as that inside a glass container for instance. Rather, one has to
estimate the section of the streamline which interacts with a solid probe

at a given time, or in optical methods - the size of the focal volume,
a somewhat arbitrarily defined area of maximum intensity. It can be
seen that even in general terms, the exact extent of the "measured" volume
(i.e. the volume being interacted with in the measurement process), will
be very difficult to evaluate, even though its order of magnitude is readily apparent. This difficulty, incidentally, explains why instrument calibration (i.e. the ratio of measurements for an unknown condition and for
a known one, but with the same geometry) is always preferred to single
"absolute" measurements, where the volume must be calculated to extract
the property of interest.

Time resolution is also a simple concept: it is the time interval
when a meaningful measurement can be made. As will be seen in the next
chapter, some methods are limited to near-steady operations (e.g. sampling probes) while others (e.g. pulse optics, hot wires) give a fine time
resolution of unsteady processes. Also, a meaningful measurement may require a long observation time (continuous or a series of pulses), due to
the weakness of the signal (e.g. Raman).

From the standpoint of both time and space, many practical situations
arise, where the control volume or the observation time cannot be made
small enough to measure a sample of uniform properties. What is measured
then is an average, and signal interpretation may be delicate. Conversely,
the instrument (e.g. pulse lasers, probe rakes) or the phenomenon (eddies,
seeding particulates in LDV) may be intermittent in space or time and assumptions must be made as to how representative these sample signals are
of the whole flow itself. These two aspects of averaging and sample bias
are discussed below.

Averaging: In turbulent combustion, temperature can vary by as much
as 500°K and composition of a species by a factor of two over distances of
the order of 0.1 mm and in a time of a few microseconds or less, both of
these variations happening in the same experiment. There is no measurement
technique with the possible exception of LDV that can approach this requirement for simultaneous fine scale resolution in both space and time. Several
techniques can obtain adequate spatial resolution but require long averaging
times; others can obtain adequate time resolution but must average over
large distances in space. If the average obtained is a true time or spatial
average of the property of interest then this is still a meaningful measurement. Often, however, composition measurements are a function of temperature and the average obtained will be the average of a complex function of
both composition and temperature. We note that:

$$\overline{\phi(c, T)} \neq (\bar{c}, \bar{T})$$

although it is a common error to assume this equality. (The equality is only true for a linear function $\phi = ac + bT$ and such linear behaviour is far from being approached in most systems of interest here.) In the above the overbar can indicate either a time or space average.

Depending on the nature of the function the average obtained may be more or less useful. Bilger (1975a) indicates that an isokinetic sampling probe will obtain a Favre averaged composition $\overline{\rho Y_i}/\overline{\rho}$ which is a useful average as it can be predicted by theory. Setchel (private communication) has shown that in Raman measurements the slit width can be chosen such that ϕ has the form

$$\phi(c,T) = (a + bT)c$$

so that $\qquad\qquad \overline{\phi} = (a + b\overline{T})\overline{c} + b\overline{T'c'}$

Some theories predict $\overline{T'c'}$ as well as $\overline{c}$ and $\overline{T}$ and so this result will be useful even if it is not directly meaningful (note: in the above it is usually possible to have b = 0 at very large slit widths but then the background radiation and interference of other lines drastically reduce accuracy).

Sampling Bias: Many candidate measurement systems do not obtain the complete time dependent variation of the quantity being measured but effectively sample this signal either by using a pulse measurement as in pulsed laser methods or through some inherent feature of the technique as in single particle LDV. All the usual features and problems of sampled data will be present. If the sampling is not random then biasing will occur. In LDV measurements for example the particle concentrations may be correlated with the velocity and biasing will result. Measurements obtained at regular intervals by pulsing a laser for example will show a bias in the results if there is a significant peak in the spectrum of the variable at this frequency.

5.   Cost Effectiveness

Cost is often quoted in terms of the purchase of the measuring equipment. At the Squid Workshop, the point was made that a great deal of labor went into tailoring off-the-shelf equipment to the needs of the particular experiment, not to speak of the additional brackets, connections, insulation, etc...., necessary for its survival in an engine environment. Finally the cost of running the engine itself, as well as monitoring the equipment and processing the data is a substantial one. These separate cost items will be evaluated in the next chapter for the different techniques.

Effectiveness: It is clear that more money will buy better lasers, better optics, automatic signal processing, etc... which will improve in turn: accuracy, versatility, dependability, etc... Those tend to make for "effective" measurements. Time to perfect the installation, to make more measurements, is also a factor of effectiveness...and cost.

In this study, as was done at the Squid Workshop, we shall assume the context of a "bright , well trained, non-specialized experimentalist," given the basic environment and money support, and interested in the kind of measurements which are needed in combustion, as listed in Part II of this report. This should set a reasonable common basis for our evaluation and comparison of the different measurement techniques.

It must be noted of course that the same level of effectiveness is far more expensive in a real engine environment, where it "shakes, rattles and rolls...", than in a quiet laboratory situation. Also the bulk and complexity of the engine prevents the use of the more accurate measurement techniques which can be arrayed around a laboratory flame (forward scattering, fixed optics and movable flame, etc...). As a result, thresholds of 10 ppm and accuracies of 1% or better are never mentioned in engine tests, whereas they are often accomplished in the laboratory. On the other hand, one should not fall into a fatalistic acceptance that "real life" measurements are necessarily bound to poor accuracies. There has been successful transposition of laboratory techniques into the field (e.g. Leonard 1974), and higher measurement standards are both feasible and necessary to obtain the knowledge which is needed for the design of future combustors. Still, at any given time, there are important differences between test cell and laboratory measurements and we shall distinguish between these. We shall also evaluate separately implementation times and the cost of the equipment, of its installation and of its use.

It is quite clear, after this quick overview of measurement criteria that they are very interrelated. Hence, each situation can be expected to be a trade-off between a number of advantages and disadvantages. Rarely will the same instrument be suited to obtain all the information desired in a given test.

B. MEASUREMENT TECHNIQUES

A review of some combustion measurement techniques is presented here.
The principles involved in each technique are assumed to be familiar to the
reader. Its purpose is to evaluate their relative merits in a combustion
environment, to indicate their strengths and weaknesses and to point out
the areas of expected improvement. Quantitative assessments of the criteria
listed in the previous chapter are given for typical combustor conditions,
with apologies for the brashness of such generalization.

The terms which will appear later in this chapter will use the following nomenclature:

For implementation times: short means less than one year
                           intermediate   "   from one to three years
                                   long   "   from three to five years

For costs: low means less than $50,000
                           intermediate   "   from $50,000 to $150,000
                                   high   "   from $150,000 to $500,000
                           out of sight   "   more than $500,000

Finally, when reading this chapter, it might be useful to keep in mind
the different interests and requirements of the combustion community as they
are discussed in Part II and schematized in Fig. 1.

1. Probes

The capabilities of probes for measuring concentration, temperature and
velocity in turbulent combustion have been reviewed by Bilger (1975a). Material probes inherently interfere with the flow but with careful design
and the use of small probes this interference can often be kept down to an
acceptable level. particularly in streaming flows that do not involve recirculation or strong swirl. A good rule of thumb is that the probe diameter
should be less than 0.1 L where L is a characteristic length scale of the
flow (such as the integral length scale of the turbulence), and the axis of
the probe should be aligned with the mean flow direction for at least 10
probe diameters; interference effects should then only be of the order of
3 percent.

Probes sampling for concentration will be as specific for any species as is the analysis instrument used. Usually this is no problem. For some species reaction within the probe will change the composition; particular problems exist with NO and NO_2. Thermocouple probes are primarily specific for temperature but show secondary dependence on local composition, temperature and radiant flux. Pitot probes for measurement of velocity are also dependent on the density.

The sensitivity of composition measurements is determined largely by that of the analytical instrument and these can be made to have sensitivities to the ppm range for most species. Sensitivity for thermocouple and velocity measurements is also limited by the detection and is usually not a problem.

The accuracy of probe measurements is largely determined by various sorts of biasing that enter the averaging process rather than by errors of a random nature. If sampling is carried out under isokinetic smapling conditions in a streaming flow, Bilger (1975a) estimates that compositions can be measured to within 10 percent of the Favre average $\tilde{Y} \equiv \overline{\rho Y_i}/\rho$, where Y_i is the mass fraction of species i and ρ is the density. There is no way of checking the accuracy of measurements made in the highly turbulent flow patterns within a gas turbine combustor; repeatability of the data indicates that some species can be measured to an accuracy approaching ten percent but what sort of average is obtained and what sort of biasing may be involved is an unknown. (Hopefully comparison with optical methods will put these measurements on a firmer foundation.) Bilger (1975a) estimates that temperatures can be measured to within 5 percent of the time mean and velocities to within 10 percent of the Favre mean in streaming flows where high radiant fluxes and droplets are absent. Inside a gas turbine combustor these figures must be doubled at least.

The only probe that purports to give any semblence of adequate time resolution is the cooled film probe which unfortunately is sensitive to both temperature and velocity and to a lesser extent composition. A frequency response to several kHz can be obtained (Ahmed, 1971). Time resolution of sampling measurements is of the order of several seconds and for pitot tubes and thermocouples, can be made as small as several milliseconds with appropriate design although such response is usually not desired. Spatial resolution for sampling, thermocouple and pitot probes is of the order of 2 mm in directions normal to the flow; in the streamwise direction, it is limited by the time resolution and the flow velocity.

The cost of probe measurement techniques is determined largely by the cost of the detector. This varies from a few dollars for a thermocouple galvanometer to about $2,000 per species for gas analysis equipment. Every probe must be custom designed for the application and there is usually a few thousand dollars of labor involved here. Probes are an effective and proven technique.

2. Laser Velocimetry

Chigier (W) presented a strong case for dual scatter (fringe) laser velocimetry as the only system which can measure a velocity field in a combustion environment. Its interference with the flow is clearly better than that of probes although its need of a port window makes it applicable to research combustors rather than to development combustors.

The specificity is also much better in the sense that it can pick all the components of the velocity vector, positive or negative, an essential feature in the recirculation zone or in the highly turbulent primary zone. This, of course, is impossible for either pitot tube or hot wire anemometer. It also measures velocity directly (by a count of particle crossing rate), and does not need additional measurements of properties [such as density for the pitot tube ($p_0 = p + \frac{1}{2} \rho v^2$)].

Accuracy is excellent, because it is based on a frequency measurement (Chigier). However, most practitioners at the Workshop did not claim better than 1% for either u_x, $\overline{u_x}$ and $\overline{u_x u_y}$. Sensitivity is good since jet entrainment velocities ($\simeq$ mm/sec) have been measured, as well as supersonic speeds.

Spatial resolution is excellent in theory since the local beam of a laser has a theoretical size limit of the order of tens of microns. However in a flame environment no one claims much better than one mm^3. Time resolution is limited by the number of scattering particles which cross the control volumes in any given interval of time (Self, W.). For real time turbulence measurements for instance, one needs near continuous reading. The proper injection and distribution of seeding particles (Al_2O_3, MgO, etc.) presents interpretation problems (drag/mass problems, sampling bias) and the possibility that the flow chemistry will be affected. Most experimenters keep seeding to a minimum and some eliminate it altogether (Lennert, W.). In short, one can make instantaneous measurements of u, but not continuously (Owen, W.). Hence the need to average over time for such quantities as turbulence. Also one can establish enough correlations over a period of time, by double pulsing for instance, to be able to reconstruct the turbulence spectrum.

Another difficulty has to do with the refraction index fluctuations on the path of the laser beam. Although it is talked down by some, some others consider it a serious obstacle in trying to measure two components simultaneously (i.e. keep two fringe systems at the same spot). Again sequential measurements seem to give some answer, at least to the analytical modelers who search - so far - for streamline functions and such steady state values only.

A great deal of the quality of LV measurements is limited by combustor constraints. Forward scattering which offers a signal gain of two or three orders of magnitude over the back scattering mode, requires an optical path through the flame (2 ports). Many combustors do not give this flexibility. Also the presence of "natural" particulates (soot, droplets) creates both a measurement and a background problem, which has been discussed extensively (Chigier, W; Eckbreth, W.). A number of these shortcomings can be overcome by larger laser powers, but these are limited by the elctrical breakdown threshold (ionization), which is especially worrisome in dusty flows (Bershader, W.).

The cost of LV measurements was rated low, if one wishes to concentrate on simple average velocities. Any attempt at measuring velocity vectors or real time correlations runs into much larger amounts of money (several hundred thousands of dollars). Signal processing equipment is a large fraction of this cost. Implementation time would run from low on average properties to intermediate on vectors and real time measurements.

3. Absorption - Emission

It was pointed out in Part II that absorption methods have the significant advantage of much larger cross sections than those found in scattering (see Table 1). Conversely, a significant limitation is that they are "path" methods which integrate all the absorption effects along the line of sight, whereas scattering methods give information at a point (see Fig. 3). Therefore, it is expected that absorption emission methods will be mostly useful in 2 dimensional geometries such as may be encountered in fundamental studies or some research combustors (Case D and E of Fig. 1). Their application to axisymmetrical (or near axisymmetrical) geometries, (Case C and D of Fig. 1) will be discussed later in this section.

Interference with the flow, as for the other optical techniques, is absent and since this technique is not applicable to the highly three-dimensional interior of development combustors the problem of access does not present itself. Indeed, most of the available engine experience with this method has been gathered in "open" exhaust measurements (Herget 1967, Lennert W., McGregor 1972).

Specificity is not always remarkable when measurements are made across a whole band, since bands of other components usually overlap (see paragraph A2b). However, since the signal is not weak, and especially for steady flows, it is possible to filter out all but a "clean" single line or group of lines, by correlation for instance. At the limit, one may wish to use tunable lasers to pick only the center of one line (Sulzmann 1973). This highly specific technique might be necessary to separate strongly over-lapping lines such as those of the various hydrocarbons (CH band). Another aspect of specificity is the dependence of absorptivity on temperature (see paragraph A2a). There seem to exist (Neer 1974, Chen 1975) lines or com-binations of weighted lines which eliminate this dependence. This proper-ty which is very useful in engine exhaust pollutant measurements (Chen 1975), acquires a special importance in all optical measurements, even including scattering, where there might be wide property variations inside the small control volume (see paragraph A4).

Accuracy in a combustor environment was given at 7%. In a careful "quiet" laboratory experiment, it could be much higher, since the noise statistics of detectors are favorable (many photons) and stable sources can be found. However, it is the sensitivity which is pursued most of the time with these methods, and very low thresholds can be reached with this method such as 10 ppm-m in an engine flow and as low as 1 ppb-m in a "quiet" condition.

Space resolution would be difined here as a practical cross section for a scanning beam: 1 mm^2 seems easy to accomplish. No practical time resolution seems to exist since absorption is not Poisson-statistics limi-ted: they are standard methods in shock tube research (< 1 μsec).

The equipment is standard (except for the new tunable sources) and cost was rated low and implementation time intermediate. Again the cost of actually running the test on an engine was not included.

Emission techniques have not been discussed extensively at the Work-shop as measurement techniques in jet engine burners. However, they might well become important, if only as a correction factor to the absorption measurement, as temperatures and soot concentration increase (Setchell 1974). Indeed, they play an important part in the analysis of industrial burners and kilns (Hottel, Beer 1973) and the new aromatic fuels currently considered would increase this radiation contribution to the heat transferred to the combustor walls.

The mathematical "inversion" of such data in the case of axisymmetrical flow is common practice (Abel, onion-peeling, etc....) and it seems to
be giving good average profiles [5% accuracy at the edge and 25% in the
middle (Ludwig, W)], where the rough turbulent detail is eliminated.
With proper optical or algorithmic precautions, it seems that turbulence
does not degrade the transmission of the average information carried by
the signal (Reynolds 1975, Rhodes W., Daily W.).

It seems also that in general the errors generated by numerical procedures can be kept to a minimum, if these experimental data are accurate
(Blair 1974). Inversion techniques have been considered in emission on
the basis of frequency scanning (Krakow 1963, Chao 1974). Inversion is
still an art and a frustrating one at that, but it is the only way as yet
to estimate the concentration profiles of some minute traces in nonhomo-
geneous flows. Note in passing that some information of average nature
(e.g. EPA engine pollution fluxes) can be acquired from absorption measurements, without having to resort to inversion.

4. Scattering

Raman scattering is a concentration and temperature technique which
has been developing rapidly in the last few years. It is now beginning
to be applied to flames (Lederman, Lapp, Setchell, W.).

Non-interference with the flow, as with any optical system, is avoided but access to the flow requires windows for the beam to come in and
out. This restricts the application to port-equipped research combustors
(although fiber optics have been suggested). Specificity is the great
advantage of the Raman process since each molecule and its excited states
carry its own distinctive signatures, i.e. its Raman frequency shift.
There a few exceptions: CO overlap with heated N_2 as it was discussed
earlier. Fluorescence lines also overlap with Raman lines: OH with H_2O
(Lapp), NO_2 with NO (Chang) and NO with hydrocarbons (Leonard 1974). But
in general, most species, including low concentration pollutants, have a
distinct signal.

Accuracy is given by Eq. 4 in paragraph A3. Since the main error
contributions are often the Raman scattering statistics themselves, the
signal to noise expression does not include - in first approximation -
the terms E_b and E_d. If one assumes a quantum efficiency of 25%, the

number of photons E_m necessary to produce an accuracy of 10% is
$$E_m = \frac{1}{\eta} \left(\frac{S}{N}\right)^2 = \frac{1}{0.25} \left(\frac{1}{0.1}\right)^2 = 400.$$ On the attached figure (Fig. 4), the ratio

$$X_{min} \equiv \frac{q^4}{\eta} \frac{E_m}{E_s} \tag{6}$$

is shown as abscissa, where η is the receiver quantum efficiency and q^4 is
a nondimensional frequency correction to compare lasers of different wavelengths on the same basis (Goulard 1973). Physically, X_{min}^{-1} is the number
of sets of 400 photons in a given laser pulse. A 1J Ruby laser corresponds
to $0.85 \times 10^{+13}$ sets. The probability that any such set will be Raman
scattered into the receiver is determined by the experimental conditions,
as given by the right side of Eq. 2. It is:

$$Z \equiv Y N_i Q_{\nu i} \Omega e \tag{7}$$

If the product $X_{min}^{-1} \times Z$ is larger than unity, at least 400 photons
will have been scattered into the detector and a detection measurement of
10% accuracy or more will have been effected. On Fig. 4, it amounts to
having log Z plotted on the same abcissa scale as log X_{min}: if $Z > X_{min}$,
the 10% accuracy measurement is possible, if $Z < X_{min}$, it is not.
For a typical flame, the value of Z is shown for various concentrations
c_i in a 1900K flame, where

$$N_i = c_i \frac{273}{1900} N_{STP}$$

It shows that a concentration unity is measurable by one Ruby laser
pulse in 1 mm^3. If one looks at Eq. 7 for the possible trade offs of the
situation, it is clear that a concentration of 10% could be obtained for
a 1 cm^3 sample (for our particular purpose, the increase of control volume Y^3 increases only the path Y of the photons and not their number since
all we do is to defocus a given flow of photons when we go from 1 mm^3 to
1 cm^3). The only way to go to lower concentrations would be either to
increase the power of the laser pulse or to install it in a cavity. We
could go then perhaps to 1% or even 10^{3}ppm.

The most common solution for pollutant levels (i.e. less than 10^3
ppm), is to go to repetitive pulses where eventually enough photons are

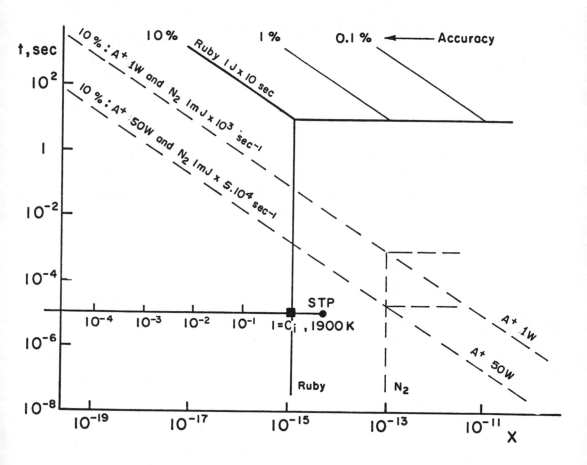

Fig. 4 - Feasibility of Laser Raman Measurements

generated to pick up small pollutant levels with off-the-shelf lasers, such as the two lasers mentioned above. On Fig. 4, one sees that 100 ppm of NO can be picked up in a 1 cm³ volume by a N_2 1mJ laser fired 10^3 times a second for 100 sec, which is precisely what Leonard did in his jet engine exhaust measurements (1974). Similarly successful results were obtained by Setchell (1974) for 1% of CO over a period of 100 sec. In short then, it seems that instantaneous (or say less than 10^{-3} sec) measurements cannot yield pollutant levels in small volumes ($\simeq$ mm³). Raman will probably <u>not</u>

be useful for pollutant turbulence studies or even reciprocating engines studies, the way LDV might.

However, it can very well measure major constituents ($c_i > 10\%$), including N_2 and their excited states, and thus obtain both density and temperature, instantaneously (Lederman 1974). Double pulsing would then provide turbulence correlations.

In a dust-free environment where Mie scattering is not interfering, Robben (W) has shown that a <u>Rayleigh</u> scattering measurement (i.e. at the laser frequency itself) would bring a relative gain of 40 in accuracy for density because of its cross section improvement of three orders of magnitude over vibrational Raman. For temperature, since Rayleigh scattering does not distinguish between excited states, the Boltzmann ratio technique has to be replaced by a Doppler profile analysis which yields an improvement of accuracy of 7 only over Raman. This gain of signal strength makes turbulence spectra more precise.

At the Squid Workshop, Raman practitioners confirmed these results. They claimed 5% accuracy for 1 mm^3 spatial resolution for the major constituents and the temperature, as intermediate cost and implementation time. It is hoped that unforseen problems will not degrade these levels of accuracy and resolution in practical combustors. One which could possibly stand in the way of small volume measurements, is the ionization threshold (<u>electric breakdown</u>) of gas submitted to highly focused laser beams, especially pulsed ones. Some rough first estimates (Penney, Boni, 1975) mention a threshold as low as 1 Jmm^{-2} for a 20 nanosecond pulse. Dust (soot? LDV tracers?) tends to lower the threshold. The usefulness of the 1 J Ruby laser might become questionable for making reasonably detailed profiles in real time. New knowledge is critically needed in this area.

The potential of scattering for the measurements of <u>particulates</u> (concentration and size distribution) was reviewed at the Workshop by Self. The outlook seems to be brighter for droplets than it is for soot ($< 1\mu$). Classical probe methods are reviewed on p. 46 of this report.

Finally, one should mention briefly a method of great apparent promise: the Coherent Anti-Stokes Raman Scattering method (CARS), which is a non-linear interaction technique. It has been demonstrated by Taran (Lapp 1973) in a hydrogen jet and it is actively investigated in this country as well. It divides the light of a laser in two beams, shifts the frequency of one by a dye laser by the exact Stokes Raman shift and mixes the two beams again at some desirable focal volume. The non-linear result of this interaction is an <u>Anti</u>-Stokes beam, i.e. frequency shifted <u>away</u> from possible laser-generated fluorescence. Best of all, about 1% of the applied power is scattered into this beam, as compared to 10^{-8} in "ordinary" Raman

(Harvey, W). Thus a limited amount of power can produce enough scattered
photons to insure good statistical accuracy without risking electric breakdown. Extremely high sensitivity (less than 1 ppb) has been claimed and
Taran (W) predicted - if one tunes in a near-resonance frequency for the
species of interest - an accuracy of 1% and a one microsecond response
time, for a laser input of 1 watt (!). Species turbulence could then be
handled at a cost comparable to ordinary Raman or slightly higher. However,
the technique is far from being developed.

5. Fluorescence

It can be seen from Table 1, that fluorescence (i.e. the near instantaneous reemission of the radiation energy absorbed by a particle) has a
cross section of the same order of magnitude as absorption but is also a
"point" emission process, like scattering. Thus, it presents the measurement advantages of the two methods. One disadvantage is its lack of specificity: the deexcitation of excited states also occurs by collisional processes (quenching). Therefore there is no direct proportionality between
the number of particles in the ground state and the fluorescent signal:
one must allow for the quenching rates, which depend in turn on the concentration of the other species which collide with the species of interest,
their quenching cross section, etc...

A promising set of particles seems to be the diatomic radicals, such
as NO or OH, which have large quenching times ($\tau_c \sim 10^{-7}$ sec) compared to
their radiative lifetime ($\tau_R \sim 10^{-8}$ sec). In this case, errors in evaluating the quenching rates are not vital and quenching could be ignored
with good accuracy. There are a number of situations where the characteristic times of the two processes are more equivalent and at least qualitative observations have been made in IC and Diesel engines on OH, CH, C_2,
NH,...(Lavoie 1970, Rodig 1969). [See also Hartley (1974)].

To exploit the potential of this highly sensitive method, two advances are necessary: the availability of reliable tunable lasers and the
accumulation of a large amount of quenching information (cross sections).
Given these advances, the fluorescence technique could make a major contribution to the 3-dimensional instantaneous mapping of flames. At the Workshop, Fontijn discussed these prospects and foresaw no problem in adapting
this technique to combustor work at low cost and intermediate implementation time.

6. Imaging

It is often useful to measure and display the properties of the whole flow field instantaneously, especially if an understanding of its overall features and of their time evolution is needed. One can think of optical systems which will produce the point measurements discussed above at a number of points on a line, on a surface or in a volume. This can be done instantaneously by using an array of sensors, or by rapid scanning, using vidicon tubes and other modes of information storage (Schildkraut 1974). Similarly, the frequency range of interest - in Raman, for instance - can be scanned in as short a time as 20 nanoseconds (Bridoux 1974).

However, the most common form of imaging is based on the path measurement of the refraction properties of the flow. Such are the well known techniques of interferometry, schlieren and shadowgraphy. These methods give a signal proportional to the density, the density gradient across the path, and its double derivative, respectively (Liepmann 1957, Merzkirch 1974).

In their more trivial photographic form, shadow methods yield pictures of liquid or solid objects (such as sprays, droplets or particulate clouds) at a given focal plane. This plane can be moved with time to cover the whole depth of field (Chigier W) or this information can be seized instantaneously by a hologram (Matthes 1974, Trolinger 1974, Rhodes W). The accuracy is good but the automation needed to eliminate the tediousness of manual data reduction is quite expensive.

Single phase flow fields do not present singular points of reflection in space as particulates do. But two dimensional flows can be inspected usefully by shadowgraph techniques (Roshko 1974) for qualitative eddy structure and intermittency.

Interferometry has been used extensively in two dimensional flows for the mapping of density profiles and by implication of temperature profiles. (Hauf and Grigull 1970, Weinberg 1963, El Wakil W). Methods seem to exist to eliminate extraneous effects, by double exposure (Sandhu 1972) or defocusing (Matthews 1974). The double exposure techniques (as compared to the double path Mach Zender techniques) allow also for the investigation of time dependent phenomena: the turbulent spectrum of a whole flow can be reconstructed by a set of double pulses at different time intervals, just as it is done in LDV (Swithenbank 1975).

Two promising directions of interest to combustion studies were discussed at the Squid Workshop. They have to do with frequency scanning and angular scanning.

Bershader (W) brought out the possibility that by using the refraction resonance <u>frequency</u> of the gases present in the flow, one would multiply the signal strength by orders of magnitude (6 for instance), even when the proper broadening mechanisms are taken in account. Since such resonance happens within 1 Å of the absorption resonance, very narrow line tunable lasers seem necessary to the success of this approach. In principle, it would be then possible to pick the refractive resonance frequency of each chemical component of interest and to map its density independently. At the moment, nonresonant techniques can be applied to two components only (e.g. fuel vapor and air: Panknin W).

As for the absorption path techniques, interferometry cannot be extended to three dimensional fields, unless <u>angular scanning</u> is performed. However, whereas absorption scanning has been limited to one direction and therefore to axisymmetrical flows ("onion peeling"), a more general <u>multidirectional</u> scanning method has been developed for reconstructing arbitrary three-dimensional density fields. This method was discussed briefly at the Workshop by Sweeney and Vest (see also their 1973, 1974 papers). A scanning range of 180° is necessary to obtain all the information needed for an accurate mathematical treatment. A holographic camera built on this principle has been proposed (Heflinger 1966). Appreciable errors (up to 5%) occur if the viewing angles are limited, say to 45°, although some techniques alleviate this problem (e.g. frequency plane restoration - Sweeney (1973)). Also a poor choice of series expansion for the profile under observation leads to numerical errors. Even if the choice is a happy one, there might be considerable error amplification (say from 1% to 8%) if there are fluctuations in the experimental measurements themselves (Matulka 1971). Furthermore, sharp density gradient complicate the interpretation of fringes, especially near the solid boundaries.

Finally, interferometry is restricted at the moment to density mapping. The time resolution, if one uses beam-split laser sources, is a few nanoseconds and Sweeney (W) considers 1 mm ^{3} a reasonable space resolution. Experimental accuracy appears to be less than for absorption measurements. Numerical accuracy is as bad.

In conclusion, it seems that for each of the methods reviewed in this report, there are features which make it the proper choice for one particular combustion measurement but not at all for the others. In many cases, it is a combination of techniques which will give the best answers for the least cost and effort.

II. COMBUSTION MEASUREMENTS

NEEDS OF THE ENGINEER

1. Introduction

High efficiency, low emissions, good exit plane temperature profile, and good flame stabilization are the prime concerns of the combustion engineer in gas turbine liner design and development. Measurements have tended to involve straight-forward instrumentation capable of rapidly giving information on the results of changes in the combustor design on overall performance parameters, such as combustor exhaust plane temperature profile, and mass- or area-averaged exhaust emissions of unburned hydrocarbons (THC), CO, NO, NO_2, and smoke. For aircraft engines, design is usually focused on steady-state operation at the various operating points from low speed ground idle through takeoff or military power, with little attention given to transients involved in altering the power setting, so that time averaged measurements at these various power settings are sufficient. Most designers would prefer more detailed information on other points, such as the flow field and compressed air velocity vectors in the upstream diffuser and casing surrounding the liner, the fuel spray pattern and penetration in the primary zone, the size and position of the flame holding recirculation zone(s), dilution jet flow trajectories, and local equivalence ratio distribution throughout the primary and secondary zones; but since this type of data cannot generally be obtained (with combustion in the burner) unless time permits detailed point-by-point probing for temperature, concentrations, and velocity vector, the engineer is forced to rely on semi-empirical correlations and/or cold flow tests in for example water tables (see Gradon and Miller, 1968). Qualitative optical field (imaging) measurements (schlieren or interferometric, perhaps utilizing holoraphy) would prove of particular interest here (assuming optical access is available*).

2. Demonstrated Probing Techniques

We have suggested that <u>time averaged</u> measurements and flow visualization are sufficient for combustor development, but that <u>point measurements</u> within the combustor (and at least for temperature in the exhaust plane) are desirable as well. The latter need is based on the three dimensional character of the flow in a real combustor. Spatial resolution should depend on the gradients which are encountered and test time which is available, but the size of material gas sampling probes currently in use precludes sampling from volumes much less than 1 cm^3. To minimize disturbance of the flow, probing from the downstream end of the combustor is preferable, and the frontal cross-sectional area of the probe should be less than 0.5% of that of the combustor. Due to random turbulent fluctuations in the mean flow, and the presence of reverse flow zones (most important in the primary zone), isokinetic <u>sampling</u> has little meaning, and the aerodynamic design of the exterior of the probe nose may not strongly influence the results. The assumption that samples withdrawn are not biased is not well proven in engine or laboratory experiment environments, however. For gas sampling probes the internal aerodynamic design and coolant flow should be selected for maximum quenching rate, but probe fouling and plugging due to fuel droplet and/or soot ingestion may dictate this design for the high temperatures and pressures characteristic of real combustors. Mellor et al. (1972) discuss in detail the development of one long-lived gas sampling probe.

Time averaged <u>velocity</u> magnitude and direction measurements utilize multihole pitot-static tubes (Hiett and Powell, 1962), which require cooling and can plug, and thus are subject to some of the same design problems as gas sampling probes. In addition, three degrees of freedom in probe positioning and the accurate determination of small pressure differences

---

*Optical access to the primary and secondary zones of a tubular combustor under test on a development stand is not available, but the dilution zone could be probed optically from downstream. However, for segment testing of an annular combustor, windows could be placed on the sides of the sector which are normally blocked off.

at very high absolute pressures are obligatory. We are not aware of any
successes via such probing techniques at mapping velocities and thus flow
fields in actual gas turbine liner hardware with combustion proceeding
which have been reported in the open literature. Limited results are,
however, available (see for example Clarke et al., 1963). Clearly this
is a most fruitful area for the LDV provided that mean velocities in all
three directions can be obtained.

Point-by-point _temperature_ estimates are perhaps the easiest to obtain, using appropriate thermocouples, but very difficult to accurately
quantify due to measurement errors (see for example Fristrom and Westenberg, 1965). Mellor et al. (1972) found that the smallest Pt/PtRh thermocouple which could survive within a J-33 combustor was overall 3.2 mm
diameter, with 0.64 mm diameter wires insulated with MgO and encased in
inconel. Thermocouples of this size give good cross-stream spatial resolution, but have large radiation and conduction losses. Free radical
recombination is catalyzed by an uncoated bead, and impingement of liquid
fuel droplets leads to erroneous measurements (but thermocouples can
thus be used to sense the liquid fuel spray cone and penetration: see
Hunter et al., 1974). For these various reasons Mellor et al. (1972)
used the thermocouple output simply as a qualitative indication of local
temperature. In subsequent work with different combustors at higher inlet temperatures and pressures they encountered frequent thermocouple
failure and were unable to obtain even these crude estimates (Shisler et
al., 1974).

LaPointe and Schultz (1972) avoided the annoyance of thermocouple
failure by use of a thermocouple calibrated, critical flow orifice,
water cooled temperature and gas sampling probe. The pressure downstream
of the orifice is monitored to assure critical flow, and the mass flow
of sample gas through the probe is measured, thus allowing calculation
of the local static temperature in the sampling volume upstream of the
probe. LaPointe and Schultz (1972) note the advantage of accomplishing
gas concentration and temperature determination from the same sample
volume. However, the spatial resolution is undoubtedly poorer than that
of a thermocouple, and the other problems with physical probes (such as
sample bias and plugging in the region of the fuel spray) occur.

3. Anticipated Range of Parameters

For assistance in _design and development_ studies with conventional
gas turbine combustors we must recognize that the flame zone generally

resembles a recirculation stabilized turbulent diffusion flame fed by rapidly vaporizing fuel droplets. Thus ranges in the mean concentrations
and temperatures are wide: the latter will take on all values between
the combustor inlet temperature and the stoichiometric adiabatic flame
temperature, depending on the location within the liner. Similarly, mean
equivalence ratios from zero to infinity must be anticipated. For this
type of study a relative accuracy of ten percent is probably sufficient.

Other mean concentrations of interest as a rule include CO, CO_2, NO,
and NO_2. The carbon oxides can take on values up to several percent by
volume, whereas one to several hundred parts per million of the nitrogen
oxides can be encountered. Low ranges for CO are several hundred parts
per million and for CO_2 one tenth of one percent. Here again a relative
accuracy of ten percent is adequate, and we emphasize the need to know
both NO and NO_2 concentrations, not merely NO_x: the ratio of NO to NO_2
has a pronounced effect on the induction time for photochemical smog
formation, of particular importance in the vicinity of large airports.
Unfortunately, accurate (even to ten percent) quantitative determination
of NO and NO_2 is beset by both gas sampling and instrumentation problems,
so that NO_2 formation in gas turbine engines is still subject to considerable controversy (Tuttle et al., 1973).

Particulate emission indices have been reported along the centerline of a model combustor in the secondary and dilution zones by Giovanni
et al. (1972); they also discuss engine and exhaust plane measurements
available in the literature. Mass concentrations from 1.0 to 50.0 mg/m^3
can be anticipated both within and at the combustor exhaust plane. Giovanni et al. (1972) used a cooled isokinetic sampling probe connected to
sintered bronze tortuous path filters for mass concentration determinations or to Nuclepore filters for particle size analysis via electron
micrography. The latter showed agglomerates of about 0.7 μm diameter,
which obviates the necessity of isokinetic sampling. We are not aware
of any other published study in which local particulate mass concentrations and size distributions have been obtained within combustors: Toone
(1968) and Cornelius et al. (1957) report concentrations of particulate
obtained by probing within combustors: Norgren (1971) uses spectral radiance measurements to obtain a concentration averaged across the width
of the primary zone. Occasionally other species such as aldehydes are
also of interest (see for example Cornelius et al., 1957).

In velocity determinations as noted previously both magnitude and
direction are important. The time-mean flow pattern thus obtained still
most likely need be accurate to only ten percent, however, for comparison
with results obtained in water tables.

4. Problem Areas

As was noted by Professor N. Chigier at the recent OSR/NSF/SQUID Workshop on combustion diagnostics, many of the probe techniques considered above have been proven, even in the high temperature and pressure environments of gas turbine engines. There are many limitations on their utility, such as quantification of thermocouple temperature measurements; spatial resolution and fouling of velocity and concentration probes; the need for three directional movement of velocity probes; failure of concentration probes to quench reactions in their samples; and interference with the flow within the combustor. Few if any studies within liners have determined how important these limitations are upon the magnitude and accuracy of the respective parameters the probes purport to measure: probe calibration by optical diagnostic means in well defined combustion flow fields (not gas turbine combustors) may prove, with the exception of LDV applications, the most meaningful and useful interface with the combustion engineer (see Section C below).

B. NEEDS OF THE COMBUSTOR MODELER

1. State of the Art

Only four calculational models of the combustion process in gas turbine liners have been compared with experiment (Mellor, 1975); three of these utilize modules to model various regions within the combustor, while the last is a detailed solution of the governing elliptic equations. All of the models have, to various degrees, been compared with internal composition (and occasionally temperature) measurements made with probes and of the type discussed in Section A (see Table 2).

Hammond and Mellor (1973) tested their numerical results with mass-averaged axial profiles of temperature, CO, and NO, since their model takes the primary zone as a pair of perfectly stirred reactors with common flow, and the secondary and dilution zones as many perfectly stirred reactors in series. Exhaust plane averages were also compared. All such averages were obtained from radial profiles of probe sampled concentrations and temperature estimates (Mellor et al., 1972). It is primarily the collation of internal measurements and calculations which revealed the deficiencies of the model.

TABLE 2

MODELS AND COMBUSTORS UPON
WHICH THEY HAVE BEEN TESTED

Model	Combustor	Result
Hammond and Mellor (1973)	Allison J-33	Could not match exhaust plane or internal NO and CO axial profiles simultaneously
Fletcher et al. (1971)	"Combustor A" "Combustor B"	Exhaust NO good except at idle Exhaust NO good
Heywood and Mikus (1973)	GM GT-309 NASA Swirl Can	Exhaust NO good except at idle No mid-power range NO
Lefebvre and Fletcher (1973)	Experimental with Staged Fuel Addition	Exhaust NO good except at increased equivalence ratio
Mosier and Roberts (1974b)	P and W REB-2	Exhaust NO and CO good; exhaust THC poor
	P and W JT8D	Exhaust NO and THC good; exhaust CO poor
	P and W JT9D	Exhaust NO good; exhaust THC and CO poor
	Allison T-56	Exhaust NO, THC, and CO poor
	Allison T-63	Exhaust NO, THC, and CO poor
	GE J-79	Exhaust NO, THC, and CO good
Anasoulis et al. (1974)	Berkeley Model Combustor	Internal equivalence ratio and temperatures contours poor; centerline NO profile poor
	Experimental Annular Combustor	Exhaust temperature, O_2 and NO poor; exhaust CO_2 fair
Altenkirch and Mellor (1975)	Nearly Axisymmetric Simplified Automotive Combustor	Internal THC, CO, and NO profiles poor

452

The code of Fletcher et al. (1971), Heywood and Mikus (1973), and Lefebvre and Fletcher (1973) utilizes a partially stirred reactor to model the primary zone, followed by a mixing section to obtain the mean primary zone equivalence ratio; a plug flow reactor with air addition represents the downstream regions of the combustor. Model verification has been attempted for the most part only in terms of exhaust plane parameters, which limits model development and understanding. Note however that the use of a partially stirred reactor makes unclear the need of internal measurements within the primary zone.

Mosier and Roberts (1947b) undertook a more sophisticated (but still modular) streamtube analysis. Here again most testing of the model has been with exhaust plane temperatures and compositions (Table 2), but detailed internal measurements are now available (Shisler et al., 1974) for one of the combustors which they modeled (the Allison T-56 liner). In view of the many adjustable parameters involved in all of the models, further measurements must be obtained within combustors for continued testing, particularly of the Pratt & Whitney model which shows the most promise of becoming a useful design tool for gas turbine liners (Mellor, 1975).

In contrast to the above modular studies, the Gosman-Spalding-type solutions of the elliptic differential equations are usually compared with internal probe measurements of temperature and concentration (Anasoulis et al., 1974; Altenkirch and Mellor, 1975). Although to date these comparisons are not encouraging (see Table 2), they should point the way toward future efforts of the modeler and scientist who hope for a better fundamental understanding of the elements that go into the models (heterogeneous effects associated with spray combustion, homogeneous chemical kinetics of practical transportation fuels, highly turbulent recirculating gas flows, and the crucial interactions between all of these which determine the eventual exhaust plane output of the combustor). Note also that with the exception of Altenkirch and Mellor (1975) who probed a simplified combustor (Table 2), all of the other models have been compared with production or development burners. Clearly more effort with the models should be devoted to less complex combustor geometries which retain the key elements (e.g., recirculation zones) used in actual engine hardware.

2. Desirable Additional Information

In comparisons of model output with probe temperature and composition data, confidence in the measurements is limited due to the many uncertain-

ties discussed in Section A.   Redundant measurements using optical methods
where appropriate or probes of differing designs will alleviate such uncertainties; Bilger (1975a) discusses consistancy tests which should also be
performed whenever possible.

The Gosman-Spalding-type calculations and the streamtube analysis of
Mosier and Roberts (1974b) provide local gas velocity vector estimates as
well as temperatures and concentrations.   Thus if pitot probe and/or LDV
measurements are available to the modeler even more fruitful flow field
comparisons can be made.   Again, not only real combustors, but also simpler geometries should be the subject of such studies; in the next section
we suggest experiments in the latter category which will yield meaningful
data for model verification and development.

C.   FUNDAMENTAL COMBUSTION STUDIES

1.   Introduction

As noted, the most important part of developing adequate combustion
models is founding them on the correct phenomenology.   Real combustion
systems use such a complex interaction between turbulent fluid mechanics,
heat transfer, thermodynamics and chemical kinetics that modeling of the
basic processes must be carried out if a quantitative description of the
system is to be achieved.   Many of the sciences on which combustion theory is built are themselves incomplete and the development of combustion
theory must go hand in hand with research in turbulent fluid mechanics,
hydrocarbon kinetics, and other sciences.   There is thus no one set of experiments which will yield all the answers desired and a broad front attack
such as that being carried out under Project SQUID and the NSF and AFOSR
programs is very necessary.

Much combustion research is carried out on systems which are delibe-
rately simplified so as to isolate the phenomenon under investigation and
make accurate measurements possible; laminar flame studies, and turbulent
mixing studies without chemical reaction are examples here.   On the other
hand, systems such as gas turbine combustors are studied in which all the
phenomena are present and interacting but the. system is too complex to allow accurate measurements to be made.   The new optical measurement technology represents a considerable improvement in measurement capability and,
while it is not yet suitable for completely detailed measurements in gas

turbine combustors, it is hoped that it will allow the detailed study of
relatively simple but realistically interacting systems. Here we propose
some fundamental combustion studies of outstanding interest and state the
sorts of measurements that will be required. This is done without regard
at this stage for any limitations in measurement capability due to laser
power or the possibility of gaseous breakdown. Actual experiments are then
proposed and the factors involved in the design of the experiments discussed.

2. Studies of Outstanding Interest

The objectives of the experiments chosen should be to obtain complete
and detailed measurements on some relatively simple combustion systems designed to address particular problems of outstanding importance and interest. These should include:

1. Turbulent Mixing with Chemical Reaction (see Murthy, 1974)
    Fundamental questions to be addressed include: turbulent trans-
    port in flows with large density gradients, unmixedness, fine
    scale flame structure, local reaction rates, reactive species
    correlations, turbulence kinetics interactions, combustion ge-
    nerated turbulence, coherent structures, influence of heat re-
    lease on coherent structures, recirculation zone structure and
    dynamics. These should be studied in flow situations which are
    a little more complex than the self-preserving two-dimensional
    mixing layer but not as complex as a gas turbine combustor.
    Fuels used should also be kept simple. Diffusion and pre-mixed
    situations should be studied.

2. Hydrocarbon and Pollutant Kinetics in Turbulent Combustion
    Global kinetic models of hydrocarbon pyrolysis and oxidation,
    soot formation, and pyrolysed product oxidation under turbulent
    diffusive conditions are needed. Mechanisms determinig OH and
    O radical concentrations in turbulent combustion need to be de-
    termined.

        Experiments should be conducted in simple combustion
    systems as in (a) above using fuels of interest.

3. Spray Combustion
    Studies of droplet dynamics and evaporation in polydisperse non-
    dilute sprays and the interaction with flame and turbulence in
    simple axisymmetric flow systems are appropriate.

4.   Acoustic Wave Interaction with Turbulent Combustion (see
     Fulkerson, 1975)
     The fundamental problem of rumble and screech instabilities in
     ramjets and afterburners is the motivation. Interest should
     center on interaction of acoustic waves on coherent structures,
     recirculation zone dynamics, mixing rates, flame microstructure,
     droplet dynamics and evaporation.

     Although such studies have been carried out for a number of years they
have either been on systems which are much too simplified or with instrumentation which was completely inadequate to obtain the detail in terms of
spatial and temporal resolution necessary to allow adequate definition of
the phenomenology. Some examples follow.

     Kent and Bilger (1973) have obtained a detailed mapping of mean tem-
perature and concentrations (measured by a sample probe) in a turbulent
hydrogen/air diffusion flame taking care to accurately measure the boundary
conditions of the flow (initial profiles and axial pressure gradient).
While these data are of value to modelers as a final comparison they do
not allow resolution of many of the fundamental questions. Thus the measured concentrations give some information on the concentration fluctuation
level but nothing about the shape of the p.d.f. (probability density function)
of the fluctuations. The nitric oxide levels measured strongly suggest
(Bilger, 1975) that the main species chemistry is far from equilibrium,
but there is no direct measure of this. Concentration measurements of high
spatial and temporal resolution would be of great value in improving models
of such a system. Detailed measurements of velocity, turbulence kinetic
energy and sheer stress would also be of great assistance in determining
whether the models can adequately describe turbulent transport under conditions of large density gradients and chemical reaction. Baker et al. (1975)
have made such measurements in a model combustor using a laser Doppler velocimeter. These measurements would be of much greater value if high resolution information on temperature and species concentrations was also available.

     Fundamental chemical kinetic studies (Baldwin and Walker, 1973) which
attempt to identify the reaction mechanism and individual reaction rates
are far too complex to be considered in modeling except for fuels such as
H_2, CO and perhaps methane. Even then, molecular diffusion will strongly
affect radical concentrations in practical combustion systems and the coupled diffusion/kinetics problem becomes far too complex to treat a priori.
The modeler must resort to global kinetic models which have in the past
been developed from experiments with well-stirred reactors (Longwell and
Weiss, 1955) or plug-flow reactors (Dryer and Glassman, 1973). While the
measurement techniques used are probably adequate for these systems the

systems themselves give a gross oversimplification of the processes actually
involved in a real combustion system where turbulent mixing and diffusion
processes are likely to provide radical concentrations considerably different
to those existing in the laboratory flow reactors. The new optical instrumentation available gives us the opportunity of studying reaction rates under turbulent mixing conditions with enough ancillary information to enable
the kinetic rate to be sorted out from the mixing rate. Models of radical
concentration levels should also become possible.

The combustion of fuel sprays has been reviewed by Beer and Chigier
(1972), Williams (1973), Chigier and McGrath (1974), and Mellor (1973, 1975).

3. Measurements Required

The type of measurements required include:

A. Discrete (but Simultaneous) Measurements of Temperature,
    Velocity and Concentration
    Spatial resolution to 0.1 mm, temporal resolution to 10 μs
    (100 KHz), and accuracy to 5 percent or better are desirable.
    Data accumulation at each point sufficient so that means, va-
    riances, convariances, p.d.f.s and joint p.d.f.s can be de-
    termined to better than 5 percent. Sufficient simultaneous
    information must be obtained so that instantaneous density
    can be computed and temperature or other corrections applied.
    Good coverage (mapping) of the combustion field is requred.
    (Required for studies (1) to (4) above).

B. Time and Space Correlated Data
    Instantaneous temperature, velocity and concentration data
    separated by variable time and spatial amounts is desirable
    to determine fine scale flame structure turbulence length
    scales, local reaction rates etc. Resolution and accuracy
    as in A above but only limited coverage of the combustion
    field is required. (Required for study (1) only.)

C. Spray Characteristics
    Droplet sizes, velocities and frequencies at one point and
    Lagrangian measurements of droplet velocity and size as a
    function of time are needed. Instantaneous spatial distri-
    bution of droplets (holograms?) are necessary. (Required for
    study (3) only).

D. Imaging
   Schlieren, shadowgraph and other integrated path imaging, as
   well as Ramanograph or other single scattering plane imaging
   are useful. Instantaneous and sequential (movie) images are
   required. Spatial resolution to 0.1 mm and temporal resolution to 10 μs are needed but information does not have to be
   quantitative as it is required only as a diagnostic for conditional sampling work on coherent structures and recirculation zone. [Required for studies (1) to (4)].

E. Coherent Structure Investigation
   If data under A and B are obtained in a continuous analog
   fashion or the equivalent by rapidly pulsing it is possible
   to do conditional sampling on the data at a later time; simultaneous imaging with accurate cross-referencing is highly
   desirable. If data under A and B are obtained in a more or
   less random discrete manner, then simultaneous imaging and
   or CW signal (Raman N_2?) is essential if quantitative investigation (conditional sampling etc.) of coherent structures
   is to be carried out; the data processing problem is then
   enormous but the measurements are not dependent on the experimenter. Alternatively the obtaining of discrete data
   (A & B) can be triggered by some continuous monitoring
   device (e.g. CW Raman N_2) with appropriate time delays,
   the selection of the trigger criterion making the measurements experimenter subjective. [Required for studies (1)
   and (4)]

The same data obtained should be made available for many workers to
analyze and model.

4. Experiment Design

Figure 5 shows three experimental configurations in which the studies
(1) to (4) mentioned above can be carried out.

Jet Diffusion Flame
The configuration is axisymmetric and the flame is stabilized
by a fuel/O_2 pilot. For studies (1) and (4), CO + 2% H_2
should be used as fuel and in studies (2) and (4) prevaporized JP-4 should also be used. In study (4) an acoustic
perturber would be used. Inlet temperature and to some extent pressure is varied to vary chemical kinetic rates.

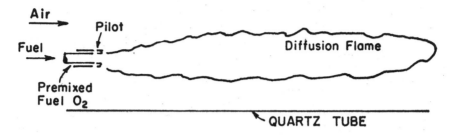

(a) JET DIFFUSION FLAME

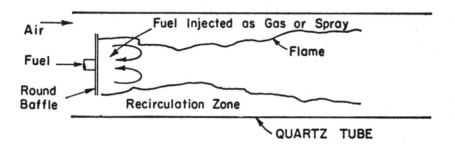

(b) BAFFLE STABILIZED DIFFUSION FLAME

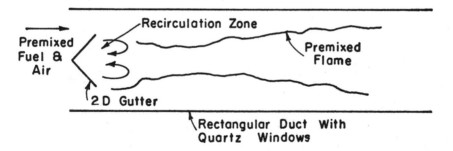

(c) 2D PREMIXED FLAME

Fig. 5 - Suggested Experimental Configurations

(b)   Baffle-Stabilized Diffusion Flame
The configuration is axisymmetric. CO with 2% H_2 is used
as fuel in study (1) with JP-4, used in prevaporized form
in study (2) and as spray in studies (3) and (4). An acoustic perturber is used in study (4).

(c)   2D Premixed Flame
The configuration is planar two-dimensional with a gutter of
aspect ratio of at least 6. Fuel and air are premixed using
either CO plus 2% H_2 or prevaporized JP-4. This experiment
is for studies (1), (2) and (4) only. An acoustic perturber
is used in study (4).

Sizes and velocities must be chosen carefully so that the fluid mechanic scales are realistic while at the same time easing as much as possible
the measurement problem. Figure 6 plots some of these fluid mechanic
scales against combustor flow velocity U and integral length scale L; this
latter can be taken as the flame diameter or width at the measuring station.
The definition and meaning of these scales and dimensionless groups is as
follows.

(i)   Turbulence Reynolds Number $Re_L \equiv \dfrac{u'L}{\nu} \sim 0.2\dfrac{UL}{\nu}$

We take $\nu P \simeq 2 \times 10^{-4}$ m²/s the value for air at
1400°K, P being the pressure in atmospheres. Typically
$Re_L \sim 10^5$ in gas turbines but should be kept above about
2×10^3 if turbulence is to be realistic.

(ii) Froude Number U^2/gL
The Froude number is a measure of the effects of buoyancy on
the flame. Keep $U^2/gL > 10^2$ to remove effects of buoyancy
on fluid mechanics. May not be necessary in kinetic studies.

(iii) Kolmogoroff Microscale $\eta \equiv (\nu^3/\epsilon)^{1/4}$
This microscale is the size of the smallest "waves" in the
flow or viscous cut-off size (ϵ is the rate of dissipation
of turbulence kinetic energy, $\epsilon \sim U'^3/L$.) Measurement volume should be kept less than 3η if adequate spatial resolution is to be obtained.

(iv) Microscale Convection Time $\tau_U \equiv \eta/U$
This is the time for the Kolmogoroff microscale to be con-
vected past a point in the flow. The measurement time
should be kept less than $3\tau_U$ if adequate temporal resolution
is to be obtained.

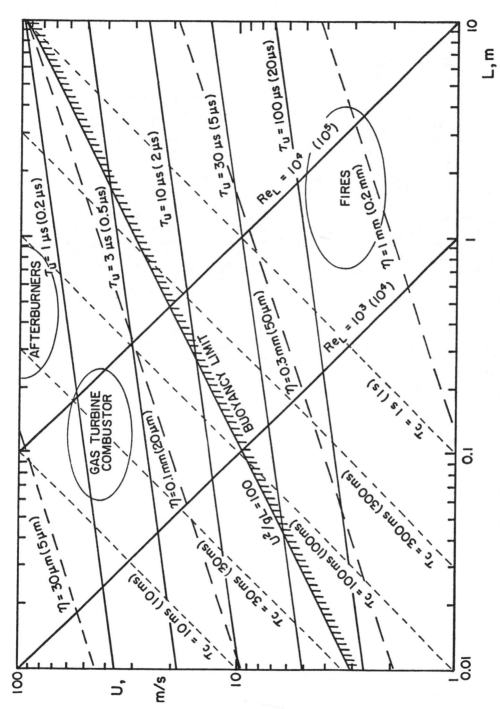

Fig. 6 - Combustion Parameters, in Terms of Characteristic
Flow Velocity U and Combustor Size L

461

(v)   Corrsin Time $\tau_c \equiv \lambda^2/6\nu$
_____

λ is the Taylor microscale $\lambda \sim (15 \nu U'^2/\epsilon)^{1/2}$ and thus $\tau_c \sim 2L/U' \sim 10L/U$. This is the Lagrangian time scale for dissipation of concentration fluctuation and is the appropriate time scale for comparing with chemical kinetic times (which vary approximately as $P^{-0.8}$ in hydrocarbon oxidation).

The figure gives values of these parameters for atmospheric pressure and in parenthesis for 10 atm. pressure.

From the discussion in Part I it is evident that the new optical techniques are going to be hard pressed to achieve adequate temporal and spatial resolution for the studies contemplated. The optimum choice appears to be U = 12 m/s. L = 15 cm, P = 1 atm which gives 3η = 0.5 mm, $3\tau_U$ = 40 μs, $Re_L \sim 1600$, τ_c = 120 ms, U^2/gL = 100. The turbulence Reynolds number is rather low and τ_c is rather large; however, the ratio of τ_c to chemical kinetic times will be similar to that in a gas turbine combustor at about 5 atm if the $P^{-0.8}$ behavior applies.

Where sufficient resolution is not possible it is estimated from the equation for the averaging error (see p. 433) that the errors will be less than 5% if the averaging is done over less than 30η and $30\tau_U$. Hence the practical lower limit of the control volume size, 1 mm^3, should not be a problem, for average values at least.

REFERENCES

Ahmed, A. A. (1971), "Application of the Cooled Film to the Study of Premixed Turbulent Flames," Ph.D. Thesis, McGill University, Montreal, Canada.

Altenkirch, R. A. and Mellor, A. M. (1975), "Continuum Flow Predictions of Measured Species Concentrations within a Prevaporizing Combustor," Rep. PURDU-CL-75-01, School of Mechanical Engineering, Purdue University.

Ansoulis, R. F., McDonald H., and Buggelin, R. C. (1974), "Development of a Combustor Flow Analysis. Part I: Theoretical Studies," AFAPL-TR-73-98, Part I.

Baker, R. J., Hutchinson, P., Khalil, E., and Whitelaw, J. H. (1975), "Measurements of Three-Velocity Components, and their Correlations, in a Model Furnace with and without Combustion," Fifteenth Symposium (International) on Combustion, The Combustion Institute, Pittsburgh, in press.

Baldwin, R. R. and Walker, R. W. (1973), "The Role of Radical-Radical Reactions in Hydrocarbon Oxidation," Comb. Flame 21, 55-67.

Beer, J. M. and Chigier, N. A. (1972), Combustion Aerodynamics, Wiley-Interscience, New York.

Beer, J. M. (1974), "Methods for Calculating Radiative Heat Transfer from Flames in Combustors and Furnaces," from Heat Transfer in Flames, Afgan and Beer, Editors, Scripta Book Co., Washington, D.C., p. 29-45.

Bilger, R. W. (1975a), "Probe Measurements in Turbulent Combustion," Rep. No. PURDU-CL-75-02, School of Mechanical Engineering, Purdue University.

Bilger, R. W. (1975b), "Turbulent Jet Diffusion Flames," Pollutant Formation and Destruction in Flames, Vol. 1 of Progress in Energy and Combustion Science, Chigier, N. A., Editor, Pergamon, Oxford, in press.

Blair, D. W. (1974), "Abel Inversion and Error Propagation" JQSRT, 14, p. 325.

Boni, A. A., SAI, Inc., Private Communication, 1975.

Bridoux, M., Chapput, A., Delhaye, M., Tourbez, H. and Wallart, F. (1974), "Rapid and Vetra Rapid Raman Spectroscopy," from "Laser Raman Gas Diagnostic" Lapp & Penney, Eds., Plenum Press, pp. 249-258.

Burch, D. E. And Gryvnak, D. A. (1974), "Infra Red Gas Filter Correlation Instrument for in situ Measurement of Gaseous Pollutants," Philco-Ford Corp., Newport Beach, CA, 92663 (Report EPA-650/2-74-094), December 1974.

Chao, C. M. and Goulard, R. (1974), "Nonlinear Inversion Techniques in Flame Temperature Measurements" from <u>Heat Transfer in Flames</u>, Afgan and Beer, Editors, Scripta Book Co., Washington, D. C., pp. 295-337.

Chao, C. M. (1974), "Retrieval of Concentration Profiles by Emission-Absorption Measurements and Inversion Techniques," a Ph.D. Thesis, Purdue Univesity.

Chen, F. P. and Goulard, R. (1975), "Optical Measurements of Jet Engine Exhaust Pollutant Levels," (to appear).

Chigier, N. A. and McCreath, C. G. (1974), "Combustion of Droplets in Sprays," Acta Astronautica <u>1</u>, 687-710.

Clarke, A. E., Gerrard, A. J., and Holliday, L. A. (1963), "Some Experiences in Gas Turbine Combustion Chamber Practice Using Water Flow Visualization Techniques," <u>Ninth Symposium (International) on Combustion</u>, Academic Press, New York, pp. 878-891.

Cornelius, W., Burwell, W. G., and Turunen, W. A. (1957), "Progress Report on Fundamental Gas Turbine Combustion Studies," General Motors Research Memorandum, 1957-1959.

Davis, M. G., McGregor, W. K. and Few, J. D., "Spectral Simulation of Resonance Band Transmission Profiles for Species Concentration Measurements: NO α-Bands as an Example, AEDC, Rept. AEDC-TR-74-124, January 1975.

Dryer, F. L. and Glassman, I. (1973), "High-Temperature Oxidation of CO and CH_4," <u>Fourteenth Symposium (International) on Combustion</u>, The Combustion Institute, Pittsburgh, pp. 987-1003.

Fletcher, R. S., Siegel, R. D., and Bastress, E. K. (1971), "The Control of Oxides of Nitrogen Emissions from Aircraft Gas Turbine Engines. I. Program Description and Results," Rep. No. FAA-RD 71-111, 1.

Forsyth, J. M. (1970), Editor "Development in Laser Technology" SPIE Seminar Proceedings, Vol. 20 (Society of Photo Optical Instrumentation Engineers).

Fristrom, R. M. and Westenberg, A. A. (1965), <u>Flame Structure</u>, McGraw-Hill, New York.

Fulkerson, G. D., et al. (1975), "Low Frequency Combustion Instability Seminar," Aero-Propulsion Laboratory, Wright-Patterson Air Force Base.

Giovanni, D. V., Pagni, P. J., Sawyer, R. F., and Hughes, L. (1972), "Manganese Additive Effects on Emissions from a Model Gas Turbine Combustor," Comb. Sci. Tech. 6, 107-114.

Goulard, R. (1974), "Laser Raman Scattering Applications," JQSRT, Vol. 14, pp. 969-974.

Gradon, K. and Miller, S. C. (1968), "Combustion Development on the Rolls-Royce Spey Engine," Combustion in Advanced Gas Turbine Systems, Pergamon, Oxford, pp. 45-76.

Gregory, G. L. (1974), "Evaluation of Chemiluminescent Hydrogen Chloride and a NDIR Carbon Monoxide Detector for Environment Monitoring," The JANNAF Propulsion Meeting, October 1974, San Diego, California.

Hammon, D. C., Jr. and Mellor, A. M. (1973), "Analytical Predictions of Emissions from and within an Allison J-33 Combustor," Comb. Sci. Tech. 6, pp. 279-286.

Hartley, Ed., "The Role of Physics in Combustion," An APS Study in Technical Aspects of Efficient Energy Utilization," July 1974 [See M. Lapp-Section 2-"Diagnostics for Experimental Combustion Research"].

Hauf, W. and Grigull, U. (1970), "Optical Methods in Heat Transfer," Advances in Heat Transfer, J. P. Hartnett and T. F. Irvine, Eds., Academic Press, N.Y.

Heflinger, L. O., Wuerker, R. F., and Brookds, R. E. (1966), "Holographic Interferometry," Journal of Applied Physics, Vol. 37, pp. 642-649.

Herget, W., "Temperature and Concentration Measurements in Model Exhaust Plumes Using Inversion Techniques," from the "Specialists 'Conference on Molecular Radiation and its Applications to Diagnostic Techniques" R. Goulard, E., pp. 359-379, NASA TM X-53711, Oct. 1967.

Heywood, J. B. and Mikus, T. (1973), "Parameters Controlling Nitric Oxide Emissions from Gas Turbine Combustors," Atmospheric Pollution by Aircraft Engines, AGARD CP No. 125.

Hiett, G. F. and Powell, G. F. (1962), "Three Dimensional Probe for Investigation of Flow Patterns" The Engineer 213 (1), pp. 165-170.

Hinkley, E. D. and Kelley, P. L., (1971), "Detection of Air Pollutants with Tunable Diode Lasers," Science, 171, pp. 635-639, Feb. 1971.

Hottel, H. C. (1974), "First Estimates of Industrial Furnaces Performance-
The One-Gas-Zone Model Reexamined," from Heat Transfer in Flames, Afgan
and Beer, Editors, Scripta Book Co., Washington, D. C., pp. 5-28.

Hunter, S. C., Johnson, K. M., Mongia, H. C., and Wood, M. P. (1974),
"Advanced, Small, High-Temperature-Rise Combustor Program. Volume I:
Analytical Model Derivation and Combustor-Element Rig Tests (phases I
and II)," USAAMRDL Tech. Rep. 74-3A.

Jachimowski, C. J., Private Communication, 1975.

Kent, J. H. and Bilger, R. W. (1973), "Turbulent Diffusion Flames,"
Fourteenth Symposium (International) on Combustion, The Combustion Institute, Pittsburgh, pp. 615-625.

Kildal, H. and Byer, R. L. (1971), "Comparison of Laser Methods for the
Remote Detection of Atmospheric Pollutants," Proceedings of the IEEE, Vol.
59, No. 12, pp. 1644-1663, Dec. 1971.

Krakow, B. "Spectroscopic Temperature Profile Measurements in Inhomogeneous Hot Gases," Applied Optics, Vol. 5, No. 2, p. 201, February, 1966.

LaPointe, C. W. and Schultz, W. L. (1972), "Measurement of Nitric Oxide
Formation within a Multi-Fueled Turbine Combustor," Emissions from Continuous Combustion Systems, Plenum Press, New York, pp. 211-242.

Lapp, M. and Penney, C. M. (1973), "Laser Raman Gas Diagnostics," Plenum
Press.

Lapp, M., Penney, C. M. and Asher, J. A. (1973), "Application of Light
Scattering Techniques for Measurements of Density, Temperature and Velocity in Gas Dynamics," A General Electric Co. Report SRD-72-085, January
1973.

Lavoie, G. A., Heywood, J. B., Keck, J. C. (1970), "Experimental and
Theoretical Study of Nitric Oxide Formation in Internal Combustion Engines," Comb. Sci. and Tech., Vol. 1, p. 313.

Lederman, D., Bloom, M. H., Bornstein, J., Khosla, P. K. (1974), "Temperature and Specie Concentration Measurements in a Flow Field," International Journal of Heat and Mass Transfer, Vol. 17, pp. 1479-1486.

Lefebvre, A. H. and Fletcher, R. S. (1973), "A Preliminary Study on the
Influence of Fuel Staging on Nitric Oxide Emissions From Gas Turbine
Combustors," Atmospheric Pollution by Aircraft Engines, AGARD CP No. 125.

Leonard, D. A., "Field Tests of a Laser Raman Measurement System for Air-craft Engine Exhaust Emissions," The AVCO Everett Research Laboratory, Inc., Report AFAPL-TR-74-100, October 1974.

Lewis, J., ARO, Inc., Private Communication, 1975.

Longwell, J. P. and Weiss, M. A. (1955), "High Temperature Reaction Rates in Hydrocarbon Combustion," Ind. Eng. Chem. 47, pp. 1634-1643.

Ludwig, C. B., et al., "Remote Measurement of Air Pollution by Non Disper-sive Optical Correlation," AIAA Paper No. 71-1107, November 1971.

Ludwig, C. B., Private Communication (A proposal to EPA Applications of Remote Monitoring Techniques in Air Enforcement and Regulatory Programs), June 1974.

McGregor, W. K., Seiber, B. L. and Few, J. D. (1972), "Concentration of OH and NO in YG 93-GE-3 Engine Exhausts Measured in situ by Narrow Line UV Absorption," Second Conference on the Climatic Impact Assessment Program, Cambridge, Mass., November 1972.

Matthews, B. J. and Wuerker, R. F. (1969), "The Investigation of Liquid Rocket Combustion Using Pulse Laser Holography," AIAA Fifth Joint Pro-pulsion Specialists Conference, Colorado Springs, Colorado, June 9-13, 1969.

Matthews, B. J., Wuerker, R. F., Chambers, H. F. and Hojnacki, J. (1974), "Holography of JP-4 Droplets and Combusting Boron Particles," from Instrumentation for Air Breathing Propulsion, A. E. Fuhs, Ed., Progress in Aeronautics and Astronautics Vol. 34, the MIT Press, pp. 296-313.

Matulka, R. D. and Collins, D. J. (1971), "Determination of Three Dimen-sional Density Fields from Holographic Interferograms," Journal of Applied Physics, Vol. 12, pp. 1109-1119.

Mellor, A. M., Anderson, R. D., Altenkirch, R. A., and Tuttle, J. H. (1972), "Emissions from and within an Allison J-33 Combustor," Rep. No. CL-72-1, School of Mechanical Engineering, Purdue Univesity.

Mellor, A. M. (1973), "Simplified Physical Model of Spray Combustion in a Gas Turbine Engine," Comb. Sci. Tech. 8, 101-109.

Mellor, A. M. (1975), "Gas Turbine Engine Pollution," Pollution Formation and Destruction in Flames, Vol. 1 of Progress in Energy and Combustion Science, Chigier, Editor, Pergamon, Oxford, in press.

Mosier, S. A. and Roberts, R. (1974a), "Low-Power Turbopropulsion Combustor Exhaust Emissions. Vol. II: Demonstration and Total Emission Analysis and Prediction,: AFAPL-TR-73-36, Vol. II.

Mosier, S. A. and Roberts, R. (1974b), "Low-Power Turbopropulsion Combustor Exhaust Emissions. Vol. III: Analysis," AFAPL-TR-73-36, Vol. III.

Murthy, S. N. B., Editor (1974), <u>Proceedings of SQUID Workshop on Reacting and Nonreacting Turbulent Flows</u>, Project SQUID Report, Purdue University.

Neer, M.E., "Numerical calculations of UV Emission and Absorption Spectra of OH," Technology, Inc., Report ARL-TR-74-olo9, August 1974 (see also AIAA Paper 73-1320).

Norgen, C. T. (1971), Determination of Primary-Zone Smoke Concentrations from Spectral Radiance Measurements in Gas Turbine Combustors," NASA TN D-6410.

Parts, L., "An Assesment of Instrumentation and Monitoring Needs for Significant Air Pollutants Emitted by Air Force Operations and Recommendations for Future Research on Analysis of Pollutants," Monsanto Research Corp., Report ARL-TR-74-0015, February 1974.

Penney, C. M., General Electric Co., Private Communication, 1975.

Platt, W. K. (1969), "Laser Communication Systems," John Wiley.

Reynolds, G., Tech Ops, Private Communication (1975).

Roshko, A. (1974), "Proceedings of the SQUID Workshop on Reacting and Nonreacting Turbulent Flows," S. N. B. Murthy, Ed., a Project SQUID Report.

RCA Phototubes and Photocells Technical Manual PT-61, Radio Corporation of America, Commercial Engineering, Harrison, N.J. 07029, (1970).

Rodig, J. and Zalud, F. (1969-70), "Some Contributions to Experimental Combustion Research," <u>Proceedings of the Institute of Mechanical Engineers</u>, p. 203.

Roquemore, W. M., Private Communication, 1975.

Salzmann, J. A., Masica, W. J. and Coney, T. A., "Determination of Gas Temperatures from Laser-Raman Scattering," NASA TN D-6336, May 1971.

Sandhu, S. S. and Weinberg, F. J. (1972), "A Laser Interferometer for Combustion, Aerodynamics and Heat Transfer Studies," Journal of Physics E: Scientific Instruments 1972 Vol. 5, pp. 1018-1020.

Setchell, R. E., "Analysis of Flame Emissions of Laser Raman Spectroscopy" Western States Section, The Combustion Institute, May 1974.   Also Sandia Corporation Report SLL-74-5422.

Shisler, R. A., Tuttle, J. H., and Mellor, A. M. (1974), "Emissions from and within a Film-Cooled Combustor," Rep. No. PURDU-CL-74-01, School of Mechanical Engineering, Purdue University.

Smith, R. A., Jones, F. E. and Chasman, R. P. (1957), "The Detection and Measurement of Infra Red Radiation," Oxford Press.

Stewart, J. E. (1970), "Infrared Spectroscopy," Marcel Dekker, Inc., N.Y.

Sulzmann, K. G. P., Lowder, J. E. L. and Penner, S. P. (1973), "Estimates of Possible Detection Limits for Combustion Intermediates and Products with Line Center Absorption and Derivation Spectroscopy Using Tunable Laser," Comb. and Flame 20, pp. 177-191.

Sweeney, D. W. and Vest, C. M. (1973), "Reconstruction of Three-Dimensional Refractive Index Fields from Multidirectional Interferometric Data," Applied Optics, Vol. 12, pp. 2649-2664.

Sweeney, D. W. and Vest, C. M. (1974), "Measurement of Three-Dimensional Temperature Fields above Heated Surfaces by Holographic Interferometry", International Journal of Heat and Mass Transfer, Vol. 17, pp. 1443-1454.

Schildkraut, E. R. (1974), "Electronic Signal Processing for Raman Scattering Measurements," from Laser Raman Gas Diagnostics" Lapp and Penney, Eds., Plenum Press, pp. 259-277.

Swithenbank, J. (1975), "Practical Application of Turbulence Theory to the Design of Combustion Systems," AFOSR Review, July 7-10, 1975, AFAPL, WPAFB, Ohio.

Thompson, H. D. and Stevenson, W. H., Eds., Proceedings of the Second International Workshop on Laser Velocimetry, Purdue University, March 27-29, 1974 [Engineering Experiment Station, Bulletin No. 144].

Toone, B. (1968), "A Review of Aero Engine Smoke Emission," Combustion in Advanced Gas Turbine Systems, Pergamon, Oxford, pp. 271-296.

Trolinger, J. D., Belz, R. A. and O'Hare, J. E. (1974), "Holography of Nozzles, Jets and Spraying Systems," from Instrumentation for Air Breathing Propulsion," A. E. Fuhs, Ed., Progress in Aeronautics, Vol. 34, the MIT Press, pp. 249-261.

Tuttle, J. H., Shisler, R. A. and Mellor, A. M. (1973), "Nitrogen Dioxide Formation in Gas Turbine Engines: Measurements and Measurement Methods," Rep. No. PURDU-CL-73-06, School of Mechanical Engineering, Purdue.

Wang, C. P. (1975), "Laser Applications to Turbulent Reactive Flows: Density Measurement by Resonance Absorption and Resonance Scattering Techniques," to appear in Combustion Science and Technology.

Wang, J. Y. (1970), "Theory and Application of Inversion Techniques: a Review," Rep. No. AAES 70-69, School of Aeronautics & Astronautics & Engineering Sciences, Purdue University.

Weinberg, F. J. (1963), Optics of Flames Butterworths, London [see also Schwar, M. J. R. and Weinberg, F. J. (1969), Combustion and Flame, Vol. 13, pp. 335-374.

Williams, A. (1973), "Combustion of Droplets of Liquid Fuels: a Review," Comb. Flame 21, 1-31.

PART VI
Conclusions and Recommen- dations

CONCLUSIONS AND RECOMMENDATIONS

In summary, it looks as if the capability of making steady state measurements in all regions of a combustor with reasonable access for measurements (i.e. at least one port) exists already. An accuracy of 1% and space resolution of 1 mm^3 is within reach, within a few years at most.

Instantaneous measurements of velocity, temperature and density (LV, Raman) are also within reach, with the same accuracy. But species concentration below 10^{-3} are not likely to be obtained instantaneously with the present methods, but only through extended observation times (Raman), or if a favorable geometry allows for path measurements.

Special mention should be made of the imaging techniques, mostly based on photographic or interferometric principles. Although they are inherently less accurate than "point" techniques, they give a time record (e.g. in movie form) of the broad features of the flow, especially in the spray (distribution and size of droplets) and in the primary combustion region. This allows for an identification of those areas to be probed in more detailed accuracy by the "point" methods. Progress in holography and data processing will increase the information contained and cost effectiveness of these techniques.

New measurement principles seem to have the potential for instantaneous ppm readings or less, with 1% accuracy. Such possibilities as fluorescence, resonance refraction and coherent Antistokes Raman Spectroscopy should be explored vigorously, as well as their supporting technology (tunable lasers, for instance). Also in urgent need of basic work are the various phenomena which might limit the applications of measurement methods: particulate radiation, hydrocarbon fluorescence, ionization breakdown...Finally it should be kept in mind that no one measurement method is the optimal answer to all the requirements of combustion research, or even one given experiment. A broad support program will guarantee that engineers and scientists are given a choice between a wide range of up-to-date techniques when designing their experiments.

Considerations of cost may limit the availability of the more elaborate techniques to a few well-equipped centers. Hopefully, one of their
responsibilities would be to evaluate and calibrate carefully less expensive instruments in typical combustion situations. This would be a valuable input to the many smaller laboratories whose contribution to the
field is essential.

Special attention should be directed to the interpretation of the
measurement of a given property. This calls for a careful analytical
elimination of the other properties affecting the data, of the integrating effect of geometries (inversion of path measurements, 3-D imaging)
and of sampling effects (spectra, seeding, pulses).

Also important is a proper integration of each measurement, at a
given place and time, into an analytical model of the overall combustor
flow. Such a systematic synthesis should result into a more accurate
model with an increased capability for generalization. It should also
serve as a sensitivity analysis which would help determine the optimum
location of measurements, in terms of maximum information and error mini-
mization for the model.

More generally, the dialogue between measurement and combustor specialists should be amplified and maintained at a more active level than
it has been so far. It is critical that measurement efforts be focused
now on the properties needed by the combustor modelers and designers.

An attempt at such a measurement program is made in pp. 463-468
of this report. The different physical processes which govern combustor
flows are characterized by their typical lengths or times. This allows
in turn for a range of scaled experiments which remain representative of
combustors while easier to run and more accurate to measure.

WORKSHOP PARTICIPANTS

Dr. Jeffrey A. Asher
General Electric Company
Corporate Research & Development
P. O. Box 43
Schenectady, New York 12301

Professor Daniel Bershader
367 Durant Building
Stanford University
Stanford, California 94305

Dr. R. W. Bilger
TSPC Chaffee Hall
Purdue University
West Lafayette, Indiana 47907

Dr. J. W. Birkeland
Aerospace Research Labs
Wright Patterson Air Force Base
Ohio 45433

Dr. Blazowski
Room 202, Building 70
Fuels Division AFAPL-AFSC
Wright Patterson Air Force Base
Ohio 45433

Professor M. Bloom
Polytechnic Institute of New York
Route 110
Farmingdale, New York 11735

Professor Louis I. Boehman
Department of Mechanical
   Engineering
300 College Park Avenue
University of Dayton
Dayton, Ohio 45469

Dr. A. A. Boni
SAI, Incorporated
P. O. Box 2351
La Jolla, California 92037

Dr. Roland Borghi
Office National D'Etudes et
   de Recherches Aérospatiales
Chatillon, France 92320

Professor Mel Branch
Department of Mechanical Engineering
University of California
Berkeley, California 94720

Mr. Michael Chaszeyka
Office of Naval Research
536 South Clark Street
Chicago, Illinois 60605

Dr. N. Chigier
Chemical Engineering & Fuel
   Technology
University of Sheffield
Mappin Street
Sheffield S13JD, England

Mr. Larry Cooper
Graduate Assistant
University of Illinois
807 W. Illinois, Apt. 5
Urbana, Illinois 61801

Professor W. B. Cottingham
School of Mechanical Engineering
Purdue University
West Lafayette, Indiana 47907

Mr. John W. Daily
Mechanical Engineering Department
University of California
Berkeley, California 94720

Mr. Ken Daniel
Research Assistant
Purdue University
West Lafayette, Indiana 47907

Dr. J. Drewry
Aerospace Research Labs (LF)
Wright Patterson Air Force Base
Ohio 45433

Professor James F. Driscoll
University of Michigan
209 Aerospace Engineering Bldg
Ann Arbor, Michigan 48105

Dr. A. C. Eckbreth
United Aircraft Research Laboratory
United Technologies Corporation
East Hartford, Connecticut 06108

Mr. Denton W. Elliott
Air Force Office of Scientific
    Research
1400 Wilson Boulevard
Arlington, Virginia 22209

Dr. M. M. El Wakil
Mechanical Engineering Department
University of Wisconsin
Madison, Wisconsin 53706

Dr. Richard Flagan
Room 31-168
Massachusetts Institute of
    Technology
Cambridge, Massachusetts 02139

Dr. Arthur Fontijn
Reaction Kinetics Group
Aero Chem Research Labs, Inc.
P. O. Box 12
Princeton, New Jersey 08540

Dr. R. M. Fristrom
Applied Physics Laboratory
Johns Hopkins University
8621 Georgia Avenue
Silver Spring, Maryland 20910

Professor Irwin Glassman
Department of Aerospace &
    Mechanical Sciences
Princeton University
Princeton, New Jersey 08540

Professor Robert Goulard
SEAS-CMEE
George Washington University
Washington, DC 20052

Dr. Fred Gouldin
Mechanical Engineering Department
Cornell University
Ithaca, New York 14850

Dr. David Hacker
Department of Energy Engineering
University of Illinois
Box 4348
Chicago, Illinois 60680

Dr. D. R. Hardesty
Division 8115
Sandia Laboratories
Livermore, California 94550

Mr. Philip T. Harsha
R & D Associates
P. O. Box 3580
Santa Monica, California 90403

Dr. A. B. Harvey
Code 6110
Naval Research Laboratory
Washington, DC 20375

Dr. Robert Henderson
AFAPL/TBC
Wright Patterson Air Force Base
Ohio 45433

Mr. M. E. Hillard
MS.S 235A
NASA Langley Research Center
Hampton, Virginia 23665

Mr. Dale Hudson
AFAPL/TBC
Wright Patterson Air Force Base
Ohio 45433

Dr. C. J. Jachimowski
MN 419A
NASA Langley Research Center
Hampton, Virginia 23665

Dr. C. W. Kaufman
Aerospace Engineering
University of Cincinnati
Cincinnati, Ohio 45221

Dr. James J. Komar
Professor of Energy Engineering
University of Illinois
919 Science Engineering Offices
Chicago, Illinois 60680

Dr. W. Kuykendal
EPA/NERC/N115
Research Triangle Park
North Carolina 27111

Dr. Marshall Lapp
General Electric Company
Corporate Research & Development
P. O. Box 8
Schenectady, New York 12301

Professor N. M. Laurendeau
TSPC Chaffee Hall
Purdue University
West Lafayette, Indiana 47907

Professor Mel Lecuyer
Purdue University
West Lafayette
Indiana 47907

Dr. Samuel Lederman
Aerospace Research Laboratory
Polytechnic Institute of New York
Farmingdale, New York 11735

Dr. Andrew Lennert
USS AEDC-DYR
Arnold AFS
Tennessee 37389

Mr. Donald A. Leonard
Computer Genetics Corporation
18 Lakewide Office Park
Wakefield, Massachusetts 01880

Dr. R. Levine
Room B64, Building 225
National Bureau of Standards
Washington, DC 20234

Mr. T. F. Lyon
General Electric Company
Aircraft Engine Group
Mail Zone H52
Cincinnati, Ohio 45215

Dr. C. B. Ludwig
Science Applications, Inc.
P. O. Box 2351
La Jolla, California 92037

Dr. Cecil Marek
MS 60-4
NASA Lewis Research Center
21000 Brookpark Road
Cleveland, Ohio 44135

Dr. W. K. McGregor
Engine Test Facility
ARO, Incorporated
Arnold AFS, Tennessee 37389

Dr. W. J. McLean
291 Grumman Hall
Cornell University
Ithaca, New York 14850

Professor A. M. Mellor
TSPC Chaffee Hall
Purdue University
West Lafayette, Indiana 47907

Mr. E. J. Mularz
Mail Stop 60-6
NASA Lewis Research Center
21000 Brook Park Road
Cleveland, Ohio 44135

Professor S. N. B. Murthy
Chaffee Hall
Purdue University
West Lafayette, Indiana 47907

Dr. James Nash-Webber
Room 31-140
Massachusetts Institute of
    Technology
Cambridge, Massachusetts 02139

Dr. F. K. Owen
United Technologies Research Center
United Technologies Corporation
East Hartford, Connecticut 06108

Mr. Walter Panknin
Institut Fur Verfahrenstechnik
Hannover/BRD
3 Hannover
Callinstrasse 1SF
West Germany

Mr. James R. Patton, Jr.
Office of Naval Research
Power Program, Code 473
Arlington, Virginia 22217

Professor S. S. Penner
Department of Engineering Physics
University of California
Box 109
La Jolla, California 92037

Mr. Dick Peterson
Mechanical Engineering Department
Purdue University
West Lafayette, Indiana 47907

Mr. Steve Plee
Purdue University
TSPC Combustion Laboratory
West Lafayette, Indiana 47907

Mr. Edward J. Rambie
Executive Program Engineer
U. S. Army Tank Automotive
    Command, AMSTA-RGR
Warren, Michigan 48089

Professor Bruce Reese
School of Aero & Astro
Purdue University
West Lafayette, Indiana 47907

Mr. John P. Renie
Student
Purdue University
West Lafayette, Indiana 47907

Mr. David L. Reuss
University of Illinois
105 Transportation Building
Urbana, Illinois 61801

Dr. G. Reynolds
Technical Operations
20 South Avenue
Burlington, Massachusetts 01803

Mr. Robert P. Rhodes
Research Engineer
ARO Incorporated, ETF/TAB
Arnold Air Force Station
Tennessee 37389

Dr. F. Robben
Mechanical Engineering Department
University of California
Berkeley, California 94720

Dr. R. Roberts
Pratt & Whitney Division
United Technologies Corporation
East Hartford, Connecticut 06108

Dr. W. M. Roquemore
Room 202, Building 70
Fuels Division AFAPL-AFSC
Wright Patterson Air Force Base
Ohio 45433

Dr. D. Santavicca
Engineering Quadrangle D 207
Princeton University
Princeton, New Jersey 08540

Mr. Helmut F. Schacke
10610 Mautz Road
Silver Spring
Maryland 20903

Professor Steve Schmidt
Mechanical Engineering Department
Washington State University
Pullman, Washington 99163

Dr. S. A. Self
Mechanical Engineering Department
Stanford University
Stanford, California 94305

Dr. R. E. Setchell
Sandia Laboratories
Division 8115
Livermore, California 94550

Mr. Chou Shen
Naval Air Propulsion Center
Department of the Navy
Trenton, New Jersey 08628

Professor J. G. Skifstad
TSPC Chaffee Hall
Purdue University
West Lafayette, INdiana 47907

Dr. B. W. Smith
Department of Chemistry
University of Florida
Gainesville, Florida 32603

Professor W. H. Stevenson
School of Mechanical Engineering
Purdue University
West Lafayette, INdiana 47907

Dr. R. A. Strehlow
University of Illinois
101 Transportation Building
Urbana, Illinois 61801

Professor D. W. Sweeney
School of Mechanical Engineering
Purdue University
West Lafayette, Indiana 47907

Professor Tankin
Mechanical Engineering School
Northwestern University
Evanston, Illinois 60201

Mr. J. P. Taran
Office National D'Etudes et
    de Recherches Aérospatiales
Chatillon, France 92320

Professor Dwight C. Tardy
Department of Chemistry
University of Iowa
Iowa City, Iowa 52242

Professor H. D. Thompson
TSPC Chaffee Hall
Purdue University
West Lafayette, Indiana 47907

Mr. J. H. Tuttle
Graduate Assistant of Research
Chaffee Hall
Purdue University
West Lafayette, Indiana 47907

Professor M. Vanpe
University of Massachusetts
Amherst
Massachusetts 01002

Professor C. M. Vest
Department of Aero & Astro
377 Durant Building
Stanford University
Palo Alto, California 94305

Professor R. Viskanta
School of Mechanical Engineering
Purdue University
West Lafayette, Indiana 47907

Mr. W. W. Wagner
Naval Air Propulsion Test Center
Department of the Navy
Trenton, New Jersey 08628

Mr. Curtis L. Walker
Combustion & Heat Transfer Research
Detroit Diesel Allison Division of GMC
W-16 P. O. Box 894
Indianapolis, INdiana 46260

Dr. C. Wang
Aerodynamics & Heat Transfer
Aerospace Corporation
Box 92957
Los Angeles, California 90009

Dr. S. Weeks
Department of Chemistry
University of Florida
Gainesville, Florida 32603

Mr. Herbert White
Aerospace Corporation
2350 E. El Segundo Boulevard
El Segundo, California 90245

Dr. B. T. Wolfson
Air Force Office of Scientific
   Research (SREP)
1400 Wilson Boulevard
Arlington, Virginia 22209

Dr. A. D. Wood
Office of Naval Research
495 Summer Street
Boston, Massachusetts 02210

Dr. Kurt Wray
Physical Sciences Incorporated
18 Lakeside Office Park
Wakefield, Massachusetts 01880

Dr. Barbara Zilles
Chemistry Department
Old Dominion University
Norfolk, Virginia 23508

Dr. B. T. Zinn
Georgia Institute of Technology
School of Aerospace Engineering
Atlanta, Georgia 30332

INDEX

This index does not include a few key words, such as accuracy, temperature, or velocity, as they appear nearly everywhere in the proceedings, and as they are analyzed separately in the forum (Part IV) and in the review (Part V). For the same reason, the two main techniques discussed (laser velocimetry and Raman scattering) do not appear as separate entries.